Concentrated Emulsion Polymerization

Functional and Modified Polymeric Materials Two-Volume Set

Eli Ruckenstein
Hangquan Li
Chong Cheng

Volumes in the Set:

Concentrated Emulsion Polymerization (ISBN: 9780367134556)

Solution and Surface Polymerization (ISBN: 9780367134563)

Concentrated Emulsion Polymerization

Eli Ruckenstein
Hangquan Li
Chong Cheng

CRC Press is an imprint of the
Taylor & Francis Group, an informa business

CRC Press
Taylor & Francis Group
6000 Broken Sound Parkway NW, Suite 300
Boca Raton, FL 33487-2742

Printed on acid-free paper

International Standard Book Number-13: 978-0-3671-3455-6 (Hardback)

Library of Congress Cataloging-in-Publication Data

Names: Ruckenstein, Eli, 1925- author. | Li, Hangquan, author. | Cheng, Chong, author.
Title: Concentrated emulsion polymerization / Eli Ruckenstein, Hangquan Li, Chong Cheng.
Description: Boca Raton : Taylor & Francis, a CRC title, part of the Taylor & Francis imprint, a member of the Taylor & Francis Group, the academic division of T&F Informa, plc, 2018. | Includes bibliographical references and index.
Identifiers: LCCN 2018047976 | ISBN 9780367134556 (hardback : alk. paper) | ISBN 9780429026577 (ebook)
Subjects: LCSH: Emulsion polymerization. | Polymer colloids.
Classification: LCC TP156.P6 R83 2018 | DDC 660/.294514--dc23
LC record available at https://lccn.loc.gov/2018047976

Visit the Taylor & Francis Web site at
http://www.taylorandfrancis.com

and the CRC Press Web site at
http://www.crcpress.com

Printed in the United Kingdom
by Henry Ling Limited

Contents

Preface

With the current annual worldwide production of over 400 million tons, polymers are indispensable materials that have been broadly used in everyday life and in almost every industry. As the chemical process to convert small molecule precursors (i.e., monomers) to polymers, polymerization can be performed in homogeneous systems in bulk or solution, as well as heterogeneous systems in emulsion or suspension, at interface, or on surface. Although polymers have versatile mechanical properties, and are commonly produced for general applications such as plastic, rubber and fiber, their structures can be tailored to target new and enhanced properties and applications. Accordingly, the preparation and comprehensive studies of functional and modified polymeric materials represent an increasingly important research direction in polymer field. This book set of two volumes summarizes the research work performed by Professor Eli Ruckenstein and co-workers on functional and modified polymeric materials over the past decades. Volume 1 depicts the research studies using emulsion polymerization, especially concentrated emulsion polymerization, as the synthetic method. The corresponding polymeric materials include conductive polymer composites, core-shell latex particles, enzyme/catalyst/herbicide carriers, and toughened and compatibilized plastics. Volume 2 presents the research studies mainly using solution or surface polymerization in material synthesis. The corresponding polymer-based materials include polymers with special functional groups, degradable and de-crosslinkable polymers, pervaporation membranes, soluble conductive polymers, composites with conductive surfaces, and so on.

Emulsion polymerization is a unique process with emulsions as reaction media, and it has several distinct advantages. Among all types of commonly used polymerization processes, it is the only process in which polymerization rate and polymer molecular weight can be increased simultaneously. It allows for easier control of heat transfer than bulk polymerization. It is also generally more environmentally benign than solution polymerization. The product of emulsion polymerization, commonly referred to as latex, can be used directly or re-processed readily in typical applications, such as surface modification and blending.

Emulsion polymerization can be classified based on specific emulsion types, which can be obtained by changing emulsion components and the preparation parameters. As a special type, concentrated emulsion is highly attractive as a reaction medium for applications in polymer synthesis. A concentrated emulsion differs from a conventional emulsion in that the volume fraction of the dispersed phase is higher than that of the most compact arrangement of mono-spheres (0.74), and can be as high as 0.99. The droplets of the dispersed phase are compactly packed and squeeze each other. As a result, they are no longer spherical, but polyhedral in shape, and separated by a thin film of the continuous phase. With such a structure, there is no presence of micelles, and the site of polymerization is in either the droplets or the continuous phase, or both. When polymerization occurs in the droplets alone, the product is colloidal particles; when in the continuous phase alone, the product is a porous material; when in both phases, a hydrophilic/hydrophobic composite is obtained.

Ruckenstein and co-workers have performed seminal studies on the preparation of functional and modified polymeric materials via (concentrated) emulsion polymerization. The corresponding research papers, after further selection and classification, are collected in the four chapters within Volume 1.

Chapter 1 reports the synthetic approaches for the preparation of conductive polymeric composites via (concentrated) emulsions. Conductive polymers are an important class of functional materials, but generally have low processability due to their conjugated main-chains. Relative to conductive polymers, conductive polymeric composites can possess improved processability, and therefore, have remarkable applicability. These composites are prepared by either using concentrated emulsion-derived cross-linked porous polymers as the synthetic substrate of conductive polymers, or by the synthesis of conductive polymers in (concentrated) emulsions in the presence of dissolved non-conductive polymers.

Chapter 2 presents the (concentrated) emulsion-based methods for the synthesis of polymeric core-shell particles. As structured composite particles consisting of different materials at core and shell domains respectively, core-shell particles may possess new and enhanced properties as compared to typical particles. They may find applications in coating, material modification, controlled release, and other areas. These polymeric core-shell particles are synthesized by either seeded polymerization in (concentrated) emulsions, or polymerization using initiator-embedded particles obtained via concentrated emulsions.

Chapter 3 describes the preparation of porous polymeric materials or latex particles as carriers of a variety of functional cargoes. The overall functional efficiency of these cargoes can be enhanced with the presence of polymeric carriers, mainly because of improved recovery or controlled release behavior. Specifically, the immobilization of enzyme or chemical catalyst on porous polymeric materials, as well as the incorporation of a phase-transfer catalyst or herbicide with latex particles, are studied. In principle, the technology of encapsulation and controlled release using concentration emulsion-derived polymeric carriers can be expanded by using other interesting cargoes, such as therapeutic agents, to target other emerging applications.

Chapter 4 illustrates plastics toughening and compatibilization using concentrated emulsion approaches. Toughening of brittle plastics can significantly broaden their applicability. Compatibilization of immiscible polymers also leads to polymer blends with stabilized morphologies that are required for common applications. Such modifications are achieved either by using polymerization in a single concentrated emulsion with dissolved polymeric modifiers, or by mixing two partially polymerized concentrated emulsions for continuous polymerization.

This book set will be of considerable value to material scientists in universities and industry, and to graduate students. In particular, this book set will be of interest to researchers working in emulsion-assisted polymer preparation, polymeric composites, conductive polymers, core-shell particles, encapsulation and controlled release, plastics toughening, and polymer compatibilization.

Authors

Eli Ruckenstein is a Distinguished Professor at the State University of New York (SUNY) at Buffalo. He has published more than 1,000 papers in numerous areas of engineering science and has received a large number of awards from the American Chemical Society and the American Institute of Chemical Engineers. Dr. Ruckenstein has also received the Founders Gold Medal Award from the National Academy of Engineering and the National Medal of Science from President Clinton. He is a member of the National Academy of Engineering and of the American Academy of Art and Sciences.

Hangquan Li received his PhD in polymer science and engineering from Beijing University of Chemical Technology in 1990. He was a visiting scientist at the State University of New York (SUNY) at Buffalo, working with Dr. Eli Ruckenstein from 1993 to 1996. He has been a Professor at Beijing University of Chemical Technology since 1996. He has published over 100 papers, mainly on polymer research.

Chong Cheng received a PhD in chemistry (polymer) from City University of New York in 2003. He currently is an Associate Professor at Department of Chemical and Biological Engineering at the State University of New York (SUNY) at Buffalo. He has published over 70 papers on polymer synthesis and characterization, as well as biomedical applications of polymers.

1 Conductive Polymer Composites

CONTENTS

INTRODUCTION

Since the discovery of conductive polyacetylene in the late 1970s, conductive polymers have become a hot topic in both science and engineering communities. Today, a broad variety of conductive polymers have been developed, showing promising properties in electrical conduction, electroluminescence, chemical sensoring, electro-magnetic shielding, antistatic coating, and corrosion inhibition. One of the most exciting applications of these novel materials is compact electronic devices such as polymer-based transistors, light-emitting diodes, lasers, flat flexible television screen, and acceptors in polymeric solar cells.

However, behind these brilliant scenes, there are countless ordinary and difficult explorations. It is well known that the disadvantages of intrinsically conductive polymers are brittleness and the lack of processability. To overcome these shortcomings, various technical approaches have been investigated. Among them, a crucial approach is the preparation of conductive polymer composites. Such composites are flexible and processable, and thus can overcome the two shortcomings simultaneously. In conductive polymer composites, the polymeric materials (plastics or elastomer) play a role as a glue to bind the particles of conductive polymers together. As a result, the conductivity of the composites must exhibit a percolation behavior. For this reason, the preparation technique of the conductive polymers and the composites plays an important role in the conductivity of the composites.

This chapter collects six papers on conductive polymeric composites by Ruckenstein and co-workers. Three general types of conductive polymers are employed, including polyaniline, polypyrrole, and polythiophene. All of the three are very attractive conductive polymers because they are relatively inexpensive, easy to synthesize, and can be readily chemically modified. They are electrical insulators when un-doped; however, they become electrical conductors when oxidized. Their conductivity strongly depends on the preparation techniques and the dopants, and can be increased by two or more orders of magnitude. Not surprising, they are among the most studied conductive polymers and find many applications as aforementioned.

The conductive composites were prepared via three approaches.

The first approach is based on an imbibition technique (Sections 1.1 and 1.2). A solution of monomer (pyrrole or bithiophene) is imbibed onto a porous substrate, followed by partial drying. Subsequently, the substrate is imbibed by a solution of an oxidant. Then the occurrence of polymerization of the monomer inside the host leads to a conductive composite. Although a variety of porous substrates are tested, the cross-linked porous polymers prepared by concentrated emulsion polymerization lead to the final composite with the best overall properties.

The second approach is based on polymerization in an emulsion (Sections 1.3 and 1.4). An oil-in-water emulsion with the dispersed oil phase containing poly(alkyl methacrylate) and a monomer (aniline or pyrrole) is prepared at first. Subsequently, an aqueous solution of an oxidant is introduced into the emulsion. The monomer present in the emulsion is polymerized and doped. The polyaniline or polypyrrole is deposited on the poly (alkyl methacrylate), to form the conductive composite.

In the third approach, the emulsion configuration is inverted for the preparation of conductive composite (Sections 1.5 and 1.6). A water-in-oil emulsion, with an oxidant ($FeCl_3$) dissolved in the dispersed water phase and an elastomer dissolved in the continuous organic phase, is prepared at first. Subsequently, a solution of a monomer (pyrrole or 3-methylthiophene) is introduced dropwise to induce polymerization and the produced polymer is deposited on the host elastomer to yield the conductive composite.

1.1 New Method for the Preparation of Thick Conducting Polymer Composites*

Eli Ruckenstein and Jun Seo Park
Department of Chemical Engineering, State University of New York at Buffalo, Amherst, New York, 14260

ABSTRACT A highly porous and absorbable crosslinked polystyrene, prepared by the concentrated emulsion polymerization method, was used as host polymer for the preparation of conducting, large objects, polymer composites. The composites, whose conductivity can be as high as 0.80 S/cm, were prepared by (1) imbibing the host polymer with a pyrrole (or oxidant) solution, (2) partially drying the imbibed host polymer, and (3) imbibing again with an oxidant (or pyrrole) solution for polymerization to take place. The electrical conductivity of the composite and the penetration of polypyrrole in the host polymer are influenced by the polymerization conditions (i.e., the concentrations of oxidant and pyrrole and the nature of the solvents used for the oxidant and pyrrole), the order in which the two imbibing solutions are introduced, and the drying time used after the first imbibation. The mechanical properties of the host polymer are improved with the incorporation of polypyrrole. Scanning electron micrographs of the composites indicate that the polypyrrole coats uniformly as a film inside of the porous host polymer.

1.1.1 INTRODUCTION

Since the discovery of electrical conducting polymers,[1,2] they have drawn considerable attention as possible substitutes for metallic conductors or semiconductors in a wide variety of electrical and electronic devices.[3–5] Potential advantages of conductive or semiconductive polymers lie in their light weight and in the versatility with which their synthesis and fabrication can be accomplished. However, most conducting polymers have at least one of the following undesirable

* *Journal of Applied Polymer Science*, Vol. 42, 925–934, (1991).

characteristics: (1) environmental instability, (2) poor processability, and (3) poor physical properties.[5] In the last few years, a considerable amount of investigation into conducting polyheterocyclic polymers has been carried out because of their good environmental stability. Among these polyheterocyclic polymers, polypyrrole (PP) has been extensively employed.[6,7] PP has been prepared either by the chemical oxidative polymerization method[6] or by the electrochemical oxidative polymerization method.[7] The electrochemical polymerization of pyrrole produces free-standing conducting films, whose conductivity at room temperature is as high as 10^2 S/cm.[8] On the other hand, the chemical polymerization method produces a finely divided, insoluble black powder, whose conductivity ranges from 10^{-15} S/cm to 10^1 S/cm, depending on the specific preparative conditions.[9] However, both the electrochemically prepared PP films and the chemically prepared PP powders are difficult to handle, and this restricts their potential for applications. A useful approach for the improvement of the mechanical properties of polymers has been the synthesis of hybrids, i.e., copolymers and polymer blends. A slight improvement in the mechanical properties has been, indeed, achieved by the copolymerization of pyrrole and styrene.[10] A better improvement in the mechanical properties can be obtained by formulating a conducting polymer blend in which the conducting polymer is well mixed with one or more conventional polymers. This polymer blending method involves the inclusion of electrochemically or chemically polymerized pyrrole in the matrix of a host insulating polymer. PP can be impregnated electrochemically into a host polymer that coats the electrode. Conducting polymer blends have been obtained by this electrochemical polymerization of pyrrole in a poly(vinyl chloride) (PVC),[11] poly(vinyl alcohol) (PVA),[12] or polyurethane (PU) matrix.[13] There are, however, limitations regarding the practical applications of the polymer blends prepared by the electrochemical method. Indeed, uniform thin films of the host polymers are necessary for this process and it is difficult to prepare them at a large scale. Although the chemical method usually produces a less conductive PP compared with the electrochemical method, it has the advantages of easier mass production and shorter reaction time. To improve the mechanical properties of PP obtained by the chemical method, conducting polymer blends were prepared by exposing an oxidant containing nonporous host polymer to pyrrole vapor.[14] Poly(methyl methacrylate) (PMMA), PVC, or PVA were employed as host polymers. Another type of conducting polymer composite was prepared by incorporating pyrrole into a nonporous polymeric matrix, followed by the exposure of the impregnated matrix to an oxidant.[15] However, the low penetration of PP into the matrix limits the utility of these procedures. The penetration of the PP in the host polymer matrix can be increased by employing a porous host polymer instead of a nonporous host polymer. A conducting polymer composite of PP–cellulose was prepared by impregnating thin porous filter papers with an aqueous oxidant solution and then contacting them with pyrrole as liquid or vapor.[16] However, in this case, the host polymer is thin and has poor mechanical properties. For the preparation of thick and large objects, the host polymer must be able to be appreciably imbibed by both the monomer and oxidant solutions, thus permitting PP to grow uniformly inside the matrix.

In this work we report a new approach to preparing thick conducting polymer composites with good mechanical properties. The concentrated emulsion polymerization method[17,18] is employed for the preparation of the host polymer. The host polymer is imbibed either with the monomer and subsequently with the oxidant or vice versa. The porous host polymer was synthesized starting from a concentrated emulsion of water dispersed in a hydrophobic continuous phase of a mixture of styrene and divinyl benzene. A concentrated emulsion has a large volume fraction of dispersed phase (about 0.8 in the present study) and the appearance of a gel. For sufficiently large volume fractions, it consists of polyhedral droplets separated by thin layers of the continuous phase.[18] This concentrated emulsion is stabilized by dissolving suitable surfactants in the continuous phase. The incorporation of PP into the porous host polymer has been achieved by the chemical oxidative polymerization method. After the well-dried porous host polymer was imbibed with a solution of pyrrole (or oxidant) and then partially dried, it was contacted with a solution of oxidant (or pyrrole). As a result, pyrrole polymerizes inside the host polymer. Both aqueous and nonaqueous solvents were employed for the oxidant, and nonaqueous solvents for pyrrole. $FeCl_3$ was used both as polymerization oxidant and dopant. The effect of the order in which the host polymer was imbibed with the solutions of monomer and oxidant, as well as the effect of the concentrations of oxidant and pyrrole on the conductivity of composite, was investigated. The relationship between the depth of penetration of the PP in the host matrix and the drying time after the first imbibation was also examined.

1.1.2 EXPERIMENTAL

1.1.2.1 MATERIALS

Styrene (Aldrich), divinyl benzene (Polysciences), and pyrrole (Aldrich) were distilled and stored in a refrigerator. Azobisisobutyronitrile (AIBN, Alfa) was purified by recrystallization in methanol. Sorbitane monooleate (Span80, Fluka), iodine (Baker Chem.), ferric chloride (Aldrich), and ferric chloride hexahydrate (Aldrich) were used as received. All solvents were of reagent grade and were used as received. Water was deionized and distilled.

1.1.2.2 PREPARATION OF THE HOST POLYMER (POROUS CROSSLINKED POLYSTYRENE)

A small amount of a mixture of styrene and divinyl benzene containing AIBN and sorbitane monooleate was placed in a flask (250-mL capacity) equipped with a mechanical stirrer and an addition funnel. A set of amounts involved is listed in Table 1.1.1.

TABLE 1.1.1
Representative Compositions in the Preparation of Porous Host Polymers

	Polymer 1	Polymer 2	Polymer 3
Continuous phase			
Styrene	5 g	5 g	5 g
Divinyl benzene	1 g	1 g	1 g
Initiator (AIBN)	0.05 g	0.05 g	0.05 g
Surfactant (sorbitane monooleate)	1 mL	1 mL	1 mL
Dispersed phase			
Water	20 mL	25 mL	30 mL

Water was placed in the addition funnel. The concentrated emulsion was prepared at ambient temperature by dropwise addition of water to the stirred mixture of styrene and divinyl benzene containing AIBN and sorbitane monooleate.[17] The polymerization was carried out in a temperature-controlled oven at 50°C for 24 h. Subsequently, the water of the dispersed phase was eliminated by keeping the polymer in the oven at 100°C for 3 days.

1.1.2.3 Preparation of the Conducting Polymer Composite

Two different procedures were used to incorporate PP in the porous host polymer. The experiments were conducted at ambient temperature.

1.1.2.3.1 The First Procedure

A well-dried porous host polymer was dipped in the oxidant solution until it saturated (about 15 min). Subsequently, the solution imbibed host polymer was partially dried by exposing it to air at room temperature. The partially dried host polymer was contacted with a large amount of pyrrole-organic solvent solution for 2 h. The latter solution penetrates the pores of the host polymer and polymerization takes place there. Because of some diffusion of the oxidant outside the host polymer, some polymerization also occurs outside. Before measuring the conductivity of the composite, it was dried in air for 1 day.

1.1.2.3.2 The Second Procedure

A well-dried host polymer was saturated with a pyrrole-organic solvent solution (about 15 min). The saturated host polymer was partially dried in air. The partially dried host polymer was contacted with a large amount of the oxidant solution for 2 h for polymerization to take place. The conductivity of the composite was measured after drying in air for 1 day.

1.1.2.4 The Absorption Test

To determine the maximum absorption capacity of the porous host polymer (0.3 × 2 × 3 cm), the host polymer was immersed in various solvents at room temperature for various time intervals. The increase in weight due to absorption was determined with a Mettler balance. The rate of desorption of the solvent from the host polymer in air with time after the first imbibation was also measured by using a Mettler balance.

1.1.2.5 Conductivity Measurements

Measurements were carried out by using the standard four-point probe method on thin sheets of composites (0.3 × 0.75 × 2.5 cm) at room temperature.

1.1.2.6 Electron Microscopy

The morphologies of the porous host polymer and conducting composite were examined by scanning electron microscopy (SEM, Amray 100A). A thin layer of gold was deposited on the cross section of the sample prior to observation.

1.1.2.7 Penetration Observation

The penetration of PP inside the composite was examined by employing a well-dried pellet of porous host polymer (diameter 1.6 cm, length 4 cm). After the formation of the composite, its cross section, obtained by cutting the sample, was subjected to observation.

1.1.3 RESULTS AND DISCUSSIONS

Three porous host polymers, whose compositions are listed in Table 1.1.1, were prepared by the concentrated emulsion polymerization method. The maximum amounts of various liquids absorbed by the porous polymer 2 are listed in Table 1.1.2. These data indicate that the host polymer is highly porous and can absorb amounts

TABLE 1.1.2
Absorption of Various Liquids in the Host Polymer[a]

	Water	Pyrrole	Acetonitrile	Methanol	Ether	Chloroform
Absorption (g liquid/g host polymer)	4.4	4.6	4.1	3.8	3.5	9.5

[a] Polymer 2 was used in these experiments.

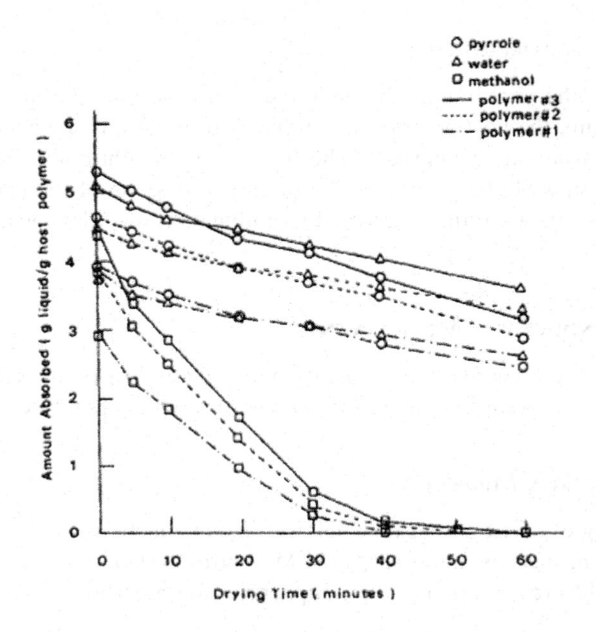

FIGURE 1.1.1 The amount of liquid against time during drying in air at ambient temperature.

of liquid greater than three times its own weight. The highest absorption is for chloroform, namely 9.5 g chloroform per gram host polymer. The desorption in air at room temperature of the absorbed liquid from its host polymer is plotted in Figure 1.1.1. As expected, the desorption rate of methanol is faster than those of water and pyrrole; most of the methanol was evaporated from the host polymer in 40 min. This figure also shows that absorption increases with increasing porosity of the polymer. The mechanical strength of the host polymer decreases, however, with increasing porosity.

The chemical oxidative polymerization method was employed to incorporate PP in the host polymer. SEM studies were carried out to investigate the morphologies of the host polymer and of the conducting composite. The host polymer, shown in Figure 1.1.2a, contains pores larger and smaller than 10 μm. According to previous studies,[17,18] the size of the dispersed phase is in the micron or submicron range. The larger pores observed here might form during drying at 100°C. These holes might constitute the tunnels through which the solutions of pyrrole and oxidant penetrate the host polymer. The PP forms in the host polymer, producing a conducting composite, which becomes stronger and harder than the host polymer. In the composite, shown in Figure 1.1.2b and c, the PP coats the inside surface area of the host polymer; in addition, some particles of PP are attached to the surface of the host polymer.

Table 1.1.3 lists the conductivity of the polymer composites, which were prepared by changing the solvent in the pyrrole solution. The value of the conductivity is not affected in a major way by the nature of the solvent and is of the order of 10^{-1} S/cm. The effect of the oxidant and its solvent on the conductivity of the composites is

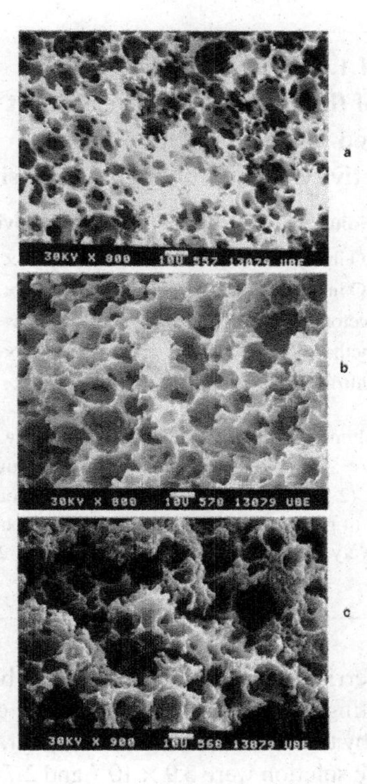

FIGURE 1.1.2 Scanning electron micrographs of the host polymer (a) and composites (b, c). The composites were prepared by the second procedure. The oxidant solution for composite (b) was $FeCl_3 \cdot 6H_2O$ (4 g) in water (10 mL) and for composite (c) $FeCl_3$ (2 g) in methanol (10 mL). The drying time of the host polymer imbibed with pyrrole-ether solution (4 g pyrrole in 10 mL ether) was 20 min.

TABLE 1.1.3
Effect of the Solvent Employed for Pyrrole on the Conductivity of Polymer Composite[a]

Solvent	Conductivity (S/cm)
Cyclohexane	2.1×10^{-1}
Chloroform	3.0×10^{-1}
Methanol	1.4×10^{-1}
Ether	3.1×10^{-1}

[a] The polymer composites were prepared by the first procedure. The drying time after imbibing with the oxidant solution (2 g $FeCl_3$/10 mL methanol) was 20 min. The concentration of pyrrole was 2 g/10 mL solvent.

TABLE 1.1.4
Effect of the Oxidant and the Solvent Employed for the Oxidant on the Conductivity of the Polymer Composite[a]

Oxidant Solution	Conductivity (S/cm)
$FeCl_3 \cdot 6H_2O$ in acetone (4 g/10 mL)	3.9×10^{-3}
$FeCl_3 \cdot 6H_2O$ in water (4 g/10 mL)	4.1×10^{-2}
$FeCl_3$ in acetonitrile (2 g/10 mL)	2.7×10^{-1}
$FeCl_3$ in methanol (2 g/10 mL)	6.5×10^{-1}
I_2 in acetonitrile (0.2 g/10 mL)	5.3×10^{-4}

[a] The polymer composites were prepared by the first procedure. The drying times for the nonaqueous oxidant solution (2 g $FeCl_3$/10 mL solvent) and aqueous oxidant solution (4 g $FeCl_3 \cdot 6H_2O$/10 mL H_2O) were 20 and 60 min, respectively. The concentration of pyrrole was 2 g/10 mL ether.

shown in Table 1.1.4. The ferric chloride, which is known to be a very effective oxidant,[9] generates a better conducting composite than iodine. The conductivities of the polymer composites prepared by using a ferric chloride hexahyrate–acetone solution and a ferric chloride–acetonitrile solution were 3.9×10^{-3} and 2.7×10^{-1} S/cm, respectively. These experimental results indicate that the nature of the solvent employed for the oxidant affects the conductivity of the composite. In the literature,[9,19,20] it was already noted that the reaction medium plays an important role. Many solvent-related factors, such as the solubility of $FeCl_3$, the solvent basicity, the dielectric constant of the solvent, and the Fe^{3+}–Fe^{2+} redox couple, are expected to affect the reaction of ferric chloride with pyrrole. The texture of the PP is, therefore, expected to depend on the reaction medium, which thus affects the conductivity of the composite. The porosity of the host polymer also plays a role, since it can determine whether the growth of the PP can such as to generate a conducting pathway.

As the chemical polymerization of the pyrrole proceeds, the host polymer darkens. When pyrrole (or oxidant) solution containing host polymer is imbibed with the oxidant (or pyrrole) solution, PP films grow on the outside surface, generating a layer that retards the diffusion of pyrrole (or oxidant) in the host polymer. For this reason the host polymer imbibed with the first solution is dried before being imbibed with the second solution. By controlling the drying time, one can influence the absorption of the second solution. Figure 1.1.3 shows a poor penetration of the PP in the composite for short drying times. The poor penetration is partly due to the poor miscibility between the pyrrole and oxidant solvents employed. Figures 1.1.4 and 1.1.5 show the effect of the penetration of PP on the conductivity of the composites, prepared as in Figure 1.1.3a and 3b, respectively. In both cases the conductivity is low for short drying times, due to the low penetration and nonuniform distribution of the PP in the composites. Figures 1.1.4 and 1.1.5 indicate that the amount of PP included in the

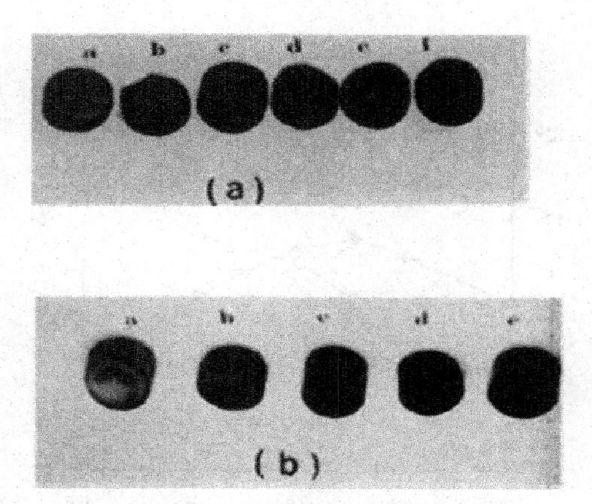

FIGURE 1.1.3 Penetration of PP in the polymer composites. Polymer composites (a) were prepared by the first procedure by changing the drying time of the host polymer imbibed with the oxidant solution. The concentrations of pyrrole and oxidant were 4 g pyrrole/10 mL ether and 4 g $FeCl_3 \cdot 6H_2O$/10 mL H_2O, respectively. The drying time was (a) 0 min, (b) 20 min, (c) 40 min, (d) 50 min, (e) 90 min, and (f) 150 min, respectively. Polymer composites (b) were prepared by the second procedure by changing the drying time of the host polymer imbibed with the pyrrole solution. The drying time was (a) 0 min, (b) 20 min, (c) 30 min, (d) 40 min, and (e) 60 min, respectively.

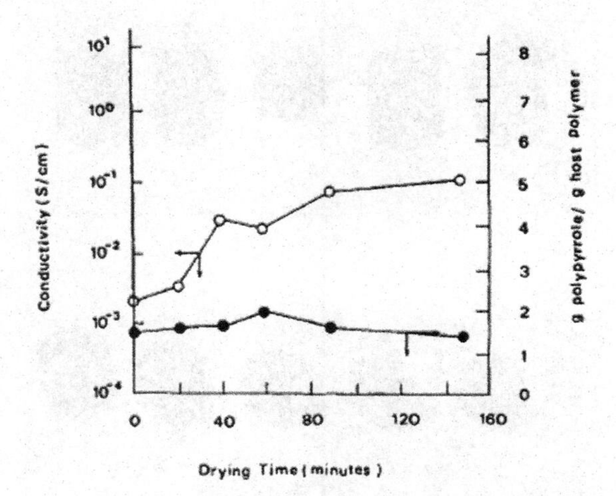

FIGURE 1.1.4 Conductivity of composite and weight ratio of PP to host polymer against drying time of the host polymer imbibed with the oxidant solution in air. The polymer composites were prepared by the first procedure with the compositions of Figure 1.1.3a.

composite is 1–2 g PP/gram host polymer for the first procedure and 3–5 g PP/gram host polymer for the second procedure.

Figure 1.1.6, in which instead of an aqueous oxidant solution a methanol solution is employed, shows a better penetration of PP in the composite, compared with that of Figure 1.1.3. This is due to the better miscibility of the oxidant solvent and pyrrole solution.

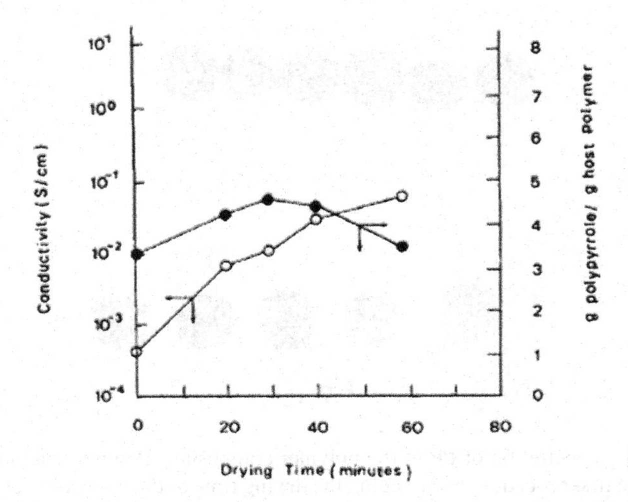

FIGURE 1.1.5 Conductivity of composites and weight ratio of PP to host polymer against drying time in air of pyrrole solution imbibed host polymer. The polymer composites were prepared by the second procedure with the compositions of Figure 1.1.3b.

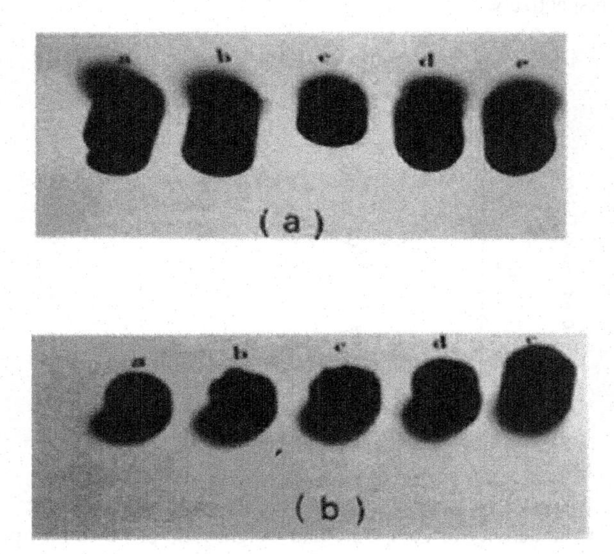

FIGURE 1.1.6 Penetration of PP in the polymer composites. Polymer composites (a) were prepared by the first procedure by changing the drying time in air of oxidant solution imbibed host polymer. The concentrations of oxidant and pyrrole were 2 g $FeCl_3$/10 mL methanol and 4 g pyrrole/10 mL ether, respectively. The drying time of the oxidant solution imbibed host polymer was (a) 0 min, (b) 5 min, (c) 10 min, (d) 20 min, and (e) 40 min, respectively. Polymer composites (b) were prepared by the second procedure by changing the drying time of pyrrole solution imbibed host polymers. The drying time was (a) 0 min, (b) 5 min, (c) 10 min, (d) 20 min, and (e) 30 min, respectively.

Comparing Figure 1.1.6a and b, one can conclude that the composite prepared by the first procedure is better penetrated by PP for short drying times than the composite prepared by the second procedure. The conductivities of the composites, prepared as in Figure 1.1.6a and 6b, are plotted against drying time in Figures 1.1.7 and 1.1.8. The conductivity of the composites in Figure 1.1.7 is by an order of magnitude higher than that of the composites in Figure 1.1.8. Both kinds of composites incorporate, however, comparable amounts of PP in the host polymer. Consequently, the solvent employed and the order in which the two solutions imbibe the host polymer affect both the penetration of PP in the composites and the conductivity of the composite. The order of contacting also

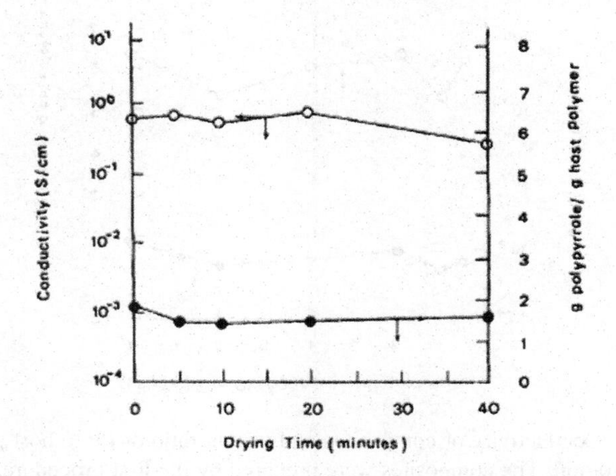

FIGURE 1.1.7 Conductivity of composites and weight ratio of PP to host polymer against drying time in air of oxidant solution imbibed host polymer. The polymer composites were prepared by the first procedure with the composition of Figure 1.1.6a.

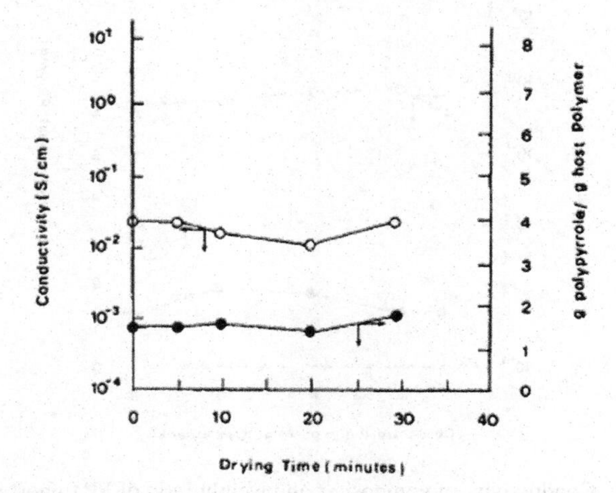

FIGURE 1.1.8 Conductivity of composites and weight ratio of PP to host polymer against drying time in air of pyrrole solution imbibed host polymer. The polymer composites were prepared by the second procedure with the composition of Figure 1.1.6b.

affects, in some cases, the amount of PP incorporated in the composite. However, the amounts of PP incorporated in the composites do not affect their conductivity.

The effect of the concentrations of oxidant and pyrrole on conductivity for both the first and second procedures and nonaqueous solvents is worth investigating. The results for the first procedure are presented in Figures 1.1.9 and 1.1.10. In Figure 1.1.9, the conductivity is about 10^{-1} S/cm and is unaffected by the concentration of the oxidant. The oxidant concentration does not affect the amount of PP formed in the

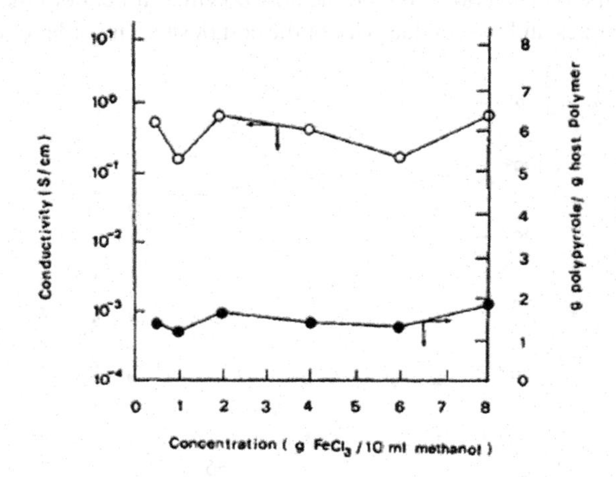

FIGURE 1.1.9 Conductivity of composites and weight ratio of PP to host polymer against oxidant concentration. The composites were prepared by the first procedure. The host polymer imbibed with oxidant solution ($FeCl_3$ in methanol) was dried for 20 min in air, followed by imbibation with pyrrole solution (4 g pyrrole/10 mL ether).

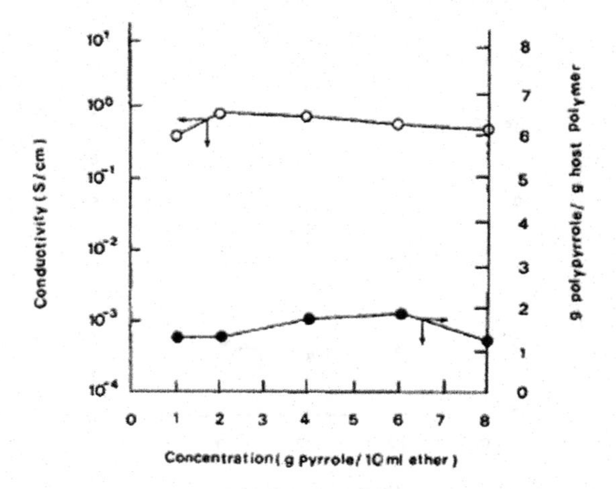

FIGURE 1.1.10 Conductivity of composites and weight ratio of PP to host polymer against pyrrole concentration. The composites were prepared by the first procedure. The saturated host polymer with oxidant solution (2 g $FeCl_3$/10 mL methanol) was dried for 20 min in air, followed by imbibation in the pyrrole solution (pyrrole in 10 mL ether).

composite either. Figure 1.1.10 shows that the monomer concentration does not affect the conductivity and the amount of PP included in the composite. Figures 1.1.11 and 1.1.12 present the conductivities of composites prepared by the second procedure. They are by an order of magnitude smaller than those obtained by the first procedure and are low at low concentrations of oxidant and pyrrole. The amount of PP included in the composite increases with the concentration of oxidant (Figure 1.1.11). However, no relationship between the amount of PP in the composite and conductivity

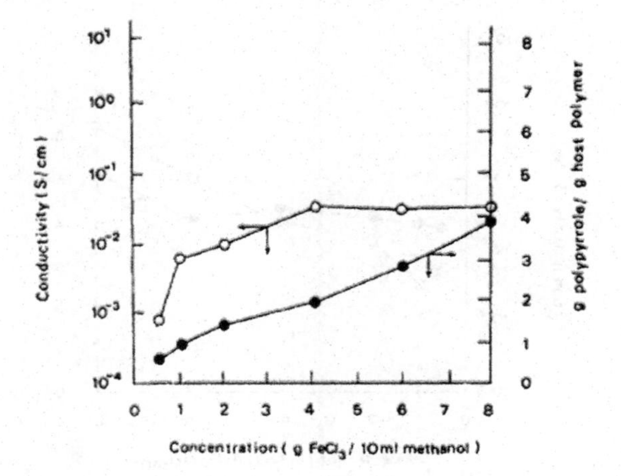

FIGURE 1.1.11 Conductivities of composites and weight ratio of PP to host polymer against concentration. The composites were prepared by the second procedure. The host polymer imbibed with pyrrole solution (2 g pyrrole/10 mL ether) was dried for 20 min in air and then imbibed with oxidant solution (2 g $FeCl_3$/10 mL methanol).

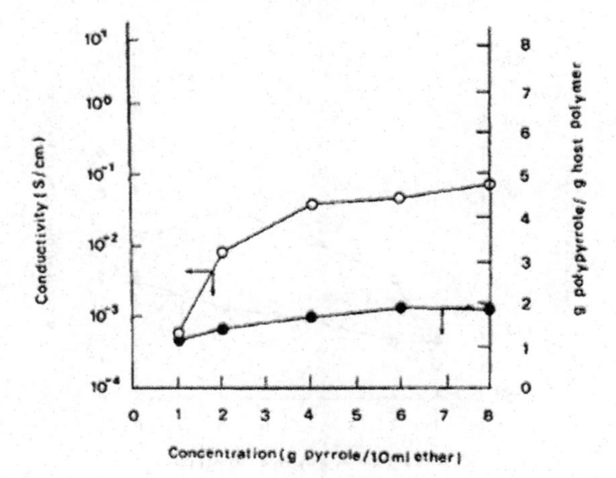

FIGURE 1.1.12 Conductivity of composites and weight ratio of PP to host polymer against pyrrole concentration. The composites were prepared by the second procedure. The host polymer imbibed with pyrrole solution (pyrrole in ether) was dried for 20 min in air and then imbibed in oxidant solution (2 g $FeCl_3$/10 mL methanol).

was found. The lower conductivities of the composites prepared by the second procedure might be due to the lower penetration and nonuniform distribution of PP in the composite, detected in the penetration experiments (Figure 1.1.6).

Figures 1.1.13 and 1.1.14 present the effect of the concentrations of oxidant and pyrrole on the conductivities of composite when an aqueous medium is employed for the oxidant. At low concentrations of oxidant and pyrrole, the conductivity has low values. In Figure 1.1.13 the conductivity increases with increasing oxidant

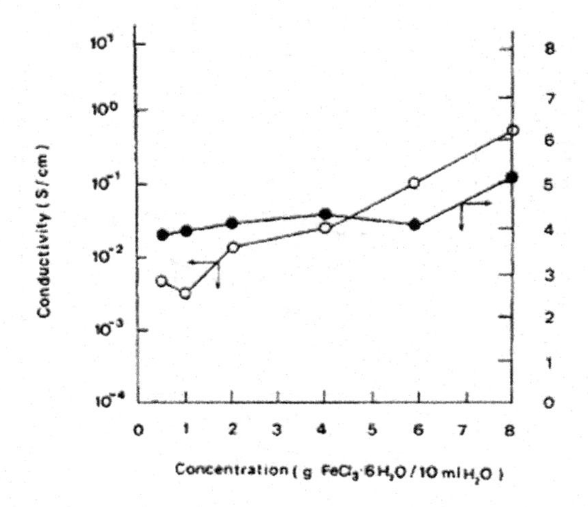

FIGURE 1.1.13 Conductivity of composites and weight ratio of PP to host polymer against oxidant concentration. The composites were prepared by the second procedure. The host polymer imbibed with pyrrole solution (4 g pyrrole/10 mL ether) was dried for 40 min in air and then imbibed with oxidant solution ($FeCl_3·6H_2O$ in 10 mL H_2O).

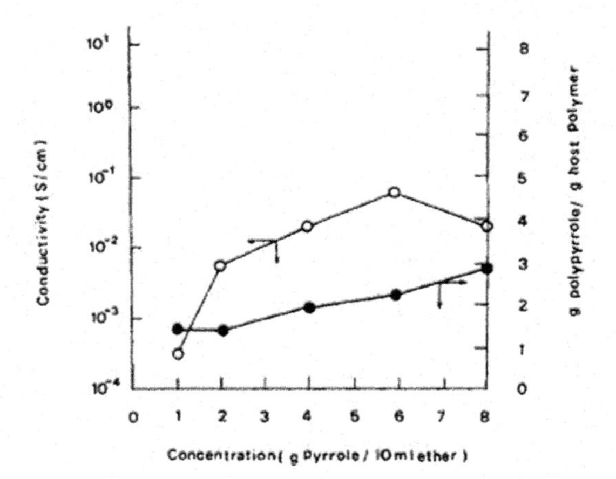

FIGURE 1.1.14 Conductivity of composites and weight ratio of PP to host polymer against pyrrole concentration. The composites were prepared by the first procedure. The host polymer imbibed with oxidant solution (2 g $FeCl_3·6H_2O$/10 mL H_2O) was dried for 50 min in air and then imbibed with pyrrole solution (pyrrole in 10 mL ether).

TABLE 1.1.5
Effect of the Porosity of the Host Polymer on the Conductivity of Polymer Composites[a]

Host Polymer	Conductivity (S/cm)	g Oxidant Solution/g Polymer	g PP/g Polymer
1	1.3×10^{-1}	0.94	1.61
2	6.5×10^{-1}	1.22	1.70
3	5.5×10^{-1}	1.42	1.77

[a] The polymer composites were prepared by the first procedure. The concentrations of oxidant and pyrrole were 2 g $FeCl_3$/10 mL methanol and 4 g pyrrole/10 mL ether, respectively. The drying time of the host polymer imbibed with the oxidant solution was 20 min.

concentration, without appreciable changes in the amount of PP. In Figure 1.1.14 the conductivity does not vary with the increase in the concentration of pyrrole.

As shown in Table 1.1.5, the amount of oxidant in the host polymer increases with increasing porosity of the host polymer. For the range of porosities investigated, there is, however, no major change of the conductivity with porosity.

1.1.4 CONCLUSION

A new method for preparing thick, conducting polymer composites is suggested. In the first step a highly porous host polymer is prepared. The concentrated emulsion method is employed to prepare a porous crosslinked polystyrene host polymer. In the second step the host polymer is imbibed with a solution of pyrrole or a solution of oxidant ($FeCl_3$). In order to facilitate the subsequent uniform distribution of a second solution, the host polymer thus imbibed is subjected to drying. After drying, the second solution (solution of oxidant or pyrrole) is allowed to imbibe the host polymer. As a result, pyrrole polymerizes to polypyrrole, with the latter uniformly distributed inside the host polymer. Morphology studies with SEM indicate that the polypyrrole grows on the inner surface of the host polymer in the form of polypyrrole films. Some polypyrrole particles are also attached to the surface of the host polymer. A higher penetration and more uniform distribution of polypyrrole in the composite are obtained by employing nonaqueous solvents (instead of an aqueous solvent) for the oxidant. Imbibing first with the oxidant solution and later with the pyrrole solution, higher conductivities have been obtained.

We believe that the present method can be employed to fabricate large and lightweight conducting material, which can be used for electromagnetic shielding and electrostatic charge protection.

The authors are indebted to Dr. M. J. Naughton and Ms. O. H. Chung, Department of Physics, SUNY at Buffalo, for the use of their facilities for the resistivities measurements.

REFERENCES

1. A. Dall'Olio, G. Dascola, V. Varacca, and V. Bocchi, *Compt. Rend.*, **433**, 2670 (1968).
2. C. K. Ching, C. R. Fincher, Jr., Y. W. Park, and A. J. Heeger, *Phys. Rev. Lett.*, **39**, 1098 (1977).
3. H. G. De Young, *High Tech.*, **3**(1), 65 (1985).
4. G. P. Kittlesen, H. S. White, and M. S. Wrighton, *J. Am. Chem. Soc.*, **106**, 7389 (1984).
5. J. E. Frommer and R. R. Chance, "Electrically conductive polymers," in *Encyclopedia of Polymer Science and Engineering,* Wiley, New York, 1985, Vol. 5, p. 462.
6. (a) F. Hautiere-Cristofini, D. Kuffer, and L. T. Yu, C. *R. Acad. Sci. Ser. C.*, **277**, 1323 (1973); (b) K. C. Khulbe and R. S. Mam, *J. Polym. Sci., Polym. Chem. Ed.*, **20**, 1089 (1982); (c) M. M. Castillo-Ortega, M. B. Inoue, and M. Inoue, *Synth. Met.*, **28**, C65 (1989).
7. (a) A. F. Diaz, K. Keiji Kanazawa, and G. P. Gardini, *J. Chem. Soc. Chem. Comm.*, 635 (1979); (b) R. J. Waltman and J. Bargon, *Can. J. Chem.*, **64**, 76 (1986); (c) G. B. Street, T. C. Clarke, M. Krounbi, K. K. Kanazawa, V. Y. Lee, P. Pfluger, J. C. Scott, and G. Weiser, *Mol. Cryst. Liq. Cryst.*, **83**, 253 (1982).
8. K. K. Kanazawa, A. F. Diaz, R. H. Geiss, W. D. Gill, J. F. Kwak, J. A. Logan, J. F. Rabolt, and G. B. Street, *J. Chem. Soc. Chem. Comm.*, 854 (1979).
9. R. E. Meyers, *J. Electron. Mater.*, **2**, 61 (1986).
10. A. Nazzal and G. B. Street, *J. Chem. Soc. Chem. Comm.*, 375 (1985).
11. M. A. De Paoli, R. J. Waltman, A. F. Diaz, and J. Bargon, *J. Polym. Sci., Polym. Chem. Ed.*, **23**, 1687 (1985).
12. S. E. Lindsey and G. B. Street, *Synth. Met.*, **10**, 67 (1984).
13. M. Bi and Q. Pei, *Synth. Met.*, **22**, 145 (1987).
14. R. Yosomiya, *Macromol. Chem., Rapid Comm.*, **7**, 697 (1986).
15. W. P. Roberts and A. Schutz, U.S. Pat. 4,604,427 (1986).
16. R. B. Bjorklund and I. Lundstrom, *J. Electron. Mater.*, **13**, 211 (1984).
17. E. Ruckenstein and K. J. Kim, *J. Appl. Polym. Sci.*, **36**, 907 (1988).
18. E. Ruckenstein and J. S. Park, *J. Polym. Sci., Polym. Lett. Ed.*, **26**, 529 (1988).
19. A. F. Diaz and B. Hall, *IBM J. Res. Develop.*, **27**, 342 (1983).
20. J. R. Raynolds, J. C. W. Chien, F. E. Karasz, C. P. Lillya, and D. J. Curran, *J. Chem. Soc. Chem. Comm.*, 1358 (1982).

1.2 Conducting Polyheterocycle Composites Based on Porous Hosts*

Jun Seo Park and Eli Ruckenstein

Department of Chemical Engineering, State University of New York at Buffalo, Amherst, New York, 14260

ABSTRACT Conducting composites based on porous substrates (cotton fiber, non-woven polypropylene mat and porous crosslinked polystyrene) have been prepared by a two-step imbibition technique. First, the substrate was imbibed with a solution of monomer (pyrrole or bithiophene) in acetonitrile, followed by partial drying. Subsequently, the substrate was again imbibed, this time with an oxidant dissolved in a suitable solvent. The polymerization of the monomer inside the host in the presence of the oxidant and the doping of the polymer with the oxidant leads to the conducting composite. The highly hydrophobic and porous crosslinked polystyrene, prepared by the concentrated emulsion polymerization method, is the most efficient. The solvent employed for the oxidant plays a major role. A $FeCl_3$-methanol system and porous crosslinked polystyrene lead to conductivities of polythiophene- and polypyrrole-based composites of 3.63 and 0.65 S/cm, respectively. Copper perchlorate and iron perchlorate are also suitable oxidants. The environmental and thermal stabilities of polypyrrole-based composites are lower than those of polythiophene-based composites. The thermal stability of polypyrrole-based composites can be enhanced by including a small amount of an organic antioxidant, such as amides or substituted phenols, in the composites.

1.2.1 INTRODUCTION

The synthesis of polyheterocycles, such as polythiophene (PTP) and polypyrrole (PPY), has received a great deal of attention for the past decades due to their reasonably high conductivity and environmental stability.[1–3] The conductivity of polyheterocycles has reached values as high as 10^2 S/cm without a noticeable decrease

* *Journal of Electronic Materials*, Vol. 21, 205–215, (1992).

over several months of standing in atmosphere. These characteristics might lead to important applications in a wide variety of electrical and electronic devices.[4,5]

One of the inherent difficulties with the intrinsically conducting polyheterocycles is the inability to process them into useful large articles.[1–5] A number of methods have been suggested to overcome this drawback by combining strong insulating materials with conducting polyheterocycles.[6–12] First, such conducting composites have been prepared by incorporating an electrochemically or chemically synthesized polyheterocycle in a non-porous polymer matrix.[6–8] The inclusion was carried out either by exposing an insulating sheet imbibed with an oxidant to the monomer or its vapor, or by electrochemical oxidation of a monomer-swollen matrix coated as a film on an electrode. However, the low penetration of the monomer into the non-porous matrix, which can be attributed to the low diffusivity of the monomer into the matrix, can produce only a very thin film of conducting polymer composite. Subsequently, conducting polymer composites have been prepared by first synthesizing a soluble conducting polymer by incorporating appropriate side chains into the monomer and then mixing this polymer with an insulating polymer by solution casting or melt-processing techniques.[9,10] The uniform mixing of the conducting polymer with the insulating polymer at the molecular level and/or the high loading of conducting polymer to reach the percolation threshold are essential in the preparation of highly conducting polymer composites. Thin porous materials were also employed as host substrates for composites in order to facilitate the penetration of polyheterocycles in the matrix.[11,12] Conducting composites based on filter paper and fabric were prepared by impregnation with an oxidant solution and subsequently by contacting them with the monomer.

In this laboratory thick and large conducting composites, based on polyheterocycles in combination with a porous matrix, have been prepared via chemical oxidative polymerization.[13,14] The highly porous crosslinked polystyrene, which was employed as host, was prepared by the concentrated emulsion polymerization method.[15] It was synthesized by heating at 50°C a concentrated emulsion of water dispersed in a mixture of styrene and divinylbenzene containing an initiator and a dispersant. In the concentrated emulsion the volume fraction of the dispersed phase is large, as large as 0.99, and the continuous phase is in the form of a network of thin liquid films that separate the polyhedral cells of the continuous phase. The concentrated emulsion has the appearance of a gel. The stability of the gel-like concentrated emulsion is ensured by the dispersant adsorbed upon the interface between the two phases as an oriented interfacial film.[16] A porous crosslinked polystyrene was obtained after the removal of water from the polymer by heating at 100°C. Conducting polyheterocycles were deposited on the inside surface of the porous host by first saturating the host with a monomer solution, followed by partial drying, and then by imbibing the partially dried host with an oxidant solution for chemical oxidative reaction to occur. PPY- or PTP-based conducting composites were prepared by employing this imbibition technique.[13,14] The conductivity of these composites was investigated by changing the initial compositions of the reactants, the nature of the oxidant and the imbibition order.[13,14]

The scope of this article is to study the effect of the porous substrate, reaction medium and antioxidants on the conductivity of the composite in order to improve its electrical properties as well as its long-term environmental stability. As porous hosts, cotton fiber, non-woven polypropylene mat and porous crosslinked polystyrene are employed. Ferric and cupric salts are used as oxidants for the chemical oxidative reaction, because they are known to provide the best performance.[13,14,17] It is known that solvent-related factors for the ferric salt, such as solubility, solvent basicity, the dielectric constant and the Fe^{+3}/Fe^{+2} redox couple affect the morphology of the polyheterocycles deposited on the inner surface of the substrate and the conductivity of the composite.[11,13,14,17] The effect of various solvents for the oxidant, which acts both as initiator and dopant, is examined. One reason for the lack of significant technical applications of intrinsically conducting polymers is their rapid aging. Positive effects on the stability of conducting polymers to aging were obtained when oxidation was carried out in the presence of some selected antioxidants.[18] Such antioxidants are therefore examined in order to improve the stability of conducting composites.

1.2.2 EXPERIMENT

1.2.2.1 Chemicals

Styrene (Aldrich), pyrrole (Aldrich) and divinylbenzene (Polysciences) were purified before use. 2,2-bithiophene (BT, Aldrich), azobisisobutyronitrile (AIBN, Alfa), 3-hydroxybenzoic acid (Alrich), 3-nitrophenol (Aldrich), phthalic acid (Aldrich), acetamide (Aldrich), pyrogallol (Aldrich), sorbitan monooleate (Span 80, Fluka), acetonitrile (Aldrich), acetone (Aldrich), methanol (Aldrich), 1-propanol (Aldrich), ether (Aldrich), ferric chloride (Aldrich), copper perchlorate hexahydrate (Aldrich) and iron perchlorate hydrate (Aldrich) were used as-received. Water was deionized and distilled.

1.2.2.2 Porous Host Substrates

Cotton fiber and non-woven polypropylene (PP, Aldrich) were used as-received. Porous crosslinked polystyrene (PS) was prepared by polymerizing the concentrated emulsion as reported in previous papers.[13–15] A concentrated emulsion of water (25 mL) in a mixture of styrene (5 g) and divinylbenzene (1 g) containing AIBN (0.05 g) as initiator and Span 80 (1 mL) as dispersant was prepared at room temperature by dropwise addition of water under vigorous stirring to a small amount of the mixture. This gel-like, concentrated emulsion was additionally packed by mild centrifugation to remove the air bubbles trapped in the emulsion. The host porous polymer was prepared by polymerizing the concentrated emulsion sandwiched between two clean glass plates (100 × 150 cm) at 50°C for 1 day, and by drying the material thus obtained at 100°C for 3 days.

1.2.2.3 Preparation of the Conducting Composite

A well-dried host was first saturated for 10 min with a solution of the monomer (pyrrole or bithiophene) in acetonitrile and then partially dried, in order to facilitate the penetration of the oxidant solution, by exposure to air for 5 min for the fiber and 30 min for the non-woven PP mat and the crosslinked PS. The durations of the partial drying times have been selected on the basis of the experiments on drying presented later in the paper. The partially dried substrates were imbibed again, by immersing them into an excess amount of oxidant solution for 10 min. The materials thus obtained were dried in air at room temperature for 6 hr and at 50°C for 1 day before measuring the conductivity.

1.2.2.4 Absorption Test

The absorption capacity of the substrates (fiber, 0.08 (diameter) × 10 cm; PP mat and crosslinked PS, 0.15 × 2 × 3 cm) was determined by immersing them in various solvents at room temperature for 20 min. The rate of desorption of the solvent from the substrates to air was measured at room temperature. A Mettler balance was used to measure the weight changes of the samples.

1.2.2.5 Instrumentation

The standard four-point probe method was employed to measure the conductivity of the specimens (0.3 × 1.2 × 3 cm). The morphologies of the fractured porous substrates and conducting composites were investigated by scanning electron microscopy (SEM, Hitachi S-450). A thin layer of gold was deposited on the surface of the samples before investigation.

1.2.3 RESULTS AND DISCUSSION

1.2.3.1 Absorption of Liquids by the Porous Hosts and Their Evaporation

The amounts of various liquids absorbed by the hosts are listed in Table 1.2.1, which shows that the crosslinked PS is the most hydrophobic among the three hosts and can absorb a relatively large amount of toluene. The cotton fiber and the non-woven PP each absorb comparable amounts of water and organic solvents (including toluene).

The amount of absorbed liquid that remains in the host during exposure to air, plotted in Figure 1.2.1, was measured in order to evaluate the time of partial drying. It shows that the liquid absorbed by the fiber was evaporated in a few minutes and

TABLE 1.2.1
Absorption of Various Liquids by the Host Substrates. The Host Was Immersed in the Liquid for 20 Min at Room Temperature

Absorption (g liquid/ g host substance)	Water	Ethanol	Acetonitrile	Toluene
Cotton fiber	2.00	1.83	2.01	1.83
Non-woven PP	2.68	2.15	2.38	2.62
Crosslinked PS	1.88	2.94	2.56	5.06

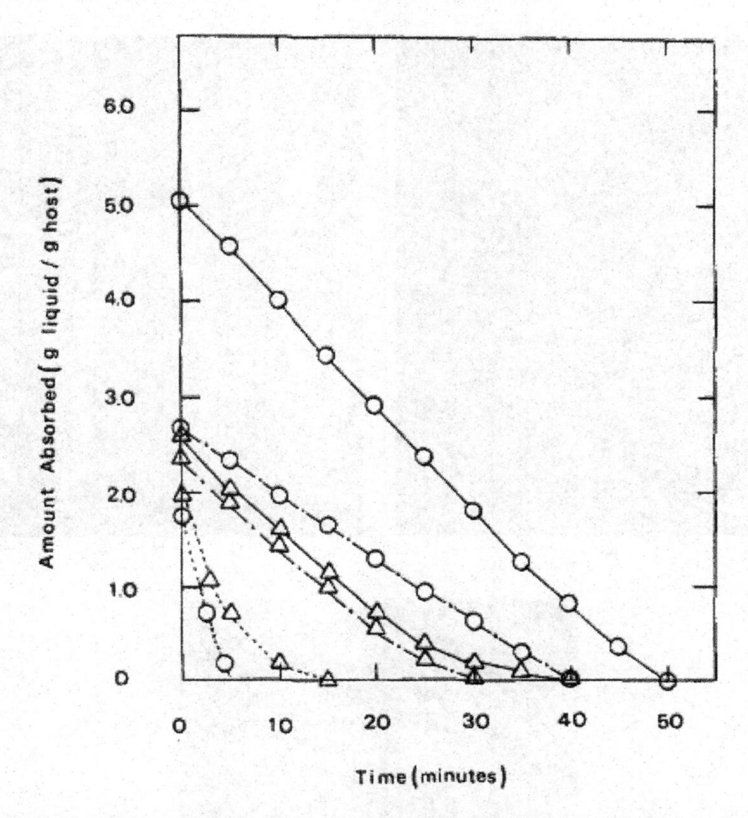

FIGURE 1.2.1 Amount of liquid against time during drying in air at ambient temperature. Δ and O refer to acetonitrile and toluene, respectively. -----, —.— and — denote cotton fiber, non-woven PP and crosslinked PS, respectively.

that the evaporation of liquids from the non-woven PP mat lasted longer than that from the fiber. The time in which crosslinked PS liberated the liquid was the longest because it has a large internal area and absorbs a greater amount of liquid.

1.2.3.2 Electron Microscopy Study

The SEM pictures (Figures 1.2.2 through 1.2.4) present the morphologies of the various hosts and composites. The morphologies of the cotton fiber (Figure 1.2.2) and non-woven PP mat (Figure 1.2.3), which are composed of many thin fibers, can be contrasted with that of the crosslinked PS. The crosslinked PS (Figure 1.2.4) contains uniform poles of about 10 μm, which were formed during the drying of the polymerized concentrated emulsion. The similarity between substrates and composites suggests that the conducting polymer coats with some uniformity the inner surface of the substrates. A close examination of the SEM pictures with higher magnification (Figure 1.2.5) reveals that the conducting polymer covers

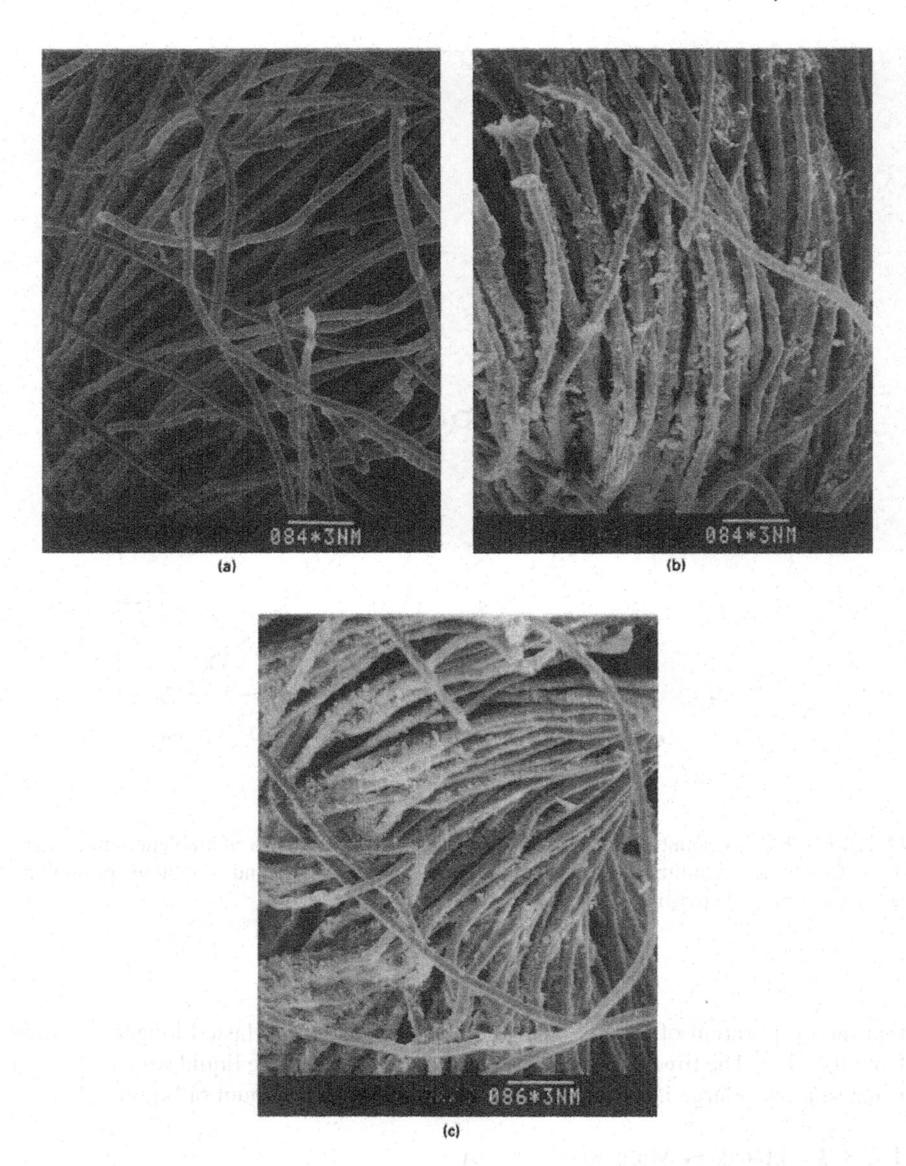

FIGURE 1.2.2 Scanning electron micrographs of (a) host cotton fiber, (b) polypyrrole-based conducting fiber and (c) polythiophene-based conducting fiber. The conducting composites were prepared by employing 2.98 mol/L of pyrrole and 1.23 mol/L of BT in acetonitrile as the monomer solution and 1.23 mol/L of $FeCl_3$ in methanol as the oxidant solution. The drying time for the monomer solution saturated host fiber was 5 min.

FIGURE 1.2.3 Scanning electron micrographs of (a) host non-woven PP, (b) polypyrrole-based composite and (c) polythiophene-based composite. The conducting composites were prepared by employing 2.98 mol/L of pyrrole and 1.23 mol/L of BT in acetonitrile as the monomer solution and 1.23 mol/L of $FeCl_3$ in methanol as the oxidant solution. The drying time for a monomer solution saturated host non-woven mat was 30 min.

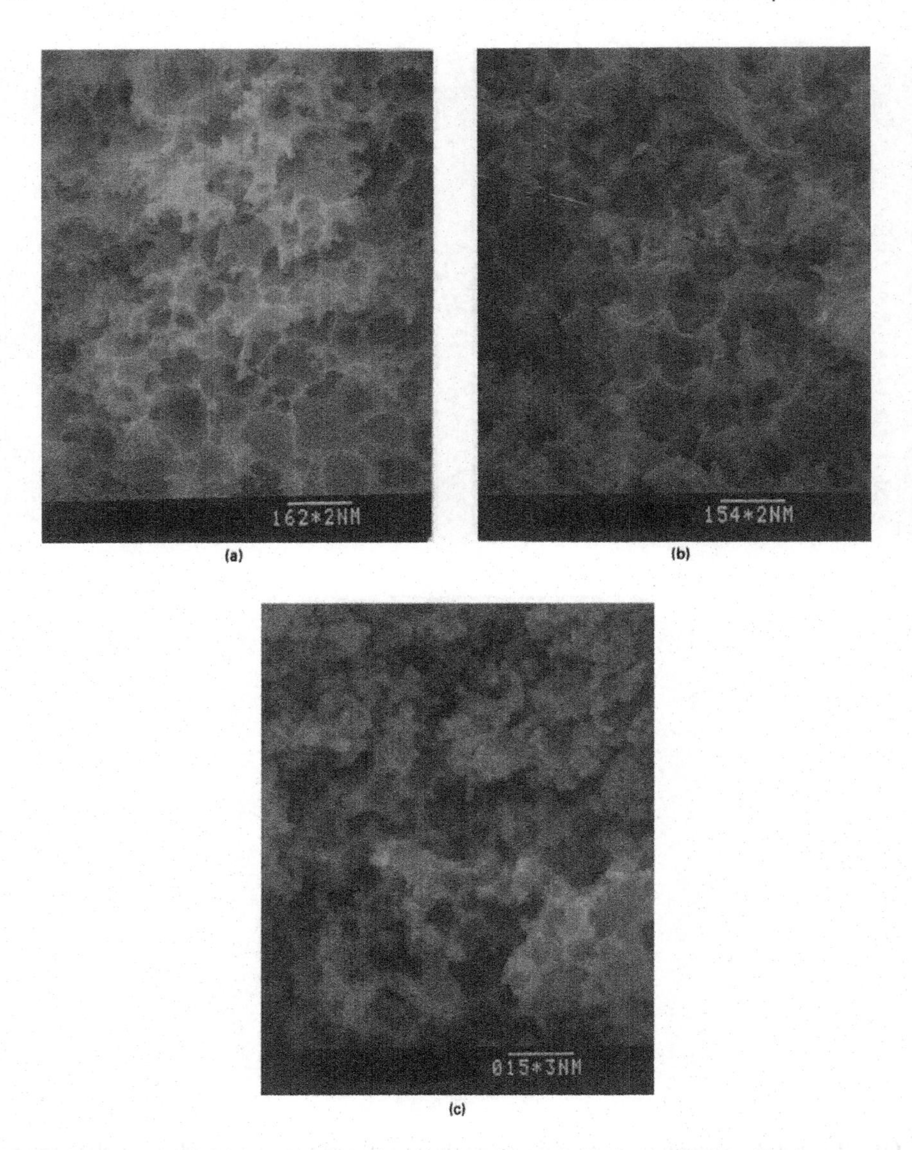

FIGURE 1.2.4 Scanning electron micrographs of (a) host crosslinked PS, (b) polypyrrole-based composite and (c) polythiophene-based composite. The conducting composites were prepared by employing 2.98 mol/L of pyrrole and 1.23 mol/L of BT in acetonitrile as the monomer solution and 1.23 mol/L of $FeCl_3$ in methanol as the oxidant solution. The drying time for a monomer saturated host crosslinked PS was 30 min.

almost uniformly the fibers of the non-woven PP mat. The substrate in the composite has a double role. It provides a macroscopic structure in which the conducting polymer generates a network of conducting thin films; it also provides mechanical strength as a reinforcing agent.

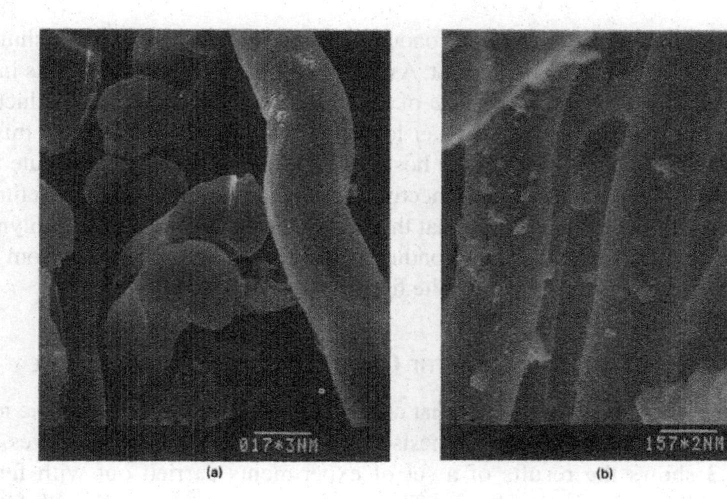

FIGURE 1.2.5 Scanning electron micrographs of (a) host non-woven PP and (b) polythiophene-based composite, (a) and (b) are higher magnifications of (a) and (c) of Figure 1.2.3, respectively.

1.2.3.3 Conductivity of the Polymer Composites

Table 1.2.2 lists the conductivities of the composites and the amount of conducting polymer loaded in the composites for the three different kinds of porous substrates employed. The initial concentrations of reactants employed in these experiments were selected (on the basis of the results reported in previous papers[13,14]) to be large enough to achieve the highest conductivity of the composites. Only above an initial concentration of monomer, corresponding to a percolation threshold, is a network generated in the host. A high conductivity is, however, achieved only when a sufficiently high ratio of oxidant to monomer concentration is also employed. The oxidant acts both as initiator and dopant. With the exception of the fiber, the polythiophene-based composites had higher conductivity than the polypyrrole-based composites. Somewhat higher conducting composites have been obtained with the crosslinked PS than with

TABLE 1.2.2
Conductivity of Composites Based on Different Porous Substrates. The Composites Were Prepared by Employing 1.23 mol/L of BT and 2.98 mol/L of Pyrrole in Acetonitrile as the Monomer Solution and 1.23 mol/L of $FeCl_3$ in Methanol as the Oxidant Solution

Host	Cotton Fiber		Non-woven PP Mat		Cross-linked PS	
Cond. Polym.	**PPY**	**PTP**	**PPY**	**PTP**	**PPY**	**PTP**
Conductivity (S/cm)	0.30	0.22	0.13	0.30	0.65	3.63
g cond. polymer per g host	1.01	0.72	0.65	0.86	0.89	1.41

the non-woven PP mat. The amount of conducting polymer loaded in the crosslinked PS-based composites was also the largest. As expected, the experimental results indicate that the morphology and the nature of the substrate affect both the conductivity and the amount of conducting polymer loaded in the composites. Perhaps this is a result of the interactions between the host and the monomer which facilitate the spreading of the latter and hence the connectivity among the formed conducting films. We have observed for the cotton fiber that the attachment of the conducting polymer to the fiber is poor. For this reason, the conducting polymer comes out easily from the fiber and the conductivity of the composite fiber decays in time.

1.2.3.4 Effect of the Solvent on the Conductivity of the Composite

In previous experiments,[13,14] we noted that aqueous solutions of ferric or cupric ions are suitable oxidant systems in the synthesis of PPT- and PPY-based composites.

Table 1.2.3 shows the results of a set of experiments carried out with ferric chloride as oxidant in various solvents. The rate of reaction of pyrrole with ferric chloride depends on the kind of solvent employed for the oxidant. The reaction of pyrrole with ferric chloride in acetone is rapid (since the color of the host changes to black rapidly). It is also associated with a high exothermic heat effect which reduces the conductivity of the composite. The lower conductivity of the composite obtained at high temperatures might be partially attributed to the lower degree of conjugation in the conducting polymer backbone.[19] The highest conductivities for PPY were obtained with alcohol or water as solvents for the oxidant. It is interesting to note that an acetonitrile-water mixture used as solvent for the oxidant increases the conductivity of the composites in comparison to that provided by acetonitrile alone. The reaction of pyrrole with the oxidant is rapid and takes place during the immersion in the oxidant solution. In contrast, the reaction of BT with the oxidant is slower and occurs particularly during the drying step at room temperature that follows the immersion in the oxidant solution. The reaction of BT with $FeCl_3$-acetone system started during immersion and continued during the drying process (as revealed by the slow change in the color of the host), and produced a low conducting composite. The oxidation reaction of BT with a $FeCl_3$-water system was very slow and occurred mostly during drying, and the conductivity was 2.84 S/cm. Composites with conductivities of 3.63 and 3.39 S/cm were obtained by employing methanol or ether as solvents for the oxidant; the reaction occurred slowly, during drying.

Copper perchlorate was reported to be a strong oxidizing agent in the one-step chemical synthesis of polythiophene.[20,21] Experimental results for conducting composites prepared with copper perchlorate or iron perchlorate as oxidants are listed in Table 1.2.4. The rate of reaction of pyrrole with copper perchlorate was strongly affected by the nature of the solvent employed. In the chemical oxidative reaction of pyrrole with copper perchlorate, only acetonitrile and acetonitrile-water mixtures were effective solvents for the oxidant, the reactions being fast and the conductivities high. In the chemical oxidative reaction of BT with copper perchlorate in various solvents, the oxidative reaction proceeded only in acetonitrile and in mixtures of acetonitrile and water. The conductivity of the polythiophene-based composite

TABLE 1.2.3

Effect of the Solvent Employed for the Oxidant on the Conductivity of the Composite. The Conducting Composites Were Prepared by Employing 1.23 mol/L of BT and 2.98 mol/L of Pyrrole in Acetonitrile as the Monomer Solution and 1.23 mol/L of $FeCl_3$ in Various Solvents as the Oxidant Solution. The Porous Crosslinked PS Is Used as Host

		Solvent for Oxidant						
Cond. Polym.		**Acetone**	**Water**	**Acetonitrile**	**90 wt% Acetonitrile in Water**	**Ether**	**1-Propanol**	**Methanol**
PPY-based	Conductivity (S/cm)	$\ll 10^{-5}$	0.26	1.7×10^{-2}	0.15	0.14	0.53	0.65
Composite	g cond. polym. per g host	1.36	1.09	1.00	1.03	1.32	1.06	0.89
PTP-based	Conductivity (S/cm)	9.4×10^{-2}	2.84	0.69	2.4	3.39	2.07	3.63
Composite	g cond. polym. per g host	1.25	1.26	1.16	1.17	1.30	1.37	1.41

TABLE 1.2.4

Effect of the Solvent Employed for the Oxidant on the Conductivity of the Composite. The Conducting Composites Were Prepared by Employing 1.23 mol/L of BT and 2.98 mol/L of Pyrrole in Acetonitrile as the Monomer Solution and 1.10 mol/L of Copper Perchlorate and 1.10 mol/L of Iron Perchlorate in Various Solvents as the Oxidant Solution. The Crosslinked PS Is Used as Host

			Solvent for Oxidant						
Oxidant	Cond. Polym.		Acetone	Water	Acetonitrile	90 wt% Acetonitrile in Water	0.1 M HCl Aqueous Solution	1-Propanol	Methanol
Copper Perchlorate	PPY	conductivity (S/cm)	7.84×10^{-4}	6.15×10^{-4}	0.11	0.14	9.75×10^{-4}	5.1×10^{-2}	1.25×10^{-3}
		g cond. polym. per g host	1.50	0.52	1.00	0.77	0.52	0.59	0.53
	PTP	conductivity (S/cm)	[a]	[a]	0.63	2.89×10^{-4}	[a]	[a]	[a]
		g cond. polym. per g host	[a]	[a]	1.16	1.16	[a]	[a]	[a]
Iron Perchlorate	PPY	conductivity (S/cm)	2.22×10^{-3}	7.14×10^{-2}	7.38×10^{-2}	0.22	0.23	9.67×10^{-2}	9.80×10^{-2}
		g cond. polym. per g host	1.81	1.24	1.17	1.27	1.13	1.03	1.13
	PTP	conductivity (S/cm)	$\ll 10^{-5}$	2.18×10^{-3}	0.78	4.4×10^{-3}	3.8×10^{-3}	3.1×10^{-3}	4.0×10^{-3}
		g cond. polym. per g host	1.81	1.42	2.16	1.78	1.39	1.81	1.26

[a] No reaction observed

prepared with a copper perchlorate–acetonitrile solution was 0.63 S/cm. The performance of various solvents used for iron perchlorate was similar to that for copper perchlorate but the reaction rate was higher. One of the highest conductivities of polypyrrole-based composites prepared with iron perchlorate was 0.22 S/cm and was achieved with a mixture of acetonitrile and water as solvent for the oxidant. Ferric ions in an aqueous acid medium were also effective, producing a polypyrrole-based conducting composite with a conductivity of 0.23 S/cm. The highest conductivity of polythiophene-based composites prepared with iron perchlorate was 0.78 S/cm and was obtained with acetonitrile as solvent for the oxidant.

As indicated by Tables 1.2.3 and 1.2.4, the conductivity of the composites is affected by the presence of water in the oxidant solution. For this reason the effect of the presence of water in acetonitrile and methanol was investigated in some detail. Table 1.2.5 summarizes the experimental results. The effect of water in the reaction medium was more pronounced on the conductivity of PPY-based composites than on that of PTP-based composites. The conductivity of PPY-based composites reached the value of 0.75 S/cm with a mixture of methanol and water as the solvent for the oxidant.

1.2.3.5 Stability of Conducting Composites

Even though very stable in comparison with polyacetylene, PPY becomes unstable after long-term exposure to air at elevated temperatures.[22] Figure 1.2.6 presents the time dependence of the conductivity of composites which were exposed to air for

TABLE 1.2.5
Effect of Water in Mixtures with Acetonitrile or Methanol on the Conductivity of the Composites. The Composites Were Prepared by Employing 1.23 mol/L of BT and 2.98 mol/L of Pyrrole in Acetonitrile as the Monomer Solution and 1.10 mol/L of Copper Perchlorate and 1.10 mol/L of Iron Perchlorate in the Solvent as the Oxidant Solution. The Porous Crosslinked PS Was Used as Host

			Concentration of Solvent in Aqueous Solution (wt%)				
Solvent	Cond. Polym.		100	90	70	50	0
Acetonitrile	PPY	conductivity (S/cm)	1.70×10^{-2}	0.15	0.49	[a]	0.26
		g cond. polym. per g host	1.00	1.03	1.08	[a]	1.09
	PTP	conductivity (S/cm)	0.69	2.40	1.58	[a]	2.84
		g cond. polym. per g host	1.16	1.17	1.12	[a]	1.26
Methanol	PPY	conductivity (S/cm)	0.65	0.61	0.75	0.44	0.26
		g cond. polym. per g host	0.89	1.13	1.10	0.97	1.09
	PTP	conductivity (S/cm)	3.63	3.96	3.39	3.60	2.84
		g cond. polym. per g host	1.41	1.29	1.19	1.23	1.26

[a] Phase separation occurred at this concentration when $FeCl_3$ was introduced in the aqueous acetonitrile solution.

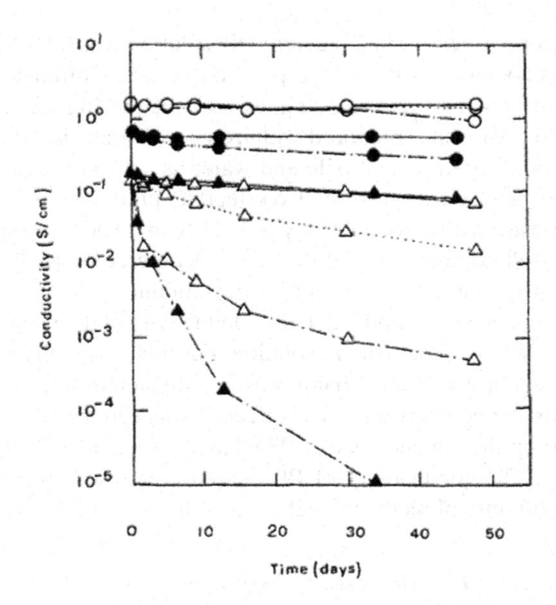

FIGURE 1.2.6 Conductivity of composites against aging time at room and elevated temperatures. Conducting composites were prepared by employing 2.98 mol/L of pyrrole and 1.23 mol/L of BT in acetonitrile as the monomer solution and 1.23 mol/L of $FeCl_3$ in methanol, 1.10 mol/L of copper perchlorate in acetonitrile and 1.10 mol/L of iron perchlorate in acetonitrile as the oxidant solution. Porous crosslinked PS was used as host. The aging samples (0.3 × 1.5 × 3.0 cm) were exposed to air. Δ and ▲ denote polypyrrole-based composites prepared by employing $FeCl_3$ in methanol and copper perchlorate in acetonitrile as the oxidant solution, respectively. O and ● denote polythiophene-based composites prepared by employing $FeCl_3$ in methanol and copper perchlorate in acetonitrile as the oxidant solution, respectively. —, ----- and —-— denote aging temperatures of 25°C, 50°C and 80°C, respectively.

a certain number of days at different temperatures. The conductivity of PPY-based composites decreases slightly at room temperature regardless of the oxidant used as initiator and dopant, while no discernable decrease in the conductivity of PPT-based composites was observed during the same period. With the increase in the aging temperature, a sharp decrease in the conductivity of the PPY-based composites was observed. This thermal instability of the conducting polymers is a result of a thermally driven undoping process in which conformational changes in the polymer backbone are activated at elevated temperature.[23] The conductivity of PPY-based composites decreases much more rapidly than that of PTP-based composites at higher aging temperatures. In PPY-based composites the rate of decrease of the conductivity of the composites prepared with copper perchlorate was faster than that of composites prepared with $FeCl_3$. This indicates that the stability of the conducting polymer can be influenced by the nature of the oxidant employed during preparation.[24]

1.2.3.6 Effect of Antioxidants on the Composite

A set of experiments was carried out to investigate the effect of an antioxidant on the stability of conducting composites. Figure 1.2.7 presents experimental results regarding the aging of composites at 80°C, when an antioxidant is introduced in the porous substrate before the imbibition with the monomer. Except for pyrogallol, which reacts with the oxidant, positive effects on the stability of composites to aging were found in the presence of antioxidants. The composite treated with pyrogallol degraded rapidly at 80°C. When the antioxidants were introduced after the preparation of the composites (Figure 1.2.8), some of them, such as 3-hydrobenzoic acid, phthalic acid and acetamide, were as effective as when they had been included before the imbibition with the monomer solution. Pyrogallol was effective when included after the formation of the composite.

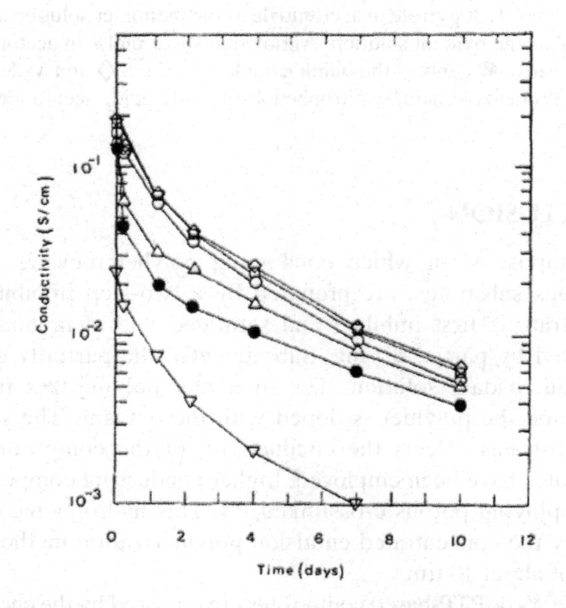

FIGURE 1.2.7 Conductivity of composite against time at 80°C. The composites were prepared by employing 2.98 mol/L of pyrrole in acetonitrile as the monomer solution and 1.23 mol/L of $FeCl_3$ in methanol as the oxidant solution. Antioxidants (0.1 mol% in acetonitrile) were used to treat the host crosslinked PS before imbibition with the monomer solution. ● denotes the blank sample. ○, Δ, □, ◊ and ∇ denote the samples treated with 3-hydrobenzoic acid, 3-nitrophenol, phthalic acid, acetamide and pyrogallol, respectively.

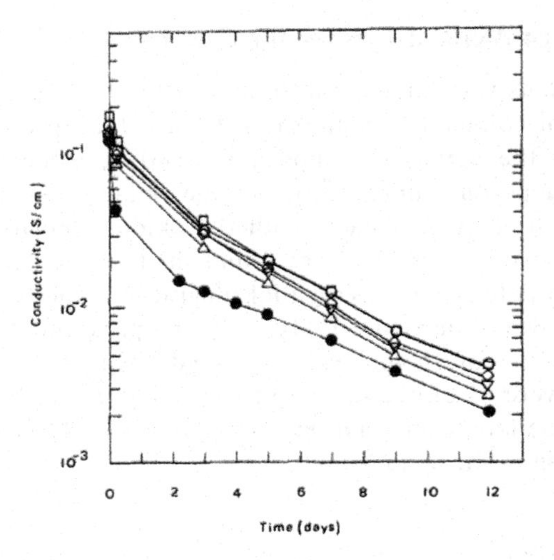

FIGURE 1.2.8 Conductivity of composite against time at 80°C. Composites were prepared by employing 2.98 mol/L of pyrrole in acetonitrile as the monomer solution and 1.23 mol/L of $FeCl_3$ in methanol as the oxidant solution. Antioxidants (0.1 mol% in acetonitrile) were used to treat the composites. ● denotes the blank sample. ○, Δ, □, ◇ and ∇ denote the samples treated with 3-hydrobenzoic acid, 3-nitrophenol, phthalic acid, acetamide and pyrogallol, respectively.

1.2.4 CONCLUSION

Conducting composites, in which conducting polyheterocycles coat the inner surface of porous substrates, are prepared by a two-step imbibition technique. A porous substrate is first imbibed and saturated with a monomer-acetonitrile solution, followed by partial drying. Subsequently, the partially dried system is imbibed with an oxidant solution. The monomer polymerizes in the presence of the oxidant and the polymer is doped with the oxidant. The structure of the porous host substrates affects the conductivity of the composite. While three different substrates have been employed, higher conducting composites have been obtained by employing porous crosslinked PS. This hydrophobic crosslinked PS was prepared by the concentrated emulsion polymerization method and contains uniform voids of about 10 μm.

Conducting PPY- or PTP-based composites are prepared by the chemical oxidative reaction of pyrrole or BT with an appropriate oxidant solution. The solvent for the oxidant affects both the rate of reaction and the conductivity of the composite. A $FeCl_3$-methanol oxidant system is very effective in obtaining a good conducting composite. Copper perchlorate and iron perchlorate are most effective with an aqueous solution of acetonitrile as solvent for PPY-based composites, and with acetonitrile for PTP-based composites. Iron perchlorate with 0.1 M aqueous HCl solution as solvent is as effective for PPY-based composites as the aqueous solution of acetonitrile. Some increases in the conductivity of PPY-based composites are achieved by employing acetonitrile-water mixtures as solvents instead of acetonitrile for $FeCl_3$.

In contrast to the PTP-based composites, the PPY-based composites are less environmentally stable at room temperature, 50°C and 80°C. The degradation of PPY-based conducting composite can be slowed down by treating the porous substrate with an antioxidant. Composites prepared with copper perchlorate are less thermally stable than those prepared with $FeCl_3$, which indicates that the stability of the conducting composites is influenced by the nature of the oxidant used in preparing the conducting polymer.

REFERENCES

1. K. K. Kanazawa, A. F. Diaz, R. H. Geiss, W. D. Gill, J. F. Kwak, J. A. Logan, J. F. Rabolt and G. B. Street, *J. Chem. Soc., Chem. Comm.* 854 (1979).
2. K. Kaneto, K. Yoshino and Y. Inuishi, *Jpn. J. Appl. Phys. 21*, L567 (1982).
3. J. E. Frommer and R. P. Chance, "Electrically conducting polymers," in *Encyclopedia of Polymer Science and Engineering*, Wiley, New York, Vol. 5, p. 462 (1985).
4. F. Tourillon and F. Gamier, *J. Electroanal. Chem. 161*, 407 (1984).
5. M. J. Davidson, "Polymers in Electronics," in *Adv. Symp. Ser., Am. Chem. Soc.* Vol. 24 (1984).
6. C. Li and Z. Song, *Synth. Met. 40*, 23 (1991).
7. T. Yoshikawa, S. Machida, T. Ikegami, A. Techagumpuch and S. Miyata, *Polym. J. 22*, 1 (1990).
8. M. Bi and Q. Pei, *Synth. Met. 22*, 145 (1987).
9. J. E. Österholm, L. Laakso, P. Nyholm, H. Isotalo, H. Stubb, O. Inganäs and W. R. Salaneck, *Synth. Met. 28*, 435 (1989).
10. D. M. Bigg and E. J. Bradbury, in *Conductive Polymers*, Plenum Press, New York, p. 23 (1981).
11. R. B. Bjorklund and I. Lundström, *J. Electron. Mater. 13*, 211 (1984).
12. R. V. Gregory, W. C. Kimbrell and H. H. Kuhm, *Synth. Met. 28*, C823 (1989).
13. E. Ruckenstein and J. S. Park, *J. Appl. Polym. Sci. 42*, 925 (1991).
14. E. Ruckenstein and J. S. Park, *Synth. Met. 44*, 293 (1991).
15. E. Ruckenstein and J. S. Park, *J. Polym. Sci. Polym. Lett. (Part C) 26*, 529 (1988).
16. E. Ruckenstein and J. S. Park, *Polym. 33*, 405 (1992).
17. R. E. Myers, *J. Electron. Mater. 15*, 61 (1986).
18. V. Bocchi, G. P. Gardini and S. Rapi, *J. Mater. Sci. Lett. 6*, 1283 (1987); I. M. Hodge, U.S. Patent 4,642,331 (1987).
19. M. Ogasawara, K. Funahashi and K. Iwato, *Mol. Cryst. Liq. Cryst. 118*, 159 (1985).
20. M. M. Castillo-Ortega, M. B. Inoue and M. Inoue, *Synth. Met. 28*, 65 (1989).
21. M. B. Inoue, E. F. Velazque and M. Inoue, *Synth. Met. 24*, 223 (1988).
22. G. L. Baker, *Adv. Chem. Ser. (Am. Chem. Soc.) 218*, 271 (1988).
23. Y. Wang and M. F. Rubner, *Synth. Met. 39*, 153 (1990).
24. J. E. Österhelm, P. Passiniemi, H. Isotalo and H. Stubb, *Synth. Met. 18*, 213 (1987).

1.3 Processable Conductive Polypyrrole/Poly(alkyl methacrylate) Composites Prepared by an Emulsion Pathway*

Eli Ruckenstein and Shiyong Yang

Department of Chemical Engineering, State University of New York at Buffalo, Box 60, Buffalo, NY 14260-4200, USA

ABSTRACT Electrically conductive polymer composites of polypyrrole and poly(alkyl methacrylate) have been prepared using a two-step procedure. First, an emulsion is generated by dispersing a chloroform solution of poly(alkyl methacrylate) and pyrrole in a small amount of an aqueous surfactant solution. The surfactant is adsorbed upon the interface between the two phases and ensures, via double-layer repulsion, the stability of the emulsion. Second, the pyrrole present in the emulsion is polymerized and doped by introducing with stirring an aqueous solution of an oxidant in the emulsion. The polypyrrole deposits on the host polymer, and the composite formed is precipitated using a suitable non-solvent. Smooth, lustrous films or other shaped objects with good mechanical properties have been prepared by hot-pressing the obtained materials. The relation between conductivity and the polypyrrole content of the composite exhibits a percolation behaviour. The electrical conductivity can reach values as high as 6-7 S/cm. The mechanical properties of the material depend on the nature of the host polymer employed and on the content of polypyrrole in the composite. The effect of the length of the alkyl chain on the mechanical properties is investigated. The polypyrrole-poly(ethyl methacrylate)

* *Polymer.* Vol. 34, 4655–4660, (1993).

composites appear to be the most suitable because they have enough flexibility and also sufficient strength. Compared to cold-pressing, hot-pressing improves the mechanical characteristics, but decreases somewhat the conductivity. Two single-step procedures have also been employed, but they are less efficient than the two-step procedure and lead to very low conductivities.

1.3.1 INTRODUCTION

Electrically conductive polymers have been the subject of considerable research effort due to their interesting and useful electronic, optical and redox properties.[1–3] These materials exhibit high conductivities and some of them have environmental stability. However, their practical use has been hampered by the fact that many of these conjugated polymers are insoluble and infusible. For this reason, the conventional methods for polymer processing, such as melt-processing and solution-casting, could not usually be applied to these materials. Numerous studies have been conducted and many methods developed for the improvement of their processability. For pyrrole-based conductive polymers these include synthesizing soluble N- or ring-substituted derivatives, i.e., introducing flexible side substituents to the main rigid chain[4] or grafting conducting polymer chains to a non-conducting polymer.[5] Polypyrrole (PPy) films have been prepared by electrochemical polymerization on the surface of an electrode,[6–8] or by the interfacial oxidative polymerization method.[9]

Of all the methods, the preparation of composites of conductive and non-conductive polymers, which was stimulated by the successful use of carbon- or silver-filled polymers in a variety of electronic devices, is considered to be a most suitable procedure. Several methods have been reported which combine strong insulating materials with conducting polymers. Initially, such conductive composites have been prepared by incorporating a chemically or electrochemically synthesized conductive polymer in an insulating polymer substrate. The process was carried out either by exposing an insulating sheet imbibed with an oxidant to the monomer or its vapour, or by electrochemical oxidation of a monomer-swollen matrix coated as a film on an electrode.[10–12] However, because of the low penetration of the monomer into the insulating substrate, only very thin conductive polymer films could be obtained. PPy-poly(methyl methacrylate) (PMMA) composite films have also been prepared via chemical polymerization by spreading a water-insoluble solvent solution of pyrrole and PMMA on the surface of an aqueous solution containing the oxidizing agent.[13,14] Another method for the preparation of composites involves chemically polymerizing the monomer of the conductive polymer inside an insulating porous polymer matrix.[15,16] The uniform distribution, with a sufficiently high loading to reach the percolation threshold, of the conductive polymer into the insulating polymer constitutes the essential factor in the preparation of highly conducting polymer composites.

Recently, a two-step pathway, involving an emulsion in the first step, has been developed in our laboratory for the preparation of composites of polyaniline and an insulating polymer. A solution of the host polymer and monomer in an organic solvent constitutes the dispersed phase, while an aqueous surfactant solution constitutes the continuous phase of the emulsion. In the second step, an oxidant solution is introduced in the emulsion. The materials thus prepared could be processed by cold- or hot-pressing, and possess excellent mechanical properties and good electrical conductivities.[17,18] In this paper, the same pathway is employed to prepare a series of processable conductive composites that combine PPy and poly(alkyl methacrylate). The effect of the length of the alkyl side chain on the mechanical properties of the composites is investigated.

The present method represents an improvement of a previous method[19] developed in this laboratory. In Ref. 19, a PPy-polyurethane composite was obtained by mixing (1) a concentrated emulsion containing polyurethane in chloroform as the dispersed phase and sodium dodecyl sulfate in water as the continuous phase, with (2) a suspension of PPy in water prepared by the chemical oxidation method. In the same year, a patent was published[20] suggesting a single-step emulsion pathway. An emulsion of PMMA, pyrrole, CH_2Cl_2, poly(vinyl alcohol), emulsifier and water was subjected to oxidative polymerization with $FeCl_3 \cdot 6H_2O$. The method is similar to method III of the present paper, which, as demonstrated later, leads to much lower conductivities than the two-step method employed by us. The conductivity obtained in the patent was indeed low (~10^{-3} S cm^{-1}). Finally, another patent[21] should be noted, in which a monomer was oxidatively polymerized in a reaction medium containing latex particles of a suitable polymer. The conductivity obtained was low.

1.3.2 EXPERIMENTAL

1.3.2.1 Chemicals

Pyrrole (98%, Aldrich) was purified by distillation in vacuum and stored in a refrigerator before use. PMMA (very high molecular weight, Aldrich), poly(ethyl methacrylate) (PEMA, high molecular weight, Aldrich) and poly(butyl methacrylate) (PBMA, very high molecular weight, Aldrich), ferric chloride (98%, Aldrich), $FeCl_3 \cdot 6H_2O$ (97%, Aldrich), $Fe(ClO_4)_3 \cdot 6H_2O$ (Aldrich), $CuCl_2$ (anhydrous, Aldrich), $Cu(ClO_4)_2 \cdot 6H_2O$ (Aldrich), sodium dodecyl sulfate (SDS, 99.5%, Polysciences), sodium dodecylbenzene sulfate (Aldrich), sodium octadecyl sulfate (97%, Aldrich), chloroform (99%, Aldrich), methanol (99% Aldrich), benzene (99%, Aldrich), 1,2-dichloroethane (99%, Aldrich) and octane (99%, Aldrich) were used as received. Water was deionized and double distilled.

1.3.2.2 Preparation of the Electrically Conductive Composites

The composites were prepared starting from an emulsion in which a solution of surfactant in water constitutes the continuous phase and a solution of the host polymer (PMMA, PEMA or PBMA) and the monomer pyrrole in an organic solvent the dispersed phase. The surfactant molecules are adsorbed upon the interface between the two phases; because of their charge these generate electrical double layers which ensure, via electrostatic repulsion, the stability of the emulsion. The pyrrole dispersed in the emulsion was polymerized and doped by introducing an aqueous solution of oxidant in the emulsion with vigorous stirring. In a typical experiment, 10 mL chloroform solution containing 0.8 g PEMA and 0.1 g pyrrole were added to a 100 mL flask containing a solution of 0.2 g SDS in 2 mL water with vigorous stirring. In the emulsion thus generated, 10 mL of an aqueous solution containing 0.5 g ferric chloride was introduced dropwise with stirring to polymerize the pyrrole and to dope the formed polypyrrole. The polymerization reaction lasted 3 h with stirring. For the precipitation of the composite formed, 25 mL methanol was added to the emulsion with stirring. The entire process took place at room temperature. The solid material was filtered, washed with methanol several times and finally dried at 40°C in vacuum for 24 h. As a result, 0.86 g of polypyrrole/poly(ethyl methacrylate) (PPy/PEMA) composite was thus obtained. The yield of the composite was 96% and was estimated using the expression

$$\text{Yield of composite (\%)} = \frac{100 \times \text{composite (g)}}{\text{pyrrole (g)} + \text{host polymer (g)}}$$

and neglecting, because of the difficulty in evaluating it, the doping of the PPy. The elemental analysis of this composite (wt%) is C = 64.24, H = 8.28, N = 1.89, Fe = 0.11 and Cl = 1.6. The values calculated from the amounts of reactants used are C = 64.26, H = 8.36, N = 2.39 and O = 24.89. The PPy/PMMA and the PPy/PBMA composites were synthesized in the same way.

The same procedure described above is subsequently referred to as method I. Two variants of method I were also employed. In the first (method II), an aqueous solution containing the oxidant was added with stirring to the organic solvent containing pyrrole and the host polymer. In the second (method III), the composite was prepared by mixing the organic solution containing pyrrole and the host polymer with an aqueous solution of surfactant and oxidant. While method I is a two-step procedure, the other two methods are single-step procedures.

1.3.2.3 Instruments

The powder was shaped into a disc (2.5 cm diameter × 0.2 cm) at room temperature, by pressing for a few minutes with a pressure of 120 MPa (cold-pressing). In addition, films were prepared by hot-pressing at 150°C for 2 h under an applied pressure of 20 MPa on the discs prepared by cold-pressing. The electrical conductivity of the discs and films was measured by the standard four-point technique. The mechanical properties of the films were determined with an Instron Universal Testing Instrument (model 1000). The morphologies of the composites were investigated by scanning electron microscopy (SEM, Hitachi S-800). The elemental analysis was carried out by Quantitative Technologies, Inc. (Whitehouse, NJ, USA).

1.3.3 RESULTS AND DISCUSSION

The conductive polymer composites were prepared using methods I–III, described above. In method I, an emulsion of an organic solvent (containing the host polymer and pyrrole) in an aqueous solution of surfactant was first formed. In the present work, the emulsion was a concentrated emulsion with a volume fraction of the dispersed phase (ϕ) equal to 0.83; it had the appearance of a gel. The continuous phase is in the form of a network of liquid films that separate the cells of the dispersed phase. The stability of the emulsion is ensured by the surfactant adsorbed upon the interface between the two phases as an oriented, charged interfacial film. The size of the cells of the dispersed phase is in the range of micrometres.[22] Consequently, the dispersed phase, which contains the pyrrole, possesses a very large surface area of contact with the continuous phase. The aqueous solution of the oxidant, which is introduced in the second step in the emulsion, can therefore be more uniformly distributed among the micrometre-sized cells. The absence of such a structure in the other two one-step procedures does not allow a sufficiently good contact between pyrrole and oxidant. As a result, lower conductivities for methods II and III than for method I are expected.

Table 1.3.1 lists the composites prepared by the three procedures and allows a comparison to be made between them. It shows that the conductivities of the composites prepared by method I are much higher than those of the composites prepared by methods II and III. The amounts of components employed are indicated in the Experimental section for method I; for method II, the 2 mL of water containing 0.2 g of surfactant is missing; for method III, the 2 mL of water is missing.

Because the host polymers employed can be moulded by melt-processing (the glass transition temperatures of PEMA and PBMA being 66°C and 27°C, respectively), the composites containing these polymers can be shaped by hot-pressing at suitable temperatures and pressures. Lustrous films, either flexible or robust, have been thus obtained. The samples prepared by hot-pressing have, however, a somewhat lower conductivity than those prepared by cold-pressing, perhaps because of the degradation of some PPy chains caused by the high temperature, which decreases the conjugation length.

All the experimental data discussed below are based on method I. The experimental data regarding the dependence of the electrical conductivity at room temperature on the weight fraction of PPy are presented in Figure 1.3.1 for the PPy/PEMA composite. They show a percolation behaviour, with reasonably high conductivities even for composites with relatively low content of conductive polymer. For example, the sample containing 10 wt% PPy has a conductivity of 0.3 S cm^{-1}. The percolation threshold occurs in the range 4–10 wt%. Similar results have been obtained in this laboratory for composites prepared from polyaniline and poly(alkyl methacrylate) or polystyrene.[17,18] The small percolation threshold indicates that the conductive polymer molecules are well dispersed in the composite. Electron tunnelling may also play a role.

Figure 1.3.2 shows the effect of the mole ratio of oxidant (ferric chloride) to pyrrole on the conductivity. There is an optimum mole ratio of oxidant to pyrrole in

TABLE 1.3.1
Comparison of the Polypyrrole/poly(alkyl methacrylate) Composites Prepared by the Three Procedures

			Conductivity (S cm^{-1})	
Method	**Composite (PPy wt%)**	**Yield (%)**	**Cold-Pressing**	**Hot-Pressing**
	PMMA/PPy (20%)	87	0.8	0.5
I	PEMA/PPy (20%)	93	0.6	0.4
	PBMA/PPy (20%)	73	0.5	0.5
II	PEMA/PPy (20%)	85	<0.01	0.04
III	PMMA/PPy (20%)	92	<0.001	0.02

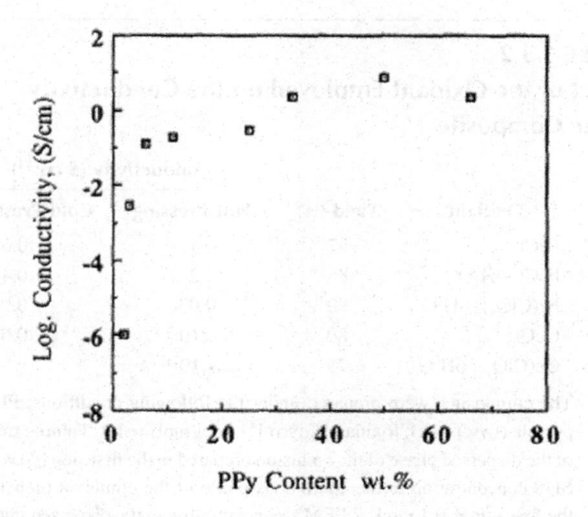

FIGURE 1.3.1 Dependence of the conductivity of the composite on the polypyrrole content. The composites were prepared under the following conditions: $[FeCl_3]$/[pyrrole] = 1:1 (mol/mol); volume fraction of the dispersed phase (ϕ) of the emulsion prepared in the first step = 0.83; surfactant concentration in the continuous phase of the emulsion prepared in the first step = 0.1 g mL^{-1} chloroform; PEMA concentration in the dispersed phase of the emulsion prepared in the first step = 0.08 g mL^{-1}; polymerization time = 3 h; polymerization temperature = 25°C. A solution of 0.14 M of oxidant in water is employed.

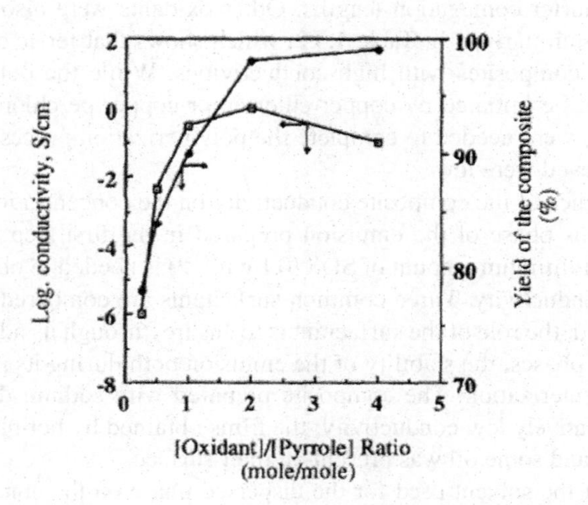

FIGURE 1.3.2 Effect of oxidant/pyrrole mole ratio on the conductivity and yield of the composite. PEMA/pyrrole = 4:1 (w/w); the other conditions are as in Figure 1.3.1.

TABLE 1.3.2
Effect of the Oxidant Employed on the Conductivity of the Composite

			Conductivity (S cm^{-1})	
No.	Oxidant	Yield (%)	Hot-Pressing	Cold-Pressing
1	$FeCl_3$	87	0.4	0.6
2	$FeCl_3 \cdot 6H_2O$	85	0.2	0.4
3	$Fe(ClO_4)_3 \cdot 6H_2O$	90	0.03	0.03
4	$CuCl_2$	80	$<10^{-3}$	0.03
5	$Cu(ClO_4)_2 \cdot 6H_2O$	75	$<10^{-4}$	

Note: The composites were prepared under the following conditions: PEMA/pyrrole (w/w) = 4:1; [oxidant]/[pyrrole] (mol/mol) = 1:1; volume fraction of the dispersed phase of the emulsion prepared in the first step (ϕ) = 0.83; SDS concentration in the continuous phase of the emulsion prepared in the first step = 0.1 g mL^{-1}; PEMA concentration in the dispersed phase of the emulsion prepared in the first step = 0.08 g mL^{-1}; polymerization time = 3 h; polymerization temperature = 25°C. A solution of 0.14 M of oxidant in water is employed.

the range of 1–2. The conductivity of the composite increases rapidly with increasing amounts of oxidant below the optimum value, and decreases with a further increase in the oxidant mole ratio. As expected, the yield of the composite increases with increasing oxidant concentration. The decrease of the conductivity beyond the maximum is probably due to the decrease in the average degree of polymerization. This implies shorter conjugation lengths. Other oxidants were also employed and the results are summarized in Table 1.3.2, which shows that ferric chloride and its hydrate lead to composites with high conductivities. While the polymerization of pyrrole can also be initiated by copper chloride or copper perchloride, long times (about 20–24 h) were needed to complete the polymerization process, and the conductivities achieved were low.

The dependence of the composite conductivity on the concentration of surfactant in the continuous phase of the emulsion prepared in the first step is presented in Figure 1.3.3. A minimum amount of SDS (0.1 g mL^{-1}) is needed to obtain a composite with high conductivity. Three common surfactants are compared in Table 1.3.3. As already noted, the role of the surfactant is to ensure, through its adsorption on the interface of the phases, the stability of the emulsion both during its preparation and during the polymerization. The composite prepared with sodium dodecylbenzene sulfate has a relatively low conductivity; the films obtained by hot-pressing were no longer lustrous and some oil was present on their surface.

The effect of the solvent used for the dispersed phase on the conductivity of the composite is examined in Table 1.3.4. Chloroform, 1,2-dichloroethane and benzene are good solvents, but the composite based on ethyl acetate exhibits a low conductivity.

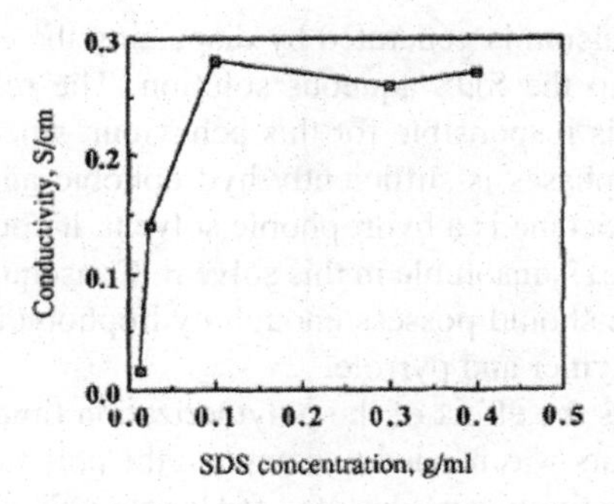

FIGURE 1.3.3 Dependence of the conductivity of the composite on the surfactant concentration in the continuous phase of the emulsion obtained in the first step. The conditions are as in Figure 1.3.2.

TABLE 1.3.3
Effect of the Nature of the Surfactant on the Conductivity of the Composite

			Conductivity ($S\ cm^{-1}$)	
No.	**Surfactant**	**Yield (%)**	**Hot-Pressing**	**Cold-Pressing**
1	Sodium dodecyl sulfate	87	0.4	0.6
2	Sodium dodecylbenzene sulfate	89	0.01	0.03
3	Sodium octadecyl sulfate	93	0.2	0.8

Note: The conditions are as in Table 1.3.2.

TABLE 1.3.4
Effect of the Solvent Used in the Dispersed Phase on the Conductivity of the Composite

			Conductivity ($S\ cm^{-1}$)	
No.	**Solvent**	**Yield (%)**	**Hot-Pressing**	**Cold-Pressing**
1	Chloroform	87	0.4	0.6
2	Benzene	80	0.6	1.1
3	1,2-Dichloroethane	85	0.4	
4	Ethyl acetate	81	0.005	0.01
5	Octane	–[a]	–[a]	

Note: The conditions are as in Table 1.3.2.
[a] PEMA does not dissolve in octane.

This is because no emulsion is generated by dispersing the ethyl acetate solution of PEMA and pyrrole into the SDS aqueous solution. The relatively low hydrophobicity of ethyl acetate is responsible for this behaviour, since an emulsion is stable only when one of the phases is sufficiently hydrophobic and the other sufficiently hydrophilic. Although octane is a hydrophobic solvent, it could not be used because poly(alkyl methacrylate) is insoluble in this solvent. Consequently, a suitable solvent for the dispersed phase should possess enough hydrophobicity but should also be a solvent for the host polymer and pyrrole.

Figure 1.3.4 presents the effect of the polymerization time on the conductivity of the composite. Two hours was needed to complete the polymerization of pyrrole, but no differences in conductivity were detected for longer polymerization times.

Scanning electron micrographs for the PPy/PEMA composites containing 11 and 25 wt% PPy are presented in Figure 1.3.5. Morphologies based on aggregated particles are observed for the samples prepared by cold-pressing (Figure 1.3.5a and c). The particle sizes are estimated to be in the range 0.05–0.1 μm. Dramatic changes in morphologies occur in the films prepared by hot-pressing (Figure 1.3.5b and d). In these films, the primary particles are replaced by completely non-particulate, homogeneous phases. A distinct PPy phase could not be detected by SEM.

Table 1.3.5 lists the mechanical properties of three typical composites: PPy/PMMA, PPy/PEMA, and PPy/PBMA. The tensile strength at the breakpoint decreases and the corresponding elongation increases as the alkyl side chains in the host polymer become larger, due to the increase in the free volume of the polymers. The stress–strain curves for these materials are presented in Figure 1.3.6. The PPy/PEMA composite appears to be the most suitable because it has enough flexibility and also has sufficient strength. The PPy/PMMA composite is somewhat brittle, while the PPy/PBMA composite does not have enough strength.

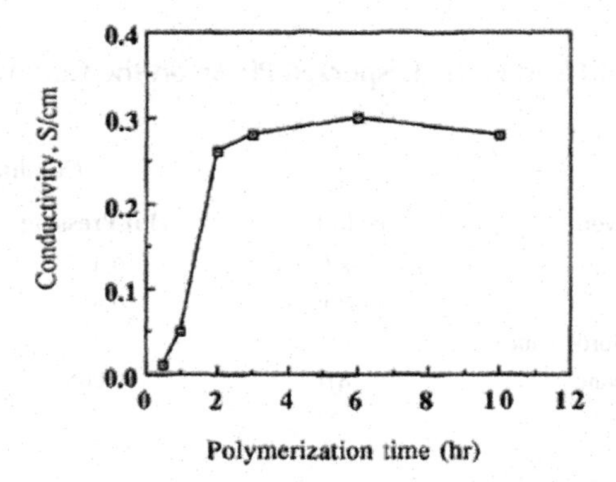

FIGURE 1.3.4 Effect of polymerization time on the conductivity of the composite. The conditions are as in Figure 1.3.2.

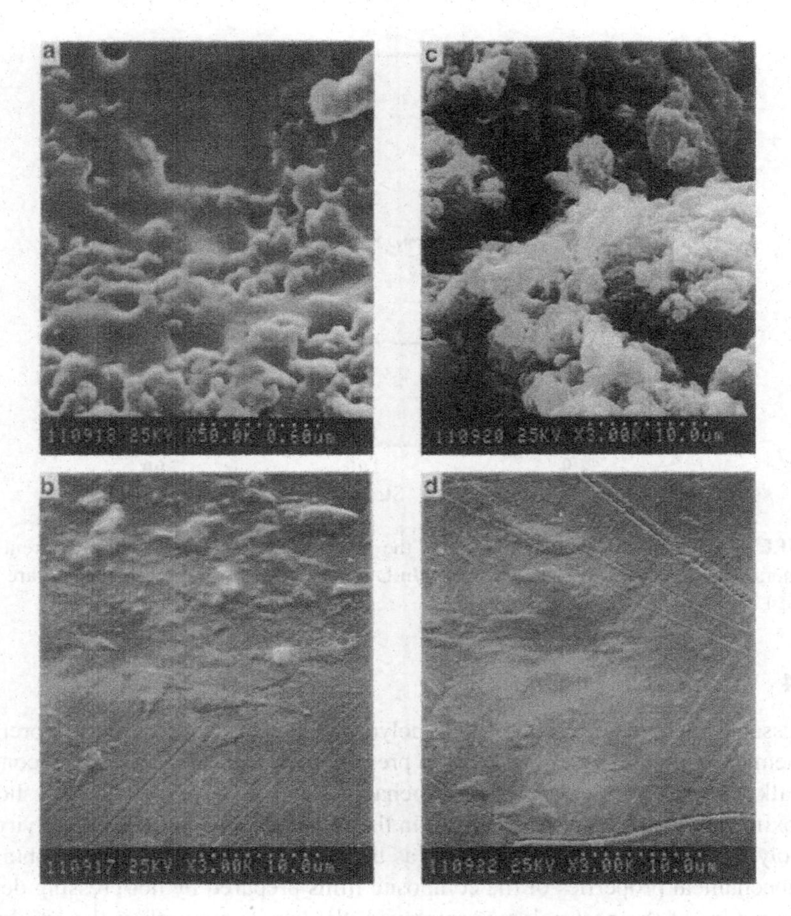

FIGURE 1.3.5 Scanning electron micrographs of PPy/PEMA composites: (a) PPy/PEMA (11 wt% PPy) prepared by cold-pressing; (b) PPy/PEMA (11 wt% PPy) prepared by hot-pressing; (c) PPy/PEMA (20 wt% PPy) prepared by cold-pressing; (d) PPy/PEMA (20 wt% PPy) prepared by hot-pressing.

TABLE 1.3.5
Mechanical Properties of Polypyrrole/Poly(alkyl methacrylate) Composites

Composite (PPy wt%)	Elongation at the Breakpoint (%)	Tensile Strength at the Breakpoint (MPa)	Conductivity (S cm^{-1})	
			Cold-Pressing	Hot-Pressing
PMMA/PPy (20)	15	25	0.8	0.5
PEMA/PPy (20)	45	13	0.6	0.4
PBMA/PPy (20)	150	8	0.5	0.5

Note: The conditions are as in Table 1.3.2.

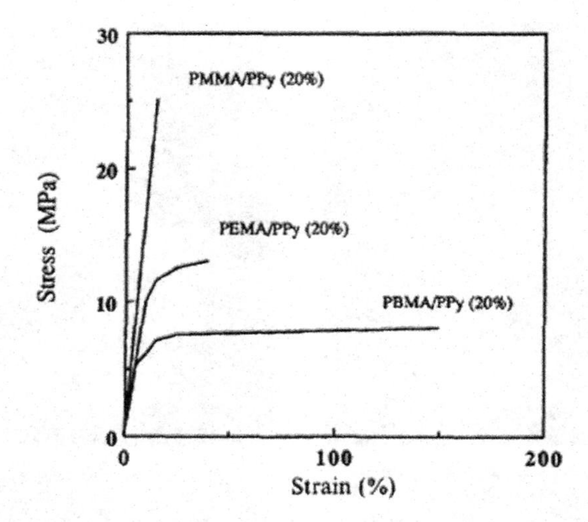

FIGURE 1.3.6 Strain-stress curves for the composites prepared with different host polymers. Polymer concentration = 0.08 g/mL chloroform; the other conditions are as in Figure 1.3.1.

1.3.4 CONCLUSION

Processable composites of polypyrrole/poly(alkyl methacrylate) have been prepared by chemically polymerizing the pyrrole present in an emulsion which also contains poly(alkyl methacrylate). A percolation behaviour with a low percolation threshold of approximately 4–10 wt% was observed in the relation between the conductivity and the polypyrrole content. Conductivities as large as 6–7 S cm^{-1} have been obtained. The mechanical properties of the composite films prepared by hot-pressing depend on the nature of the host polymer employed, the tensile strength at the breakpoint decreasing and the corresponding elongation increasing as the alkyl side chains in the host polymer become larger.

REFERENCES

1. Skotheim, T. A. (Ed.) *Handbook of Conducting Polymers*, Vol. 1, Marcel Dekker, New York, 1986.
2. Burroughes, J. H., Bradly, D. D. C., Brown, A. R., Marks, R. N., Mackay, K., Friend, R. H., Burnes, P. L. and Holmes, A. B. *Nature* 1990, **347**, 539.
3. Stevens, T. *Mater. Eng.* 1991, **108**, 21.
4. Dao, L. H., Leclerc, M., Guay, J. and Chevalier, J. W. *Synth. Met.* 1989, **29**, E377.
5. Li, S., Cao, Y. and Xue, Z. *Synth. Met.* 1987, **20**, 141.
6. Diaz, A. F. and Kanazawa, K. K. *J. Chem. Soc. Chem. Commun.* 1979, 635.
7. Kuwabata, S., Okamoto, K., Ikeda, O. and Yoneyama, H. *Synth. Met.* 1987, **18**, 101.
8. Peres, R. C. D., Pernault, J. M. and DePaoli, M. A. *J. Polym. Sci. A. Polym. Chem.* 1991, **29**, 225.
9. Nakata, M., Taga, M. and Kise, H. *Polym. J.* 1992, **24**, 437.
10. Diaz, A. F., Paoli, M. and Waltman, R. J. *J. Polym. Sci. A. Polym. Chem.* 1985, **23**, 1687.

11. Yoshikawa, T., Machida, S., Ikegami, T. N., Techagumpuch, A. and Miyata, S. *Polym. J.* 1990, **22**, 1.
12. Li, C. J. and Song, Z. G. *Synth. Met.* 1991, **40**, 23.
13. Moreta, M., Hashida, I. and Nishimura, M. *J. Appl. Polym. Sci.* 1988, **36**, 1639.
14. Chan, H. S. O., Hor, T. S. A, Ho, P. K. H., Tan, K. L. and Tan, B. T. G. *J. Macromol. Sci.-Chem.* 1990, **A27**, 1081.
15. Ruckenstein, E. and Park, J. S. *J. Appl. Polym. Sci.* 1991, **42**, 925.
16. Ruckenstein, E. and Park, J. S. *Synth. Met.* 1991, **44**, 293.
17. Ruckenstein, E. and Yang, S. Y. *Synth. Met.* 1993, **53**, 283.
18. Yang, S. Y. and Ruckenstein, E. *Synth. Met.* 1993, **59**, 1.
19. Ruckenstein, E. and Chen, J. H. *Polymer* 1991, **32**, 1230.
20. Kato, K., Mineura, Y., Tuchida, Z. and Saito, T. Japanese Patent 03.234.733, October 18, 1991.
21. Jasne, S. J. European Patent 229992, July 29, 1987
22. Ruckenstein, E., Ebert, G. and Platz, G. *J. Colloid Interface Sci.* 1989, **133**, 432.

1.4 Processable Conductive Composites of Polyaniline/Poly(alkyl methacrylate) Prepared via an Emulsion Method*

Shiyong Yang and Eli Ruckenstein

Department of Chemical Engineering, State University of New York at Buffalo, Buffalo, NY 14260 (USA)

ABSTRACT Polyaniline/poly(alkyl methacrylate) composites are prepared using an emulsion pathway which involves two steps. First, an emulsion with the appearance of a gel is prepared by dispersing a chloroform solution of poly(alkyl methacrylate) and aniline in an aqueous solution of sodium dodecylsulfate (SDS). Secondly, an oxidant (sodium persulfate) dissolved in an aqueous solution of HCl is added dropwise to the emulsion with vigorous stirring in order to polymerize the aniline and to dope the polyaniline formed. The composite precipitates spontaneously or is precipitated by the addition of methanol. The materials have been processed either by cold- or hot-pressing. Those prepared by hot-pressing can have electrical conductivities as high as 2 S/cm and good mechanical properties, while those prepared by cold-pressing have somewhat higher conductivities but somewhat inferior mechanical properties.

1.4.1 INTRODUCTION

Electrically conductive polymers have attracted attention because of their unique electrical and optical properties and, numerous potential applications [1–4]. Among the conducting polymers, polyaniline (PANI) has been extensively studied in the past decades due to its electrochromism, fast switching rate, environmental

* *Synthetic Metals*, Vol. 59, p. 1–12, 1993.

stability, low cost of raw materials and ease of synthesis, etc. [5,6]. Many industrial and academic laboratories are conducting research on polyaniline and some products (e.g., PANI/Li secondary batteries) have already been commercialized [7]. The emeraldine oxidation state of polyaniline in its base form is soluble in aqueous acetic acid or formic acid solutions, in dimethylsulfoxide, dimethylformamide and *N*-methylpyrrolidinone. Free-standing films and also fibers can be formed from these solutions [8,9]. The processability of polyaniline can be improved by various methods such as synthesizing *N*- or ring-substituted derivatives soluble in organic solvents [10,11], preparing sterically stabilized colloidal dispersions of particles of a host polymer and polyaniline [12], or electrochemically polymerizing aniline on a polymer matrix [13]. Recently Cao et al. [14] found that after doping and complexing with camphor-sulfonic acid (CSA), polyaniline becomes soluble in *meta-cresol,* chloroform, formic acid, etc. The CSA-polyaniline complex can be spin-cast from *meta*-cresol onto glass or polymer substrates to yield optical-quality transparent films with relatively low surface resistance. Flexible light-emitting diodes have been fabricated using this material, which are easily visible under room lighting and have an external quantum efficiency of about 1% [15]. A route for the preparation of polypyrrole/polyurethane composites by coprecipitation from a concentrated emulsion containing polyurethane, and an aqueous suspension of polypyrrole powder was reported from this laboratory [16]. Recently, an emulsion pathway for the preparation of processable polyaniline/polystyrene composites was described [17]. In the present paper, the latter method is employed for the preparation of a series of processable conductive polyaniline/poly(alkyl methacrylate) composites. It is shown that materials with good mechanical properties can be obtained by hot-pressing instead of cold-pressing.

1.4.2 EXPERIMENTAL

1.4.2.1 Chemicals

Aniline (Aldrich, 99.5%) was purified by distillation in vacuum and stored in a refrigerator before use. Poly(methyl methacrylate) (Aldrich, PMMA, very high molecular weight), poly(ethyl methacrylate) (Aldrich, PEMA, high molecular weight), poly(butyl methacrylate) (Aldrich, PBMA, very high molecular weight), ammonium persulfate (Polysciences), sodium dodecylsulfate (Polysciences, SDS, 99.5%), chloroform (Aldrich, 99.8%) and methanol (Aldrich, 99%) were used as received. Water was deionized and distilled before use.

1.4.2.2 Preparation of the Electrically Conductive Composites

The composites were prepared starting from an emulsion in which an aqueous solution of sodium dodecylsulfate constitutes the continuous phase and a chloroform solution of aniline and one of the host polymers (PMMA, PEMA, PBMA or their mixtures) constitutes the dispersed phase. The aniline was polymerized by introducing an oxidant dissolved in an aqueous solution of HCl to the emulsion with vigorous stirring. In a typical experiment, a solution of 0.20 g of SDS in 2 mL of water was placed in a 100 mL flask. To this solution, 5.0 mL of a chloroform solution containing 0.40 g of PEMA and 0.20 g of aniline was added dropwise with stirring. An emulsion with the appearance of a gel was thus formed. Then, 0.5 g of ammonium persulfate dissolved in 10 mL of 1 N HCl aqueous solution were introduced dropwise in the emulsion with stirring to polymerize the aniline and to dope the polyaniline formed. The hydrochloric acid assists in the polymerization of aniline and functions also as a dopant for polyaniline. The polymerization reaction lasted 3 h. The composite formed, deposited on the wall of the flask, was separated from the system, washed with 20 mL methanol, filtered, washed again with 20 mL of 1 N HCl aqueous solution and finally dried at 40°C in vacuum for 24 h. About 0.66 g of polyaniline/poly(ethyl methacrylate) composite (PANI/PEMA) was thus obtained. The yield of the composite was 83%. The yield of the composite was estimated using the expression 100 × composite (g)/(aniline (g) + host polymer (g) + SDS (g)) since the amounts of dopant and surfactant included in the composites are not known. When the amount of aniline used was smaller than about 0.1 g and other quantities remained the same, the composite could not precipitate from the emulsion and methanol had to be used for its precipitation. Polyaniline/poly(methyl methacrylate) (PANI/PMMA), polyaniline/poly(butyl methacrylate) (PANI/PBMA), PANI/PMMA/PBMA and PANI/PEMA/PBMA composites were prepared by the same method.

In addition to the above two-step procedure, a single-step procedure, in which an emulsion is generated between an aqueous solution of HCl containing the surfactant and the oxidant, and a chloroform solution containing aniline and the host polymer, was also employed. It appears that the two-step procedure has the advantage of a better distribution of the oxidant and dopant among the droplets of the dispersed phase of the emulsion generated in the first step.

1.4.2.3 Instruments

For measuring the conductivity of the composite, the powder was shaped either as a disc (2.5 cm diameter, 0.2 cm thick) at room temperature under an applied pressure of 120 MPa (cold-pressing) or as a smooth, flexible film by hot-pressing at 150°C for 2 h under an applied pressure of 20 MPa. The electrical conductivities of the disc and the film were measured by the standard four-point technique. The morphologies of the composites were investigated by scanning electron microscopy (Hitachi S-800). Fourier transform infrared spectroscopy (FT-IR) experiments were also performed on a Mattson-Alpha Centauri instrument.

1.4.3 RESULTS AND DISCUSSION

The *in situ* polymerized aniline deposits on the host polymer chains thus forming a polyaniline/host polymer composite. The host polymers investigated include PMMA, PEMA, PBMA and their mixtures. Because the glass transition temperatures of

TABLE 1.4.1
The Conductivities of Poly(alkyl methacryrate)/ Polyaniline Composites

Sample (wt./wt.)	Conductivity (S/cm) Sample A[a]	Conductivity (S/cm) Sample B[b]
PMMA/PANI composites		
1:1	3.7	1.9
2:1	0.7	0.4
4:1	0.3	0.2
PEMA/PANI composites		
1:1	3.9	0.8
2:1	0.9	1.0
4:1	0.5	0.2
PBMA/PANI composites		
1:1	1.0	0.7
2:1	0.6	0.4
4:1	0.1	0.01

Notes: The composites were prepared under the following conditions: oxidant/aniline (mol/mol) = 1:1; concentration of HCl = 0.58 mol/L of the system; 1 N aqueous solution of HCl was employed; volume fraction of the dispersed phase in the emulsion prepared in the first step = 0.71; concentration of the surfactant in the continuous phase of the emulsion prepared in the first step = 0.1 g/mL; concentration of poly(alkyl methacrylate) in the dispersed phase in the emulsion prepared in the first step = 0.08 g/mL; polymerization temperature = 25°C; polymerization time = 3 h.

[a] Sample A was prepared by cold-pressing at room temperature under a pressure of about 120 MPa.

[b] Sample B was prepared by hot-pressing at 150°C for 2 h under a pressure of about 20 MPa.

PEMA and PBMA are 66°C and 27°C, respectively, the powders of the composites containing these polymers and polyaniline have also been subjected to hot-pressing at 150°C for 2 h under an applied pressure of 20 MPa. Smooth, flexible or mechanically robust films have been obtained by this method. Table 1.4.1 lists the conductivities of the samples prepared either by cold-pressing or by hot-pressing. One can note that (i) the conductivity is affected by the content of polyaniline in the composite, those containing about 50 wt.% polyaniline reaching values as high as 4 S/cm (compared to 5–6 S/cm for pure polyaniline) and (ii) the samples prepared by hot-pressing exhibit somewhat lower conductivities than those prepared by cold-pressing.

TABLE 1.4.2
Electrical Conductivities and Mechanical Properties of PANI/Polymer (1)/ Polymer (2) Composites[a] Prepared by Hot-Pressing

Sample (wt./wt.)	Yield of the Composite (%)	Conductivity (S/cm)	Mechanical Properties
PANI/PBMA	60	0.4	Soft, flexible
PANI/PEMA	83	1.0	Robust, stiff
PBMA/PEMA			
2:1	58	0.8	Between PANI/ PBMA and PANI/ PEMA
1:2	85	0.7	
1:4	68	0.6	
PBMA/PMMA			
1:1	74	0.6	Strong, flexible

[a] The composites were prepared under the following conditions: polymer/aniline (wt./wt.) = 2:l. The other conditions are as in Table 1.4.1.

The mechanical properties of the material depend on the amount and nature of the host polymer (Table 1.4.2). For instance, the composite PANI/PEMA (2:1 wt./wt.) prepared by hot-pressing is strong and stiff while the composite PANI/PBMA (2:1 wt./wt.) prepared in the same way is soft, flexible and elastic. The butyl side chains in the host polymer increase the free volume available and are, therefore, responsible for the behavior of the latter composite. Consequently, electrically conductive materials with a variety of mechanical properties can be prepared by changing the host polymer. The composites prepared by cold-pressing have been less flexible.

The electrical conductivity of the composite versus the weight fraction of polyaniline is plotted for the PANI/PEMA and PANI/PBMA composites in Figure 1.4.1. The curves present percolation thresholds which are very low in both cases, being in the range of 4–10 wt.%. The conductivities are relatively high even when the PANI content is low. For example, the composite PANI/PEMA containing only about 6 wt.% polyaniline has a conductivity of 0.3 S/cm. Conductivities as high as 6–7 S/cm were observed for the PANI/PEMA composites. Similar results have been reported for the PANI/polystyrene composites prepared previously by the same method [17], the composite films prepared from sterically stabilized polyaniline latex particles mixed with sterically stabilized poly(methyl methacrylate-co-butylacrylate) latex particles [18] and the polyaniline/copolymer latex composites prepared by chemically polymerizing aniline in the presence of a film-forming chlorinated copolymer latex [19].

The dependence of the conductivity of the composite (PANI/PEMA) on the concentration of HCl used in the polymerization system is presented in Figure 1.4.2. The conductivity increases with increasing concentration of HCl and attains saturation for concentrations greater than 0.6 mol/L of the entire system.

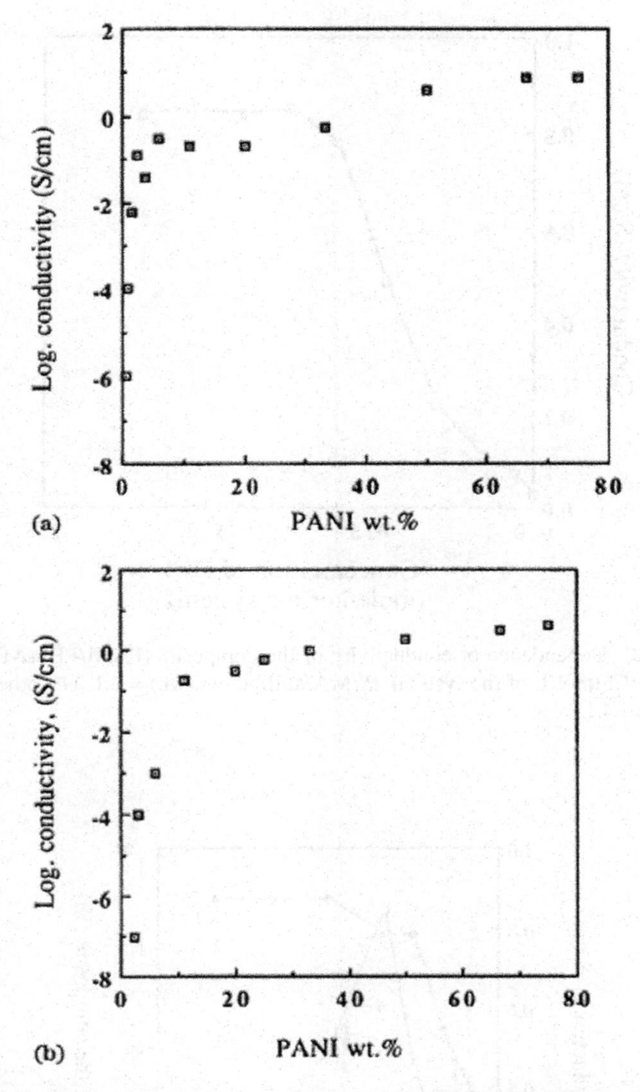

FIGURE 1.4.1 Conductivity of (a) PANI/PEMA and (b) PANI/PBMA composites vs. content of polyaniline (PANI wt.%). The conditions are as in Table 1.4.1.

Figure 1.4.3 shows that the conductivity and yield of the composite are affected by the mole ratio of the oxidant to aniline. An optimum mole ratio of oxidant to aniline (1:1) is observed. The conductivity of the composite increases rapidly below this optimum value and decreases with a further increase in the oxidant to aniline ratio. The decrease in conductivity of the composite may be attributed to the lower degree of polymerization of aniline and hence to the shorter conjugation length at higher oxidant concentrations. At high mole ratios of oxidant, the composite obtained is

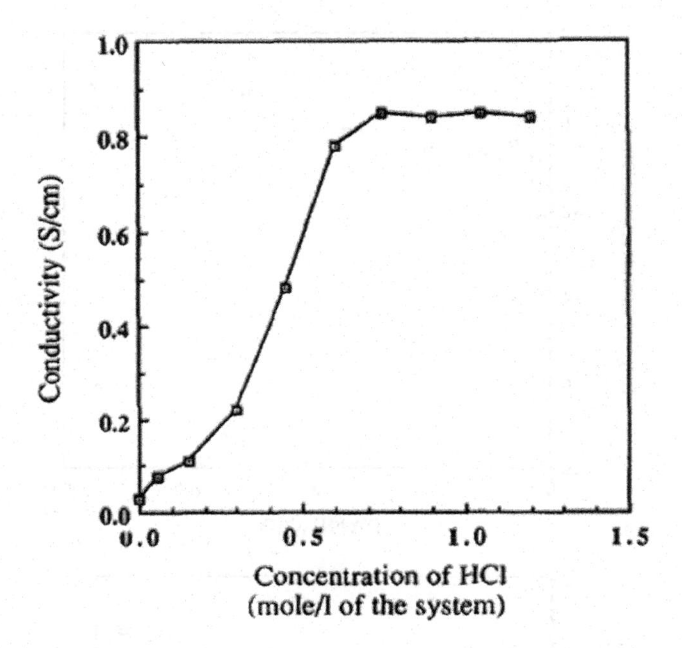

FIGURE 1.4.2 Dependence of conductivity of the composite (PANI/PEMA) on the concentration of HCl (mol/L of the system). PEMA/aniline (wt./wt.) = 2:1. The other conditions are as in Table 1.4.1.

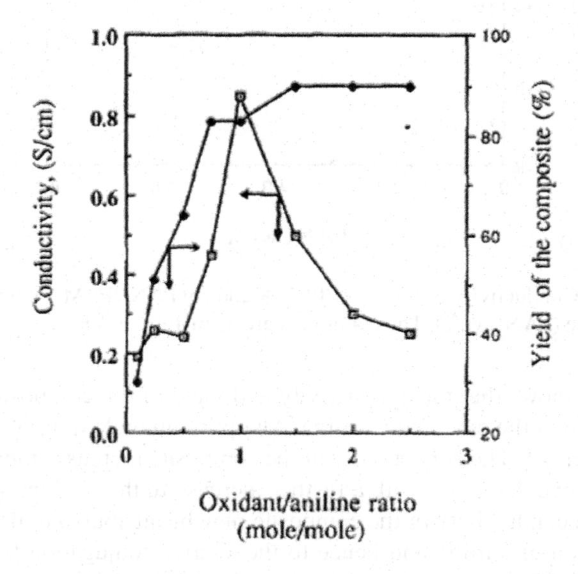

FIGURE 1.4.3 Conductivity and yield of the composite vs. oxidant/aniline mole ratio. The conditions are as in Figure 1.4.2.

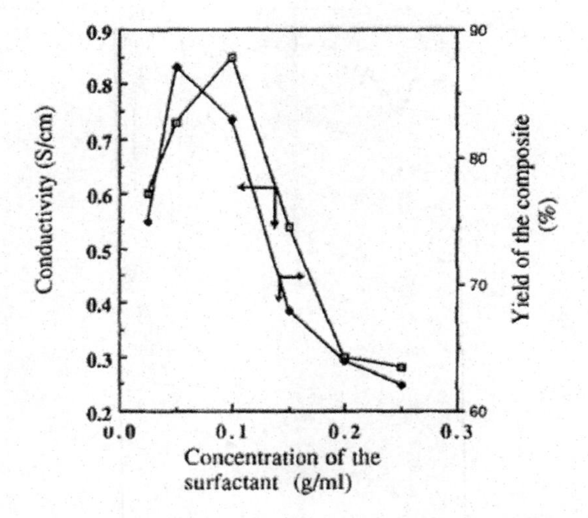

FIGURE 1.4.4 Effect of the concentration of the surfactant used in the continuous phase on the conductivity and yield of the composite: ⊡, conductivity; ◆, yield. The conditions are as in Figure 1.4.2.

a black powder while at lower oxidant mole ratios it is dark blue. The yield of the composite increases with increasing mole ratio of oxidant to aniline and attains saturation.

Figure 1.4.4 shows the effect of the concentration of surfactant in the continuous phase on the conductivity and yield of the composites. In the absence of surfactant, only a fraction of polyaniline forms a composite with PEMA and a large amount of free polyaniline is formed. The conductivity and yield of the composite pass through maxima with increasing amount of surfactant. These maxima can be probably explained as follows. The adsorption of the surfactant molecules on the surface of the droplets of the emulsion ensures, because of their charge, the stability of the emulsion. As the concentration of the surfactant increases, the contact surface between the dispersed and continuous phases has the tendency to increase, because a larger number of surfactant molecules are available for adsorption. However, the increase in ionic strength of the aqueous solution with increasing surfactant concentration is increasingly shielding the electrostatic repulsion between the globules of the dispersed phase and this stimulates their coalescence. These two opposite effects may lead to a maximum in the area of contact surface between the dispersed and continuous phases of the emulsion generated in the first step and may be responsible for the optimum surfactant concentration. Another possible explanation is the formation of an increasing number of micelles in the continuous phase in which aniline is solubilized. Finally, the most likely explanation is the formation of soluble complexes between polyaniline and surfactant which are not included in the composites.

Because of the different glass transition temperatures of PMMA, PEMA and PBMA, a series of composites with various mechanical properties can be obtained by using mixtures of PMMA and PBMA or of PEMA and PBMA as host

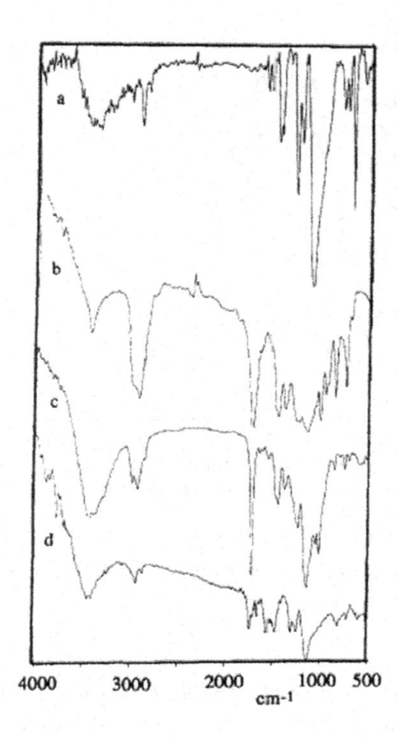

FIGURE 1.4.5 FT-IR spectra of (a) PANI, (b) PEMA and the composites of PANI and PEMA with (c) 4 wt.% PANI and (d) 33 wt.% PANI.

polymers. The conductivities and yields of these composites are listed in Table 1.4.2. However, no major differences in conductivity among these three-component composites are observed; only the mechanical properties are different.

The FT-IR spectra of the composites and of their components are presented in Figure 1.4.5. The spectra of the PANI/PEMA composites contain the main absorptions of both polyaniline and PEMA. However, the intensities of the vibrations for PEMA in the composite are much lower than in the pure polymer. For instance, the very strong intensities of the stretching vibrations of C–H in the range of 2800–3200 cm^{-1} and of C = O in the range of 1680–1760 cm^{-1} are greatly reduced by the addition of polyaniline.

Scanning electron micrographs of the PANI/PEMA and PANI/PEMA/PBMA composites are presented in Figures 1.4.6 through 1.4.8. Figure 1.4.6 shows the morphologies of PANI/PEMA composites containing 0, 4 and 33 wt.% polyaniline prepared by cold-pressing. The composites appear to have a uniform structure. However, the composites prepared by hot-pressing appear to contain two phases (Figure 1.4.7). This phase separation may be the main reason for the decrease in conductivity of the hot-pressed materials compared to that of the cold-pressed materials (Table 1.4.1). Some polyaniline probably becomes the dispersed phase during hot-pressing because, in contrast to the host polymer which has a relatively low glass transition temperature, it decomposes above 250°C before even passing through a

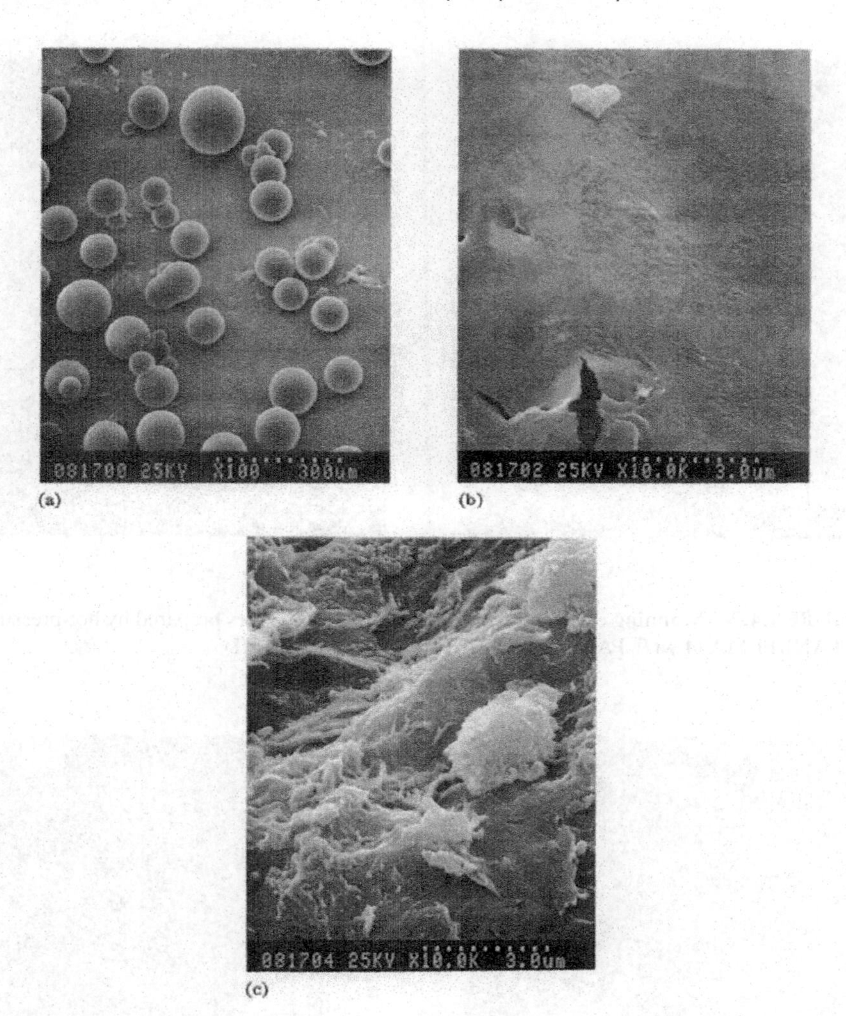

FIGURE 1.4.6 Scanning electron micrographs of the composites prepared by cold-pressing: (a) PEMA; (b) PANI/PEMA (4 wt.% PANI); (c) PANI/PEMA (33 wt.% PANI).

glass transition or a melting temperature. One can observe that the particles of the dispersed phase (estimated to be below 0.5 μm) are relatively uniformly distributed. Therefore, although some phase separation appears to occur in PANI/PEMA composites during hot-pressing, the conductive component is still well dispersed in the material. As a result, the conductivity of the material is not too much decreased. Similar observations can be made regarding the three-component composite PANI/PEMA/PBMA shown in Figure 1.4.8.

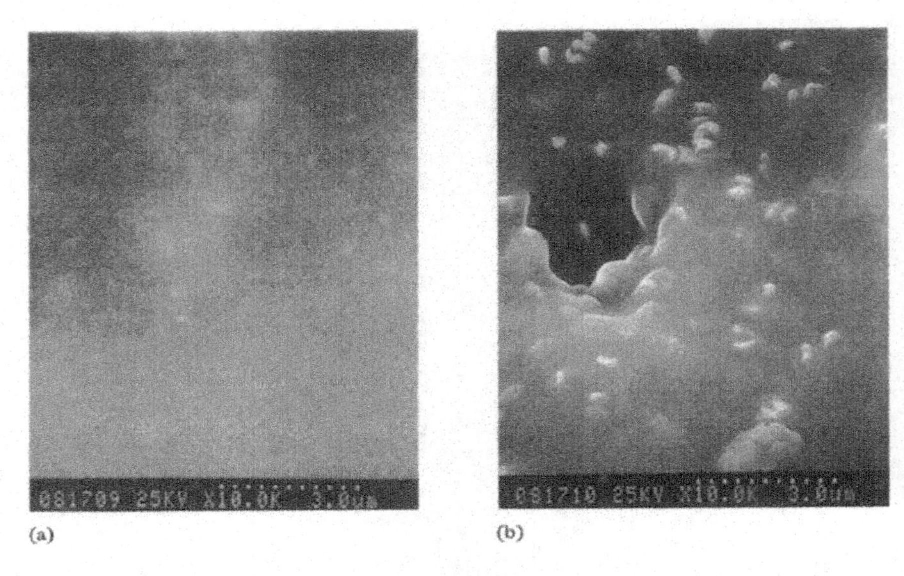

FIGURE 1.4.7 Scanning electron micrographs of the composites prepared by hot-pressing: (a) PANI/PEMA (4 wt.% PANI); (b) PANI/PEMA (33 wt.% PANI).

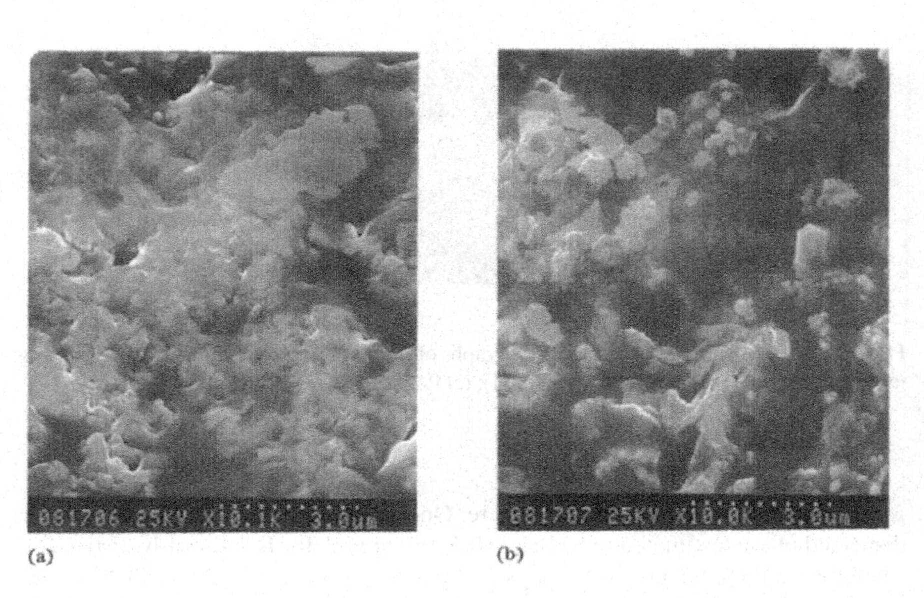

FIGURE 1.4.8 Scanning electron micrographs of the three-component composites prepared by hot-pressing: (a) PANI/PEMA/PBMA composite (33 wt.% PANI and PEMA/PBMA = 2:1 wt./wt.); (b) PANI/PEMA/PBMA composite (33 wt.% PANI and PEMA/PBMA = 1:1 wt./wt.).

1.4.4 CONCLUSIONS

Polyaniline/poly(alkyl methacrylate) composites have been prepared by the chemical polymerization of aniline in an emulsion that contains a chloroform solution of poly(alkyl methacrylate) and aniline as dispersed phase and an aqueous solution of sodium dodecylsulfate as the continuous phase. The conductivity of the samples exhibits a percolation threshold at a low polyaniline content of approximately 4–10 wt.%. The composites have been prepared by both cold- and hot-pressing: the cold-pressed composites have somewhat higher conductivities, while the hot-pressed ones have somewhat better mechanical properties.

REFERENCES

1. J. H. Burroughes, D. D. C. Bradly, A. R. Brown, R. N. Marks, K. Mackay, R. H. Friend, P. L. Bums and A. B. Holmes, *Nature (London)*, *347* (1990) 539.
2. D. Braun, G. Gustafsson, D. McBranch and A. J. Heeger, *J. Appl. Phys.*, *72* (1992) 564.
3. D. Braun and A. J. Heeger, *Appl. Phys. Lett.*, *58* (1991) 1982.
4. T. Stevens, *Mater. Eng.*, *108* (1991) 21.
5. J. C. Lacroix, K. K. Kanazawa and A. Diaz, *J. Electrochem. Soc.*, *136* (1989) 1308; C. D. Batich, H. A. Laitinen and H. C. Zhou, *J. Electrochem. Soc.*, *137* (1990) 883.
6. M. Kalaji, L. M. Peter, L. M. Abrantes and J. C. Mesquita, *J. Electroanal Chem.*, *274* (1989) 289.
7. Bridgestone Corporation and Seiko Electronic Components Ltd. Internal Rep., 1987.
8. M. Angelopoulos, A. Ray, A. G. MacDiarmid and A. J. Epstein, *Synth. Met.*, *21* (1987) 21.
9. M. Angelopoulos, G. E. Asturias, S. P. Ermer, A. Ray, E. M. Scherr, A. G. MacDiarmid, M. Akhter, Z. Kiss and A. J. Epstein, *Mol. Cryst. Liq. Cryst.*, *160* (1988) 150.
10. J. Yue and A. J. Epstein, *J. Am. Chem. Soc.*, *112* (1990) 2800.
11. S. Ni, L. Wang, F. Wang and L. Shen, *Polym. Commun.*, *31* (1989) 123.
12. S. P. Armes, M. Aldissi, S. F. Agnew and S. Gottesfeld, *Langmuir, 6* (1990) 1745; B. Vincent and J. Waterson, *J. Chem. Soc., Chem. Commun.*, (1990) 683.
13. S. A. Chen and W. G. Fang, *Macromolecules, 24* (1991) 1242; G. Bidan and B. Ehui, *J. Chem. Soc., Chem. Commun.*, (1989) 1568.
14. Y. Cao, P. Smith and A. J. Heeger, *Synth. Met.*, *48* (1992) 91; Y. Cao, G. M. Treacy, P. Smith and A. J. Smith, *Appl. Phys. Lett.*, *62* (1992) 2711.
15. G. Gustafsson, Y. Cao, G. M. Treacy, F. Klavetter, N. Colaneri and A. J. Heeger, *Nature (London), 357* (1992) 477.
16. E. Ruckenstein and J. H. Chen, *Polymer, 32* (1991) 1230.
17. E. Ruckenstein and S. Y. Yang, *Synth. Met.*, *53* (1993) 283.
18. E. C. Cooper and B. J. Vincent, *J. Phys. D: Appl. Phys.*, *22* (1989) 1580.
19. P. Beadle and P. Armes, *Macromolecules, 25* (1992) 2526.

1.5 Inverted Emulsion Pathway to Polypyrrole and Polypyrrole Elastomer Composites*

Eli Ruckenstein and Liang Hong

State University of New York at Buffalo, Department of Chemical Engineering, Clifford C. Furnas Hall, Box 604200, Buffalo, NY 14260-4200, USA

ABSTRACT A new method is proposed for the preparation of polypyrrole and polypyrrole-elastomer composites. An inverted emulsion of water containing the oxidant ($FeCl_3$) in a solution (toluene, or a mixture of toluene and isooctane) of the host polymer (styrene-butadiene-styrene (SBS) triblock copolymer), or free of the host polymer, was first generated. The bis(2-ethylhexyl)sodium sulfosuccinate (AOT) surfactant was used as dispersant. A solution of pyrrole in toluene was introduced dropwise in the inverted emulsion. Because of the large interfacial area between the aqueous solution and the continuous phase containing pyrrole, a rapid oxidation occurred and, as a result, colloidal-sized polypyrrole particles were generated. Methanol was employed to precipitate the polymers. The pure polypyrrole black prepared by using AOT as dispersant exhibited a conductivity of 34 S/cm, while that prepared without using AOT gave a conductivity of 9 S/cm. The elastomeric sheets prepared by hot-pressing possessed electrical conductivities in the range of 2–3 S/cm, a tensile strength of 7–13 MPa and an elongation at the break point of 47–259%. The factors that affect the conductivities of the composites include the content of polypyrrole, the molar ratio of H_2O to AOT, the concentration of Fe(III) salt in the aqueous solution and the nature of the solvent employed as the continuous phase of the emulsion. The effect of doping was also investigated by comparing various counter-anions of the positively charged pyrrole unit. The following sequence was observed: $Cl^- > ClO_4^- \gg p\text{-}CH_3C_6H_4SO_3^-$.

* *Synthetic Metals*. Vol. 66, p. 249–256, 1994.

1.5.1 INTRODUCTION

Polypyrrole is the most commonly used conductive material of the several known conjugated polymers because of its environmental stability to oxygen and water [1]. In some cases the conductivities of the doped polypyrroles have been as high as several hundred S/cm [2]. Polypyrrole and its derivatives have been synthesized either by electrochemical or by chemical oxidative polymerization [3]; the generated polypyrroles have been brittle powders, insoluble and infusible, and hence, unprocessable. For pyrrole-based conductive polymers, the processability could be improved, either by synthesizing N- or ring-substituted derivatives of pyrroles by introducing flexible substituents [4], or by grafting conductive side chains to an insulating polymer [5]. In order to develop pyrrole-based conductive materials for commercial applications [6], major efforts have been made toward the preparation of conductive polypyrrole-polymer composites [7] which possess enough conductivity and satisfactory mechanical properties. Several approaches have been developed: (1) An oxidant-impregnated polymer was exposed to pyrrole vapor, and thus a two-phase structure with microdomains of polypyrrole dispersed in the elastomer matrix was obtained [8]. (2) Electrodes were coated with various porous polymers and then immersed into a pyrrole solution. Diffusing through the coating layer, the pyrrole molecules contact the electrode surface and polymerize within the channels of the pores of the polymers [9]. (3) Pyrrole was polymerized electrochemically in the presence of latex particles with anionic surface characteristics and the resulting polymer composites precipitated onto the anode. The positively charged polypyrrole units become associated with the anionic charges present on the surface of the particles [10]. (4) Graft copolymers of suitable polymers containing pyrrole as grafted groups were coated onto an anode; additional pyrrole was electrochemically grafted onto the polymer [11]. (5) A porous polymer matrix, prepared by polymerizing a concentrated emulsion of water in styrene/divinylbenzene, was first impregnated with pyrrole, and then immersed into an aqueous solution of oxidant to polymerize the pyrrole [12]. In contrast to the other methods, the latter can provide thick conductive objects. In all these methods, there are a number of features which limit their uses in a variety of applications. In methods (1) and (2), hydrophilic polymers, such as poly(vinyl alcohol), are usually employed to allow the absorption of the aqueous oxidant solution in their matrixes; in methods (2) to (4), the polymerization of pyrrole is carried out on the anodes; and in method (5), the crosslinking must be sufficiently high to ensure the generation of porosity. The hydrophilic polymers employed in methods (1) and (2) cannot provide suitable mechanical properties such as those generated by elastomers and, due to the rapid oxidation, the polypyrrole formed at the entrance of the pores of the host impedes further reaction. The electro-chemical polymerization used in methods (2) to (4) is limited by the fact that the yield of the polymer composite is restricted by the area of the electrode of the electrochemical apparatus. The crosslinked polystyrene porous matrix used in method (5) is a relatively brittle material.

Two emulsion pathways to prepare conductive polymer composites were developed recently in this laboratory [13,14]. In one of them [13] a concentrated emulsion containing a solution of polyurethane in chloroform as the dispersed phase and a solution of sodium dodecylsulfate in water as the continuous phase was first prepared. Finally, the composite was obtained by blending the concentrated emulsion with an aqueous suspension of polypyrrole. The other procedure [14] also employed two steps: first, a concentrated emulsion was generated by dispersing an organic solution (toluene or chloroform was used as solvent) of pyrrole and a host copolymer (such as poly(ethylene-vinylacetate) (EVA)) in a small amount of an aqueous solution of a surfactant (sodium dodecylsulfate); secondly, the pyrrole present in the emulsion was polymerized and doped by introducing with stirring an aqueous solution of an oxidant in the emulsion. This procedure has led to a rather uniform distribution of polypyrrole inside the host polymer matrix. However, the concentrated emulsion became unstable as soon as the aqueous oxidant solution was introduced, and the coalescence of the dispersed droplets became severe because of the large volume fraction of the dispersed phase. This reduces the original high dispersion of the organic phase, thus affecting the contact of pyrrole with the oxidant. For this reason, a rather long reaction time of 2 h was needed to achieve a good conductivity. The latter procedure was also used to prepare polyaniline-polystyrene and polyaniline–poly(alkyl methacrylate) composites. An emulsion procedure, similar to that of Ref. [14a], was used by Österholm et al. [15] to prepare high molecular weight polyaniline.

In the present paper, the emulsion pathways used to prepare conductive polymer composites have been modified by oxidatively polymerizing pyrrole with an inverted emulsion system of water (containing the oxidant $FeCl_3$) in an organic solvent (containing the host polymer). The oil-soluble surfactant bis(2-ethylhexyl) sodium sulfosuccinate (Aerosol OT), which is an effective emulsifier for the formation of reversed microemulsions, was used as emulsifier. The pyrrole was dissolved in the same organic solvent as the one used as the continuous medium of the emulsion, or in an organic solvent completely miscible with the latter. The pyrrole solution was introduced dropwise in the inverted emulsion, with stirring. The main advantage of this procedure lies in the good contact between the oxidant system and the pyrrole solution, which is achieved because the solvents in the two organic systems are the same or are at least soluble in one another, the contact surface area for reaction is large and the mechanical stirring is vigorous. Since the polymerization of pyrrole is carried out under an excess of oxidant, and the chemical reaction is rapid, fine polypyrrole particles with high yields are expected to form within a short time.

Compared to other methods, the main characteristics of the present one are its simplicity and suitability for numerous host polymers and monomers. It can be easily scaled up for industrial applications.

1.5.2 EXPERIMENTAL

1.5.2.1 CHEMICALS

Styrene-butadiene-styrene triblock copolymer (SBS) (Polymer Science, styrene content 28 wt.%), pyrrole (Aldrich 99%), iron(III) chloride (Aldrich, 97%), iron(III) perchlorate monohydrate (Aldrich, 97%), sodium dioctyl sulfosuccinate (AOT, Aldrich, 98%), toluene (Aldrich, 99+%) and isooctane (Fisher, HPLC) were used as received. Distilled and deionized water was employed.

1.5.2.2 PREPARATION OF POLYPYRROLE AND SBS COMPOSITES

A typical procedure can be described as follows: either SBS (3.0 g) and AOT (13.6 g, 30 mmol) were dissolved in toluene (75 mL) or SBS was dissolved in toluene (30 mL) and AOT in isooctane (45 mL). It will be shown later that the latter solvent is more effective than the former. To the above solution, located in a 500 mL three-necked round-bottomed flask equipped with a mechanical stirrer, an aqueous solution of iron(III) chloride (9.8 g, 60.5 mmol) in 23 mL of water ($[Fe^{3+}] = 2.7$ M) was introduced with stirring (about 700 rpm). An orange-colored inverted emulsion (W/O) was thus generated. About 5 min later, a solution of pyrrole (2.0 g, 29 mmol) in 10 mL toluene (pyrrole is not soluble in isooctane) was added to the stirred inverted emulsion over 10 min. After the addition of the pyrrole-toluene solution, the stirring was continued for 5 min more and the color of the system rapidly became black. Then, 200 mL of methanol (a non-solvent for the polymers) was introduced to the stirred system to precipitate the composite. The entire process took place at room temperature. The black composite solid was filtered, then washed several times successively with methanol and with water until the filtered liquid became colorless. The wet composite material was dried in a fume hood overnight, and about 4.6 g of dry composite was obtained. The elemental analysis of the composite (wt.%) indicated: N% = 5.10 and Cl% = 3.74. The weight percentage of polypyrrole in the composite was 23.4%, while the theoretical percentage (quantitative conversion) of polypyrrole in the composite is 36%. Consequently, the yield of the oxidative polymerization was 66%. This composite is denoted as PPY-SBS (2/3), the numbers 2 and 3 in parentheses indicating that 2 g of pyrrole and 3 g of SBS were employed in the feedstock.

1.5.2.3 PREPARATION OF POLYPYRROLE

The synthesis was carried out as in the previous section, but without SBS. Five types of polypyrrole samples were prepared by changing some of the conditions in the preparation. They are summarized in Table 1.5.1. In the preparation of PPY-5, the pyrrole (2.0 g, 29 mmol) was first treated with an equivalent amount of *p*-toluenesulfonic acid in 10 mL of water. The formed organic salt is water-soluble, and its aqueous solution was introduced dropwise into an inverted

TABLE 1.5.1
Polypyrroles Synthesized by the Inverted Emulsion Method

	Oxidant	AOT (mmol)	Organic Medium	Counter-ion
PPY-1	$FeCl_3$ (62 mmol)	30	Isooctane (75 mL)/toluene (10 mL)	Cl^-
PPY-2	$FeCl_3$ (62 mmol)	0	Isooctane (75 mL)/toluene (10 mL)	Cl^-
PPY-3	$FeCl_3$ (62 mmol)	30	Toluene (85 mL)	Cl^-
PPY-4	$Fe(ClO_4)_3$ (62 mmol)	30	Isooctane (75 mL)/toluene (10 mL)	ClO_4^-
PPY-5[a]	$(NH_4)_2S_2O_8$ (31 mmol)	30	Isooctane (75 mL)/toluene (10 mL)	p-$CH_3C_6H_4SO_3^-$

Notes: The other conditions: (1) molar ratio of oxidant/pyrrole = 2; (2) 24 mL of water was used to dissolve the oxidant.

[a] In this case, pyrrole (29 mmol) was first treated with p-toluenesulfonic acid (29 mmol) in 10 mL of water.

emulsion, which was prepared by dispersing an aqueous solution of $(NH_4)_2S_2O_8$ (7.1 g, 31 mmol) in 13 mL water into an isooctane (75 mL) solution containing 13.6 g AOT.

1.5.2.4 Measurement of the Electrical Conductivity

PPY-SBS composites and PPY powders were shaped into discs of 2.5 cm diameter at room temperature by cold pressing. In addition, the discs of PPY-SBS composite were pressed at 150°C for 2–3 min and then the obtained sheets were cut to rectangular shapes (2.5 × 1.5 cm). The conductivities of cold-pressed and hot-pressed samples were measured at room temperature by the four-point technique.

1.5.2.5 The Tensile Testing

The sample sheets were cut to the size required by the ASTM D.638-58T. The tensile testing was performed with an Instron testing instrument (Model 1000) at room temperature. The elongation speed of the instrument was 20 mm/min.

1.5.2.6 The Scanning Electron Microscopy (SEM) Analysis

The SEM micrographs were obtained with a Hitachi S-800 instrument. The sample was coated with carbon before investigation. Energy-dispersive spectroscopy (EDS) was performed with PGT/TMIX field emission microscopy equipment.

1.5.2.7 The X-ray Diffraction of Polypyrrole Black

The X-ray diffraction analysis of the polypyrrole blacks (PPY in Table 1.5.1) was performed on a Nicolet powder diffractometer equipped with a Cu Kα source.

1.5.2.8 The Elemental Analysis

The elemental analysis was carried out by Quantitative Technologies, Inc. (Whitehouse, NJ).

1.5.3 RESULTS AND DISCUSSION

1.5.3.1 Pyrrole Polymerization in Reversed Emulsion Systems

Since pyrrole is water-insoluble while the commonly used oxidants, such as $FeCl_3$, are water-soluble, the chemical polymerization of pyrrole at the water/oil interface of the inverted water/oil emulsion provides a high surface area between the two immiscible liquids which allows the pyrrole molecules to rapidly approach the oxidant.

Polypyrrole was first synthesized and doped without using any elastomer. The preparation conditions and the conductivities are listed in Tables 1.5.1 and 1.5.2. For comparison purposes, a surfactant (AOT) was employed in the synthesis of PPY-1 but not in that of PPY-2; PPY-1 and PPY-3 were synthesized in different organic media; and PPY-1, PPY-4 and PPY-5 were synthesized using different oxidants and dopants.

It is interesting to note that PPY-1 has a conductivity 3.5 times larger than PPY-2. The SEM investigations revealed that the two have completely different morphologies: PPY-1 displays a sponge-like and PPY-2 a scale-like morphology (Figure 1.5.1) It is likely that the different polymerization surroundings are responsible for this difference. In the presence of AOT and under vigorous mechanical stirring, the aqueous $FeCl_3$ solution is dispersed into tiny spherical droplets suspended in the organic medium, and the polymerization of pyrrole generates a large number of fine polypyrrole particles (0.5–1.0 μm). Their agglomeration leads to a

TABLE 1.5.2
Electrical Conductivity of Polypyrrole Prepared by the Inverted Emulsion Procedure

Polypyrrole	Conductivity (S/cm)
PPY-1	34.2
PPY-2	9.8
PPY-3	29.3
PPY-4	13.3
PPY-5	negligible

Note: The experimental conditions are described in Table 1.5.1. The measurements were performed by using disc samples.

(a)

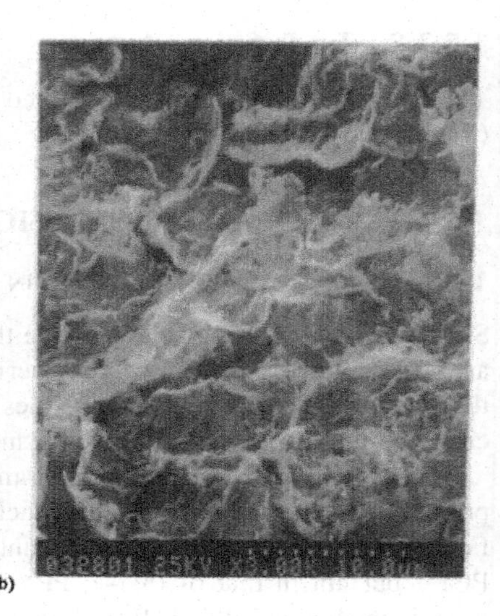

(b)

FIGURE 1.5.1 SEM micrographs of polypyrrole blacks: (a) PPY-1 (synthesized in the presence of AOT); (b) PPY-2 (synthesized in the absence of AOT).

spongelike morphology. In contrast, in the absence of AOT, the aqueous $FeCl_3$ solution is sheared into liquid stripes, and the chains of polypyrrole organize on the surface of the stripes as scale-like structures. However, the X-ray powder diffraction (Figure 1.5.2) shows that PPY-1, PPY-2 and the other three specimens have peaks at $2\theta = 18°$. This might indicate the presence of some local crystallinity in all of them. Most forms of polypyrrole reported in the literature are poorly crystalline, and X-ray diffraction patterns have been obtained only for the dimer and trimer pyrrole [1b]. This local crystallinity is very likely generated by the polymerization at the water/oil interface, and is probably due to the higher molecular weight of the formed polypyrrole. It is also worth noting that the EDS investigations show (Figure 1.5.3) that PPY-1 contains a large amount of Cl^- as dopant, a small amount of AOT (the S peak) in the polymer matrix and a trace amount of Fe.

The elemental analysis shows that the molar ratios of N(pyrrole units)/Cl(counter-ion)/S(due to AOT) = 10.71/3.46/0.44. The presence of Cl^- as dopant is essential for the high conductivity of PPY-1 (34 S/cm). Indeed, when the specimen was treated with Na_2CO_3 solution (2.5 M), thus replacing Cl^- with 1/2 CO_3^{2-}, the conductivity of the specimen became very low (0.11 S/cm).

The polypyrroles PPY-4 and PPY-5 were synthesized and doped with different oxidants and dopants than those used above (Table 1.5.1). For PPY-4, $Fe(ClO_4)_3$ was used as oxidant and dopant, and for PPY-5, pyrrole was first converted to its salt (with $p\text{-}CH_3C_6H_4SO_3^-$ as counter-ion) by its treatment with *p*-toluenesulfonic acid. In the latter case, $(NH_4)_2S_2O_8$ was used as oxidant. In the electrochemical preparation of polypyrrole, the anions ClO_4^- and $p\text{-}CH_3C_6H_4SO_3^-$ have been known [16] to be effective dopants that generate high conductivities. In contrast, in the inverted

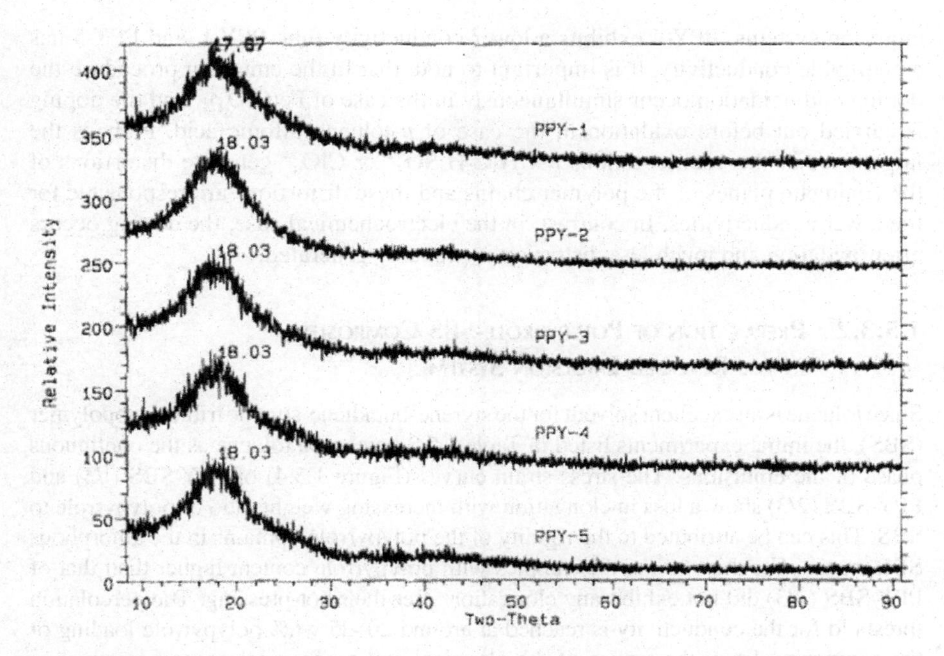

FIGURE 1.5.2 X-ray powder diffractions of the polypyrrole black samples (see also Tables 1.5.1 and 1.5.2).

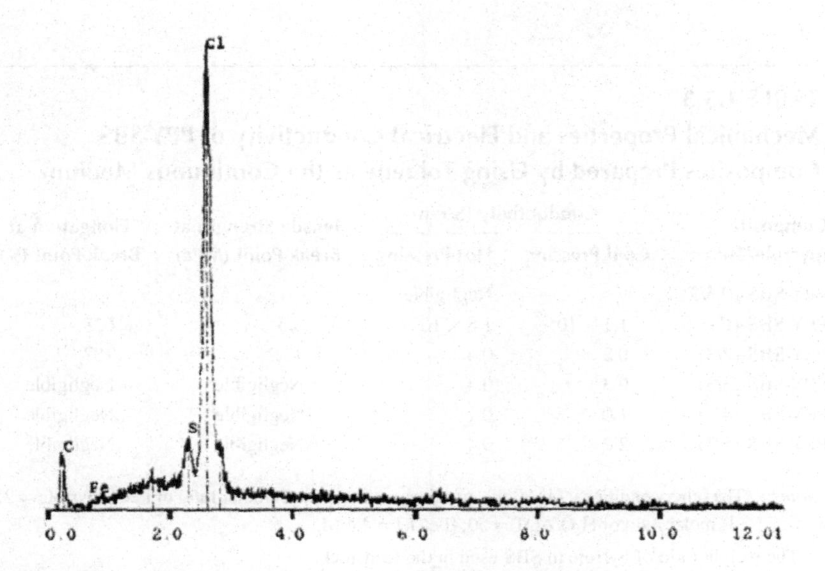

FIGURE 1.5.3 Energy-dispersive spectroscopy (EDS) of PPY-1.

emulsion systems, PPY-4 exhibits a lower conductivity than PPY-1, and PPY-5 has a negligible conductivity. It is important to note that in the emulsion procedure the doping and oxidation occur simultaneously in the case of $Fe(ClO_4)_3$, and the doping is carried out before oxidation in the case of *p*-toluenesulfonic acid. Perhaps the large sizes of the counter-anions *p*-$CH_3C_6H_4SO_3^-$ or ClO_4^- generate distortions of the conjugate planes of the polymer chains and these distortions are responsible for the lower conductivities. In contrast, in the electrochemical case, the doping occurs after oxidation and much less distortion is probably generated.

1.5.3.2 Preparation of Polypyrrole-SBS Composites in the Reversed Emulsion System

Since toluene is an excellent solvent for the styrene-butadiene-styrene triblock copolymer (SBS), the initial experiments listed in Table 1.5.3 have used toluene as the continuous phase of the emulsions. The stress-strain curves (Figure 1.5.4) of PPY-SBS (1/3) and PPY-SBS (2/3) show a loss in elongation with increasing weight ratio of polypyrrole to SBS. This can be attributed to the rigidity of the polypyrrole domains in the amorphous SBS matrix. For this reason, the samples with polypyrrole content higher than that of PPY-SBS (2/3) did not exhibit any elongation after their hot-pressing. The percolation threshold for the conductivity is reached at around 20–25 wt.% polypyrrole loading of the composite. From the values of the electrical and mechanical properties, one can conclude that the optimum weight ratio of pyrrole to SBS falls in the range of 1/3 to 2/3.

TABLE 1.5.3
Mechanical Properties and Electrical Conductivity of PPY-SBS Composites Prepared by Using Toluene as the Continuous Medium

Composite	Conductivity (S/cm)		Tensile Strength at	Elongation at
(pyrrole/SBS)[a]	Cold Pressing	Hot Pressing	Break Point (MPa)	Break Point (%)
PPY-SBS-(0.5/3)		Negligible		
PPY-SBS-(1/3)	1.1×10^{-2}	1.8×10^{-2}	8.5	628
PPY-SBS-(2/3)	0.2	0.4	12.2	187
PPY-SBS-(3/3)	0.3	0.4	Negligible	Negligible
PPY-SBS-(4/3)	1.0	0.2	Negligible	Negligible
PPY-SBS-(5/3)	2.2	0.4	Negligible	Negligible

Notes: The other conditions: (1) 80 mL of toluene was used; (2) molar ratio of $FeCl_3$/pyrrole = 2; (3) molar ratio of H_2O/AOT = 50; [$FeCl_3$] = 4.4 M.

[a] The weight ratio of pyrrole to SBS used in the feedstock.

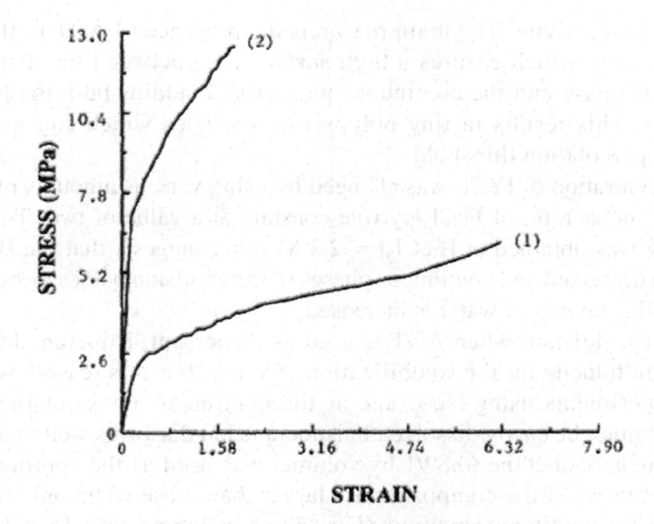

FIGURE 1.5.4 Stress-strain curves of the polypyrrole-SBS composites prepared by the inverted emulsion pathway by using toluene as the continuous phase: (1) PPY-SBS (1/3); (2) PPY-SBS (2/3) (see also Table 1.5.3).

Two parameters were found to affect in a major way the electrical conductivity: the molar ratio H_2O/AOT (Table 1.5.4) and the concentration of the aqueous solution of $FeCl_3$ (Table 1.5.5).

The ratio of H_2O/AOT was varied between a large value and 16, by changing the amount of AOT. Table 1.5.4 shows that there is a percolation threshold in the change of the conductivity, for a molar ratio of H_2O/AOT of about 50. From the weights of the obtained composites, the content of polypyrrole in the composite based on H_2O/AOT = 50 is only 8% higher than that in the composite free of AOT. It is clear that this small difference cannot explain the 400-fold

TABLE 1.5.4
Electrical Conductivity of PPY-SBS-(2/3) Composites Prepared with Various Amounts of AOT

	Conductivity (S/cm)	
Molar Ratio of H_2O/AOT	**Cold-Pressing Disc**	**Hot-Pressing Sheet**
∞	9.1×10^{-3}	9.6×10^{-4}
50	0.2	0.4
24	0.6	0.7
16	0.9	0.9

Notes: The other conditions: (1) 80 mL of toluene was used; (2) molar ratio of $FeCl_3$/pyrrole = 2; (3) weight ratio of pyrrole/SBS = 2/3; [$FeCl_3$] = 4.4 M.

increase in conductivity. The main reason is the presence of AOT in the reversed emulsion system, which ensures a high surface area between the dispersed oxidant aqueous phase and the continuous phase that contains both the pyrrole and the polymer. This results in tiny polypyrrole particles which can more readily achieve the percolation threshold.

The concentration of $FeCl_3$ was changed by using various amounts of water and keeping the molar ratio of $FeCl_3$/pyrrole constant at a value of two. The optimum conductivity was obtained at $[FeCl_3] = 2.7$ M. This suggests that the surface area between the dispersed and continuous phases of the emulsion passes through a maximum when the amount of water is increased.

In a microemulsion, when AOT is used as dispersant, isooctane has a higher capacity than toluene for the solubilization of water. For this reason, we also carried out experiments using isooctane as the continuous phase of the emulsion. Since SBS cannot be easily dissolved in isooctane but dissolves well in toluene, the mixture toluene-isooctane (0.89/1 by volume) was used as the continuous phase. The conductivities of the composites are larger than those when only toluene was employed as the continuous medium (Nos. 3 to 5 in Table 1.5.5). Even low loading composites such as PPY-SBS (0.5/3) display some conductivity. The SEM micrograph (Figure 1.5.5) of the hot-pressed sheet of PPY-SBS (2/3) composite shows that SBS constitutes the continuous medium in which polypyrrole particles are dispersed. The stress-strain curves (Figure 1.5.6) compare the mechanical properties of the composites with different contents of polypyrrole. It is worth noting that use of the mixture leads to a somewhat smaller elongation at the break point than the use of toluene (compare Figures 1.5.4 and 1.5.6). This happens because toluene is more compatible with the styrene of the SBS triblock copolymer. The composite PPY-SBS (1.5/3) of Table 1.5.6 possesses both good conductivity and mechanical properties.

TABLE 1.5.5
Effect of the Concentration of Oxidant $FeCl_3$ on the Conductivity of PPY-SBS Composites

Sample	Composite (pyrrole/SBS)	$[FeCl_3]$ (M)	Conductivity (S/cm)	
			Cold-Pressing Disc	Hot-Pressing Sheet
1	PPY-SBS (2/3)	6.0	0.4	0.3
2	PPY-SBS (2/3)	4.4	0.6	0.7
3	PPY-SBS (2/3)	2.7	0.8	1.9
4	PPY-SBS (1.5/3)	2.7	Rough surface of disc	1.6
5	PPY-SBS (1/3)	2.7	Rough surface of disc	0.8
6	PPY-SBS (2/3)	1.9	Rough surface of disc	1.1

Notes: The other conditions: (1) 80 mL of toluene was used; (2) molar ratio of $FeCl_3$/pyrrole = 2; (3) AOT (13.6 g, 30 mmol).

TABLE 1.5.6
Mechanical Properties and Electrical Conductivity of PPY-SBS Composites Prepared with 40 mL Toluene and 45 mL Isooctane as the Continuous Medium of the Inverted Emulsion

Composite (pyrrole/SBS)	Content[a] of PPY (wt.%)	Yield of Polymerization (%)	Tensile Strength at Break Point (MPa)	Elongation at Break Point (%)	Conductivity (S/cm)
PPY-SBS (2/3)	23.7	66	7.5	47	2.8
PPY-SBS (1.5/3)	20.5	87	13.4	259	2.7
PPY-SBS (1/3)	15.4	79	7.9	267	1.7
PPY-SBS (0.5/3)	6.3	48	9.7	559	2.3×10^{-2}

Notes: The other conditions: (1) molar ratio of oxidant/pyrrole = 2; (2) $[FeCl_3]$ = 2.7 M; (3) H_2O/AOT = 42.

[a] The content of polypyrrole in the composite was evaluated on the basis of the elemental analysis of nitrogen.

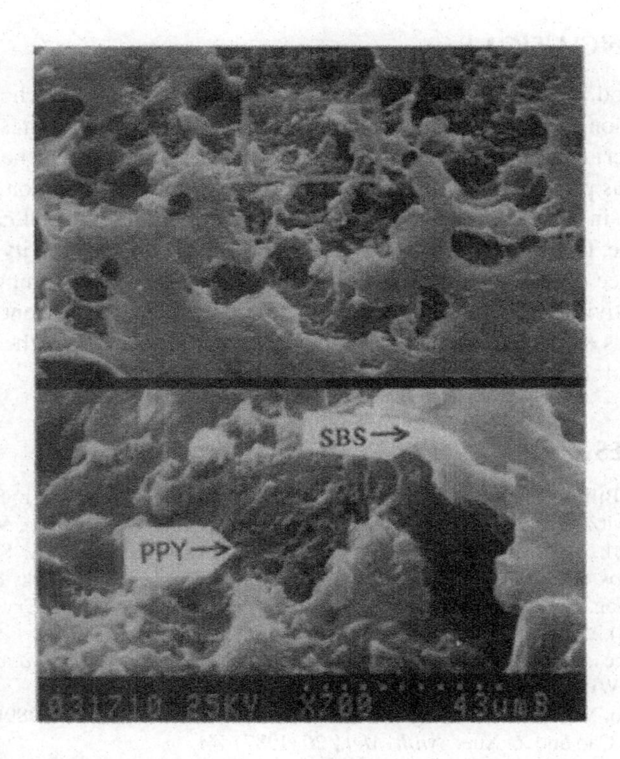

FIGURE 1.5.5 SEM micrograph of the morphology of the hot-pressed sheet of PPY-SBS (2/3) of Table 1.5.6.

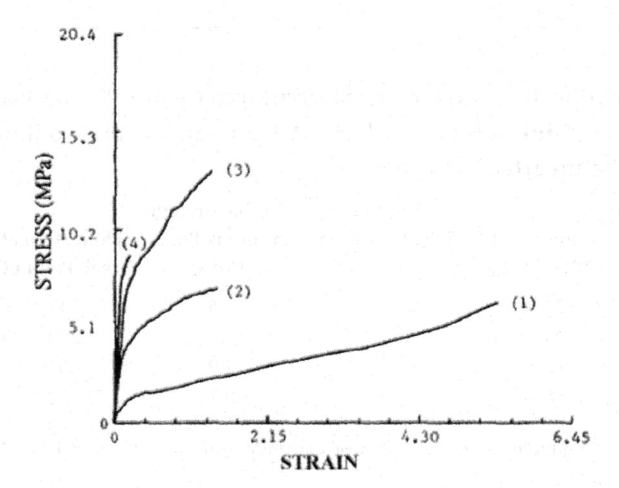

FIGURE 1.5.6 Stress-strain curves of the polypyrrole-SBS composites prepared by the inverted emulsion procedure by using toluene (40 mL)/isooctane (45 mL) as the continuous phase: (1) PPY-SBS (0.5/3); (2) PPY-SBS (1/3); (3) PPY-SBS (1.5/3); and (4) PPY-SBS (2/3) (see also Table 1.5.6).

1.5.4 CONCLUSIONS

A new method, based on inverted emulsions, is proposed for the chemical oxidative preparation of polypyrrole and polypyrrole-elastomer composites. Because of the large interfacial area between the dispersed phase containing the oxidant and the continuous phase containing the pyrrole and polymer, the reaction is very rapid (taking place in a few minutes). The dispersant (AOT in the present case) plays an important role. Indeed, when polypyrrole was prepared, its conductivity was 34 S/cm in the presence of the surfactant and 9 S/cm in its absence. The composites exhibit good conductivities and mechanical properties. When a mixed solvent (toluene and isooctane) was employed, the conductivity was as high as 2–3 S/cm, the strength was 7–13 MPa and the elongation 47%–259%.

REFERENCES

1. (a) N.C. Billingtram and P.D. Calvert, *Adv. Polym. Sci., 90* (1989) 1; (b) S. Jasne, in J.I. Kroschwitz (ed.), *Encyclopedia of Polymer Science and Engineering*, Vol. 13, Wiley, New York, 1990, p. 42; (c) P. Burgmayer and R.W. Murray, in T.A. Skotheim (ed.), *Handbook of Conducting Polymers*, Vol. 1, Marcel Dekker, New York, 1986, p. 507.
2. M. Salmon, A.F. Diaz, A.J. Logan, M. Krounbi and J. Bargon, *Mol. Cryst. Liq. Cryst., 83* (1982) 256.
3. M.B. Jones, in J.I. Kroschwitz (ed.), *Encyclopedia of Polymer Science and Engineering*, Vol. 10, Wiley, New York, 1990, p. 670.
4. L.H. Dao, M. Leclerc, J. Guay and J.W. Chevalier, *Synth. Met., 29* (1989) E377.
5. S. Li, Y. Cao and Z. Xue, *Synth. Met., 20* (1987) 141.
6. G.B. Street, in T.A. Skotheim (ed.), *Handbook of Conducting Polymers*, Vol. 1, Marcel Dekker, New York, 1986, p. 265.

7. G.E. Wnek, in T.A. Skotheim (ed.), *Handbook of Conducting Polymers*, Vol. 1, Marcel Dekker, New York, 1986, p. 205.
8. M. Makhlouki, J.C. Berne'de, M. Morsli, A. Bonnet, A. Conan and S. Lefrant, *Synth. Met., 62* (1994) 101.
9. C.J. Li and Z.G. Song, *Synth. Met., 40* (1991) 23.
10. S. Jasne and C.K. Chiklis, *Synth. Met., 15* (1986) 175.
11. A.I. Nazzal and G.B. Street, *J. Chem. Soc. Chem. Commun., 375* (1985).
12. (a) E. Ruckenstein and J.S. Park, *J. Appl. Polym., 42* (1991) 925; (b) E. Ruckenstein and J.S. Park, *Synth. Met., 44* (1991) 293.
13. E. Ruckenstein and J.H. Chen, *Polymer, 32* (1991) 1231.
14. (a) E. Ruckenstein and S.Y. Yang, *Synth. Met., 53* (1993) 249; (b) S.Y. Yang and E. Ruckenstein, *Synth. Met., 59* (1993) 1; (c) S.Y. Yang and E. Ruckenstein, *Synth. Met., 60* (1993) 249; (d) E. Ruckenstein and S.Y. Yang, *Polymer, 34* (1993) 4655.
15. J.-E. Österholm, Y. Cao, F. Klavetter and P. Smith, *Synth. Met., 55–57* (1993) 1034.
16. G.B. Street, S.E. Lindsey, A.I. Nazzal and K.J. Wynne, *Mol. Cryst. Liq. Cryst., 118* (1985) 137.

1.6 Poly(3-methylthiophene)-Rubber Conductive Composite Prepared via an Inverted Emulsion Pathway*

Yue Sun and Eli Ruckenstein

Department of Chemical Engineering, State University of New York at Buffalo, Buffalo, NY 14260, USA

ABSTRACT An inverted emulsion pathway was employed to prepare poly(3-methylthiophene)-rubber conductive composites. In the first step, an inverted emulsion containing an aqueous solution of $FeCl_3$ as dispersed phase and an organic solution of a host polymer and a surfactant as continuous phase were generated. In the second step, a toluene solution of 3-methylthiophene was introduced dropwise into the inverted emulsion where the monomer was polymerized by the oxidant $FeCl_3$ to the conductive polymer. After the completion of polymerization, the composite was obtained by precipitating the conductive polymer and the host polymer. Under suitable preparation conditions, composites with both high conductivities and good mechanical properties were obtained; these composites had a conductivity as high as 1.3 S/cm as well as a tensile strength of 14.8 MPa and an elongation of 84.0% at break point. Among thiophene, 2,2′-bithiophene and 3-methylthiophene, the latter monomer is the most suitable for preparing thiophene-containing conductive composites by the inverted emulsion method.

* *Synthetic Metals*. Vol. 74, p. 145–150, 1995.

1.6.1 INTRODUCTION

Polythiophene and its derivatives are polymers that, due to the delocalized π-bond along their chains, change from semiconductors to conductors when suitably doped with Lewis acids. This change during doping and their stability toward oxygen and moisture stimulated their use for chemical detectors [1], electrochromic devices [2], electrooptical devices [3], actuators [4], batteries [5] and in catalysis [6,7]. The conductive composites containing polythiophene and its derivatives can also be used as antistatic and electromagnetic shielding materials.

The polymerization of thiophene was first reported in the last century [8]. However, only in the last two decades extensive research was carried out regarding polythiophene and its derivatives. The thiophene units in polythiophene are linked through both the 2 and 3 positions; the irregularity caused by the units involving the 3 position decreases the conductivity of polythiophene because it interrupts the long delocalized π-bond of those linked through the 2 position. However, when the 3 position of the thiophene unit is occupied by an organic group, such as in the 3-alkylthiophene, the conductivity of the polymer increases in a major way, because of increased linkage ordering [9,10]. Polythiophene and its derivatives have been synthesized both electrochemically [11–13] and chemically [14–18]. Generally, the polymers prepared electrochemically have high conductivities; for example, poly(3-methylthiophene) has a conductivity as high as 2×10^3 S/cm [14]. However, this method can only be used for preparing films, and the conductivity depends on the thickness of the film. In contrast, by chemical polymerization, bulk polythiophene and its derivatives can be prepared, but they have relatively low conductivities [14–18]. 2,5-Dihalogenthiophene or its derivatives were transformed via dehalogenative polycondensation into conductive polymers using a Grignard reaction with Mg, in the presence of Ni catalyst, in water-free systems [14,15]. Thiophene and 2,2′-bithiophene were also polymerized to polythiophene using SbF_5 as oxidant [16], and 3-alkylthiophene and 2,2′-bithiophene were chemically polymerized using $FeCl_3$, $Fe(ClO_4)_3$ or $Cu(ClO_4)_2$ as oxidant [17,18]. Among the above three chemical oxidative polymerization methods, the latter is preferable because the restriction of a water-free system is not necessary.

Since polythiophene is insoluble, infusible and brittle, and hence unprocessable, methods were developed to increase its processability. The first consisted of introducing a long (the number of carbon atoms in the chain was at least four) alkyl [17], alkoxy [19] or alkylsulfonate [20] group in the 3 position of thiophene to generate soluble polythiophene derivatives. The second was based on processable copolymers such as polythiophene-poly(vinyl alcohol) graft polymer [21]. The third was based on composites with polythiophene or its insoluble derivative as the conductive component and another polymer as the flexible one. For instance, a poly(3-methylthiophene)–poly(methyl methacrylate) composite was prepared electrochemically using a solution of 3-methylthiophene and poly(methyl methacrylate) [22]. Another composite was prepared chemically by first polymerizing thiophene and then methyl methacrylate in a mixture of the two monomers [23]. Large objects of polythiophene-polystyrene composites were obtained by immersing a porous polystyrene material which had absorbed 2,2′-bithiophene from its acetonitrile solution in another acetonitrile solution containing Fe(III) [18].

Emulsion pathways have been developed in this laboratory to prepare conductive composites. One of them started with a concentrated emulsion containing an aqueous solution of a surfactant as the continuous phase and an organic solution of a host polymer and a monomer as the dispersed phase. The conductive composites were obtained by introducing an aqueous solution of an oxidant and a dopant with stirring into the concentrated emulsion and precipitating the conductive polymer and the host polymer with methanol after completion of the polymerization [24–27]. Another emulsion pathway was based on inverted emulsions. An inverted emulsion was first prepared, which had an organic solution of a host polymer and a surfactant as continuous phase and an aqueous solution of an oxidant and dopant as dispersed phase. Subsequently, an organic solution of a monomer was introduced with stirring into the inverted emulsion to polymerize the monomer, and the conductive and host polymers were precipitated with methanol [28–30].

In the present paper, poly(3-methylthiophene)-rubber composites with relatively high conductivities and good mechanical properties were prepared via the inverted emulsion pathway. The effects of the preparation conditions, such as the nature of the host polymer and the solvent, and the amounts of oxidant and monomer were investigated.

1.6.2 EXPERIMENTAL

1.6.2.1 Chemicals

Thiophene (Aldrich, 99+%), 3-methylthiophene (MT, Aldrich, 99+%), 2,2′-bithiophene (BT, Aldrich, 97%), poly(butyl methacrylate) (PBMA, Aldrich, high molecular weight), styrene-butadiene-styrene triblock copolymer (SBS, Scientific Polymer Products Inc., styrene content: 28%), styrene–ethylbutylene–styrene triblock copolymer (SES, Scientific Polymer Products Inc., styrene content: 28%), styrene–isoprene–styrene triblock copolymer (SIS, Scientific Polymer Products, Inc., styrene content: 14%), iron(III) chloride (Aldrich, 97%), Span 80 (sorbitan monooleate, Fluka Chemika), toluene (Aldrich, 99%+), isooctane (Aldrich, 99%+), chloroform (Aldrich, HPLC), benzene (Aldrich, HPLC), petroleum ether (Aldrich, A.C.S. reagent) and methanol (Aldrich, 99%) were used as received. Distilled and deionized water was employed.

1.6.2.2 Preparation of Poly(3-methylthiophene)-Rubber Composites

In a typical experiment, 1.5 g SBS and 0.66 g Span 80 were dissolved in a mixture containing 15 mL toluene and 15 mL isooctane placed in a 500 mL three-necked round-bottomed flask equipped with mechanical stirring. 30 g $FeCl_3$ was dissolved in 10 g water, and the aqueous solution was introduced with vigorous stirring (about 700 rpm) into the flask (as the dispersed phase) to generate an inverted emulsion. Then, a solution of 0.5 g 3-methylthiophene in 10 mL toluene was introduced at room temperature with stirring. The polymerization was allowed to proceed for about 3 h. Finally, 150–200 mL methanol was added to the system to precipitate the composite which was then filtered, washed with methanol three or four times and immersed in petroleum ether for 15–20 min and filtered. The washed composite was dried in the fume hood overnight.

1.6.2.3 Measurement of the Electrical Conductivity

The composite was pressed at 150°C for 1–2 min with a Carver Laboratory Press (model C) and the obtained sheet was cut to rectangular shape (2.5 × 1.5 cm). The conductivity was measured at room temperature by the four-point method.

1.6.2.4 Tensile Testing

The sample sheet was cut to the size required by the ASTM D.638-58T. The tensile testing was performed with an Instron Testing Instrument (model 1000) at room temperature with a cross-head speed of 10 mm/min.

1.6.2.5 Scanning Electron Microscopy (SEM) and Energy Dispersive Spectroscopy (EDS) Analyses

The hot-pressed sample was immersed in toluene for 2 h to dissolve the host polymer present on the surface. This allowed the observation of the dispersion of the conductive polymer particles in the composite with a Hitachi S-800 scanning electron microscope. The EDS analysis was performed with PGT/TMIX field emission microscopy equipment.

1.6.2.6 Elemental Analysis

The elemental analysis was performed by Quantitative Technologies, Inc. (Whitehouse, NJ).

1.6.3 RESULTS AND DISCUSSION

Table 1.6.1 presents the effect of the content of poly(3-methylthiophene) (PMT) on the properties of PMT-SBS composites. As PMT increases from 7.9 to 21.6 wt.%, the conductivity of the composite increases from 0 to 1.3 S/cm and reaches a plateau with further increase in PMT. The mechanical properties become increasingly unsatisfactory at the high content levels of PMT. The percolation threshold for conductivity is between 18.3 and 21.6 wt.% PMT. The EDS analysis (Figure 1.6.1) indicates that large amounts of S and Cl and only a small amount of Fe are present in the composites. The elemental analysis of the composite prepared from a weight ratio of MT/SBS = 0.5/1.5 indicated 7.05 wt.% S, 3.98 wt.% Cl

TABLE 1.6.1
The Effect of the Weight Feeding Ratio of MT/SBS on the Properties of PMT–SBS Composite

Weight Ratio of MT/SBS	Content of PMT (wt.%)	Conductivity (S/cm)	Tensile Strength (MPa)	Elongation at Break Point (%)
0.15/1.5	7.9	0	10.0	310
0.20/1.5	9.3	4.6×10^{-4}	10.1	250
0.25/1.5	11.8	4.1×10^{-2}	11.7	203
0.40/1.5	18.3	0.18	14.8	184
0.50/1.5	21.6	1.3	14.2	84.0
0.80/1.5	29.2	1.0	14.3	42.0
0.86/1.5	30.5	1.1	brittle	

Notes: The composites were prepared under the conditions: (1) The continuous phase of the inverted emulsion was a mixture of 30 mL toluene/isooctane (1/1 vol./vol.) which contained 0.022 g/mL Span 80 as surfactant and 1.5 g SBS as host polymer. (2) The dispersed phase was 30 g $FeCl_3$ in 10 g water. (3) 10 mL MT containing toluene was introduced into the inverted emulsion.

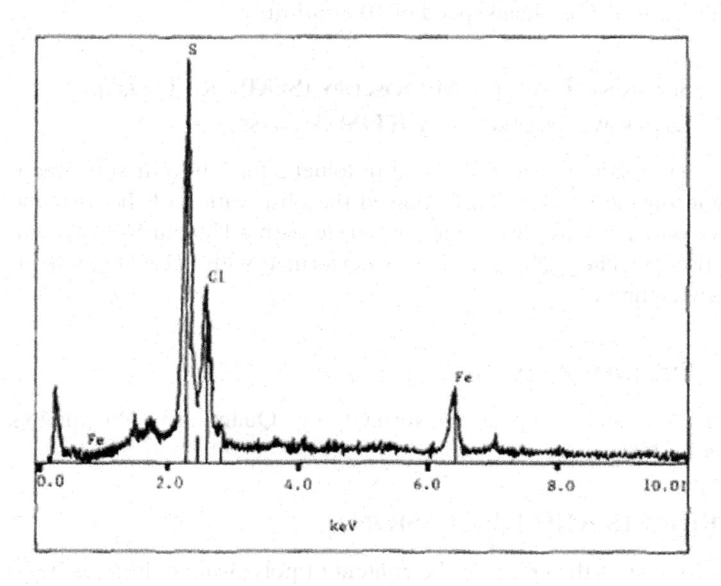

FIGURE 1.6.1 EDS spectrum of PMT–SBS composite. The composite was prepared under the conditions of Table 1.6.1 from a weight ratio MT/SBS of 0.5/1.5.

and 0.83 wt.% Fe; hence, the molar ratios of MT/Cl/Fe in the composite were 1.97/1.0/0.13. Consequently, only a moderate fraction of MT was doped and only traces of $FeCl_3$ remained in the composite, even though a large amount of $FeCl_3$ was employed. Since PMT provides rigidity to the composite and interrupts the continuity of the host polymer, which provides flexibility, it is expected that

FIGURE 1.6.2 Scanning electron micrograph of the PMT-SBS composite of Figure 1.6.1.

the elongation of the composite would decrease with a concerted increase in tensile strength as the amount of PMT increases. The optimum content of PMT in the composite is 21.6 wt.%, for which the composite has a conductivity of 1.3 S/cm with a tensile strength of 14.8 MPa and an elongation at the break point of 84.0%. Figure 1.6.2 provides some information about the dispersion of PMT particles in the composite.

For comparison purposes, PMT-SBS composites were also prepared in a single step. The procedure consisted of directly pouring the aqueous solution of 30 g $FeCl_3$ in 10 g water with vigorous stirring into the toluene/isooctane (25 mL/15 mL) solution containing 1.5 g SBS, 0.66 g Span 80 and MT. After the completion of polymerization, the composite was collected as in the inverted emulsion method. The properties of the prepared composites are presented in Table 1.6.2. Comparing with Table 1.6.1, one can see that the composites prepared in a single step have much lower conductivities than those prepared via inverted emulsions. The reason is as follows: The oxidative polymerization occurs at the interface between the aqueous and the organic phases, since the oxidant $FeCl_3$ is present in the aqueous solution and MT in the organic phase. Consequently, the larger the interface area between the two phases, the more uniform the dispersion of PMT in the composite will be, and hence the higher the conductivity of the composite. In the direct mixing (single-step)

TABLE 1.6.2
The Properties of PMT-SBS Composites Prepared by the Direct Mixing (Single-Step) Procedure

Weight Ratio of MT/SBS	Conductivity (S/cm)	Tensile Strength (MPa)	Elongation at Break Point (%)
0.20/1.5	0	4.3	73.5
0.25/1.5	1.3×10^{-3}	3.6	160
0.40/1.5	0.03	10.0	47.5
0.50/1.5	0.10	6.0	138
0.86/1.5	0.60	17.5	26.0

Notes: (1) The organic phase was a mixture of toluene/isooctane (25 mL/15 mL) containing 0.66 g Span 80, 1.5 g SBS and MT. (2) The aqueous phase was 30 g $FeCl_3$ in 10 g water.

procedure, the processes of polymerization and formation of the interfacial area start at the same time. Since the polymerization is rapid, the dispersion of the PMT in the composite is less uniform and, as a result, the conductivity of the composite is relatively low. The fact that in Table 1.6.2 the tensile strength and elongation at the break point change in an unexpected way with the increase of the weight ratio of MT/SBS also suggests a non-uniform dispersion of PMT in the composite.

Table 1.6.3 presents the effect of the concentration of $FeCl_3$ (C_{FeCl_3}) on the properties of the composite. With the increase of C_{FeCl_3} from 2.4 to 3.0 g/mL, the conductivity increases from 2.9×10^{-4} to 1.3 S/cm; however, with a further increase of C_{FeCl_3}, the conductivity decreases. Compared with the amount of MT, the amount of $FeCl_3$ employed is in high excess. Obviously, not all of $FeCl_3$ participates in the polymerization process. However, a sufficiently high C_{FeCl_3} is necessary. Indeed, the standard reduction potential of Fe(III)/Fe(II) (0.77 V) is lower than the standard oxidation potential of MT (1.48 V); therefore the oxidative polymerization cannot occur under standard conditions. However, the reduction potential can be increased by increasing the concentration of $FeCl_3$. When C_{FeCl_3} is high enough, the reduction potential of Fe(III)/Fe(II) increases to a value higher than the oxidation potential of MT and the polymerization becomes possible. For this reason, the polymerization hardly occurs with C_{FeCl_3} lower than 2.4 g/mL, but proceeds more easily for higher concentrations. As a result, the conductivity of the composite increases and reaches 1.3 S/cm at = C_{FeCl_3} 3.0 g/mL. When C_{FeCl_3} increases further, the reduction potential of Fe(III) /Fe(II) becomes so high that the oxidation proceeds faster than it takes for MT to be well mixed with the organic phase of the inverted emulsion. This leads to a less uniform dispersion of PMT in the composite and hence to a lower conductivity.

For comparison purposes, thiophene and 2,2′-bithiophene (BT) were also employed to prepare conductive composites. Since the standard oxidation potential of thiophene (1.6 V) [31] is higher than that of MT, the polymerization of thiophene does not occur even with C_{FeCl_3} as high as 3.0 g/mL. The standard oxidation potential of 2,2′-bithiophene is lower (1.2 V) [31], and therefore it can be polymerized more easily. In the latter case, the polymerization proceeds even with C_{FeCl_3} as low

TABLE 1.6.3
The Effect of the Concentration of $FeCl_3$ (C_{FeCl_3}) in the Dispersed Phase of the Inverted Emulsion on the Properties of PMT-SBS Composite

C_{FeCl_3}(g/mL)	Conductivity (S/cm)	Tensile Strength (MPa)	Elongation at Break Point (%)
3.4	0.40	17.1	75.0
3.2	0.70	18.5	83.5
3.0	1.3	14.2	84.0
2.8	0.07	14.8	104
2.4	2.9×10^{-4}	10.1	49.5

Notes: The amount of MT was 0.50 g. Other preparation conditions were as in Table 1.6.1.

as 1.5 g/mL; however, the PBT-SBS made from a weight ratio BT/SBS = 1.0/1.5 was an isolator. The conductivity of the composite increases to 7.7×10^{-4} S/cm as C_{FeCl_3} increases to 2.6 g/mL and decreases as C_{FeCl_3} increases further. When the weight ratio of BT/SBS is between 0.43 g/1.5 g and 1.0 g/1.5 g, the PBT-SBS composites prepared under the conditions of Table 1.6.1 have conductivities between 1.1×10^{-5} and 6.0×10^{-5} S/cm, which are much lower than those of PMT-SBS composites of Table 1.6.1. The reason for such a low conductivity may be the irregular linkages of BT units in 4 (4′) and 5 (5′) positions which interrupt the long delocalized π bonds. Consequently, among thiophene, 2,2′-bithiophene and 3-methylthiophene, the latter is the most suitable for preparing conductive composites.

Table 1.6.4 presents the effect of the concentration of surfactant in the continuous phase of the inverted emulsion on the properties of the composites. Without surfactant, the composite has a conductivity of 0.09 S/cm which is almost one order of

TABLE 1.6.4
The Effect of the Amount of Surfactant Span 80 in the Continuous Phase of the Inverted Emulsion on the Properties of PMT–SBS Composite

Span 80 (g/mL)	Conductivity (S/cm)	Tensile Strength (MPa)	Elongation at Break Point (%)
0	0.09	13.0	27.0
0.011	0.9	14.8	88.5
0.022	1.3	14.2	84.0
0.031	1.0	14.4	121
0.040	1.0	14.1	108

Notes: The amount of MT was 0.50 g. Other preparation conditions were as in Table 1.6.1.

magnitude lower than that of the composite prepared with the surfactant; its tensile strength and elongation at the break point are also lower. When the amount of Span 80 increases from 0.011 to 0.040 g/mL, the properties of the composite change little. The reason is as follows: As already mentioned, the higher the interface area between the $FeCl_3$ -containing aqueous phase and the MT-containing organic phase, the more uniform is the dispersion of PMT and the higher is the conductivity of the composite. Without the surfactant, the aqueous solution of $FeCl_3$ and the organic solution of the host polymer cannot generate an inverted emulsion, and the interfacial area between the two solutions is rather low. When the toluene solution of MT is introduced, the interface area between the phase containing MT and that containing $FeCl_3$ is also low. However, it becomes much higher when an inverted emulsion is generated with the surfactant. Hence, the prepared composite has a higher conductivity as well as better mechanical properties when a surfactant is employed. Since, with 0.011 g/mL Span 80, the inverted emulsion provides a high enough interfacial area for a uniform dispersion of PMT to be achieved, the further increase of the concentration affects little the properties of the composite.

Table 1.6.5 shows how the nature of the organic phase of the inverted emulsion affects the properties of PMT-SBS composites. When benzene, chloroform or toluene is employed, the composite has a relatively low conductivity (lower than 0.12 S/cm); however, when a toluene/isooctane mixture is used, the conductivity increases with the decrease of volume ratio of toluene/isooctane from 30/0 to 15/15, but changes little on further decrease. Consequently, the optimum solvent is toluene/isooctane (15/15), for which the composite has a conductivity of 1.3 S/cm as well as good mechanical properties. The reason is that the conformation of the host polymer is quite different in different solvents. In good solvents, such as chloroform, benzene or toluene, the molecular chains of SBS are extended and the PMT particles are easily surrounded separately by molecules of SBS. It is therefore more difficult for them to generate a conductive network throughout the composite. As a result, the conductivity of the composite is low.

TABLE 1.6.5
The Effect of the Nature of the Organic Solvent of the Continuous Phase of the Inverted Emulsion on the Properties of PMT–SBS Composite

Solvent (mL)	Conductivity (S/cm)	Tensile Strength (MPa)	Elongation at Break Point (%)
Benzene (30)	0.08	16.2	110
Chloroform (30)	0.04	13.7	280
Toluene (30)	0.12	9.0	300
Toluene/isooctane (25/5)	0.13	16.4	264
Toluene/isooctane (20/10)	0.16	16.3	138
Toluene/isooctane (15/15)	1.3	14.2	84.0
Toluene/isooctane (10/20)	0.9	11.7	11.0

Notes: The amount of MT was 0.50 g. Other preparation conditions were as in Table 1.6.1.

TABLE 1.6.6
The Effect of the Amount of Toluene in the Monomer Solution on the Properties of PMT-SBS Composite

Volume of Toluene (mL)	Conductivity (S/cm)	Tensile Strength (MPa)	Elongation at Break Point (%)
6.0	0.13	15.1	100
8.0	1.3	14.6	84.5
10.0	1.3	14.2	84.0
12.0	0.75	15.7	77.5
14.0	0.60	5.7	28.5
20.0	2.2×10^{-4}	brittle	

Notes: The amount of MT was 0.50 g. Other preparation conditions were as in Table 1.6.1.

Table 1.6.6 presents the effect of the amount of toluene in the solution of the monomer on the properties of the PMT-SBS composites. The conductivity of the composite increases with volume of toluene, reaches the highest value of 1.3 S/cm for volumes of 8 and 10 mL, and then decreases with a further increase of volume. The reason may be as follows: With the increase of the volume of toluene, the concentration of MT in the organic phase decreases and this results in a more uniform dispersion of PMT in the prepared composite, as well as in a lower molecular weight of PMT. The former effect increases while the latter decreases the conductivity of PMT. Due to the two opposite factors, the conductivity values shown in Table 1.6.6 pass through a maximum.

Table 1.6.7 illustrates the effect of the volume of the aqueous solution of $FeCl_3$ on the properties of the PMT–SBS composites. When the volume increases from 5 to 10 mL, the conductivity increases from 0.26 to 1.3 S/cm; when the volume increases further, the conductivity changes little, but the mechanical properties become unsatisfactory (for example, the elongation at break point decreases

TABLE 1.6.7
The Effect of the Volume of the Dispersed Phase of the Inverted Emulsion on the Properties of PMT-SBS Composite

Volume of Dispersed Phase (mL)	Conductivity (S/cm)	Tensile Strength (MPa)	Elongation at Break Point (%)
5.0	0.26	10.0	78.0
10.0	1.3	14.2	84.0
15.0	1.6	8.2	10.0
20.0	0.9	brittle	

Notes: The amount of MT was 0.50 g. The concentration of the dispersed phase and other preparation conditions were as in Table 1.6.1.

TABLE 1.6.8
The Effect of the Nature of the Host Polymer on the Properties of PMT-Rubber Composite

Host Polymer	Conductivity (S/cm)	Tensile Strength (MPa)	Elongation at Break Point (%)
SBS	1.3	14.2	84.0
SES	0.4	8.9	191
SIS	0.14	brittle	
PBMA	1.3	brittle	

Notes: The amount of MT was 0.50 g. Other preparation conditions were as in Table 1.6.1.

from 84.0 to 0%). With the increase of volume of the dispersed phase of the inverted emulsion, the interfacial area between the dispersed and the continuous phases of the inverted emulsion increases. As a result, the size of the PMT particles becomes smaller and their number becomes larger; in addition, PMT is more uniformly dispersed in the composite. When the volume increases from 5 to 10 mL, the conductivity increases. For a volume of 10 mL, the dispersion of PMT in the composite is uniform enough for a conductive network to be generated; as a result, the further increase of volume affects little the conductivity. The more uniform distribution has two effects on mechanical properties. On one hand, it has a positive effect, since the non-uniformity decreases both the elongation and the tensile strength. On the other hand, a large contact area between the increasingly smaller particles of PMT and SBS interrupts more frequently the structure of the latter. As a result, an optimum amount of solvent exists.

Table 1.6.8, which summarizes the effect of the nature of the host polymer on the properties of the composite, allows us to select SBS as the most suitable host polymer.

1.6.4 CONCLUSIONS

Poly(3-methylthiophene)-rubber conductive composites can be prepared via an inverted emulsion pathway. The procedure consists of introducing an organic solution of the monomer into an inverted emulsion containing an aqueous solution of $FeCl_3$ as dispersed phase and an organic solution of a host polymer and a surfactant as the continuous phase. The composite is precipitated with methanol after the completion of the oxidative polymerization. Under suitable preparation conditions, a composite with a conductivity as high as 1.3 S/cm and good mechanical properties (tensile strength 14.8 MPa and elongation at the break point 84.0%) was obtained. The inverted emulsion method is more efficient than the direct mixing (single-step) method, and 3-methyl-thiophene is a more suitable monomer than thiophene or 2,2′-bithiophene for obtaining conductive composites.

REFERENCES

1. H.S.O. Chan, S.C. Ng and S.H. Seow, *Synth. Met., 66* (1994) 177.
2. K. Yoshino, K. Kaneto and Y. Inuishi, *Jpn. J. Appl. Phys., 22* (1983) 157.
3. S. Destri, W. Porzio, M. Nisoli, S. De Silverstri, D. Grando and S. Sottini, *Synth. Met., 67* (1994) 293.
4. Q. Pei and O. Inganäs, *Synth. Met., 55–57* (1993) 3730.
5. K. Kaneto, K. Yoshino and Y. Inuishi, *Jpn. Appl. Phys., 21* (1982) 567.
6. T. Kawai, T. Kuwabara and K. Yoshino, *Synth. Met., 55–57* (1993) 3649.
7. G. Tourillon and F. Gamier, *Mol. Cryst. Liq. Cryst., 121* (1985) 305.
8. V. Meyer, *Ber. Dtsch. Chem. Ges., 16* (1883) 1465.
9. S. Hotta, T. Hosaka and W. Shimotsuma, *Synth. Met., 6* (1983) 317.
10. M. Sato, S. Tanaka and K. Kaeriyama, *J. Chem. Soc., Chem. Commun.*, (1985) 713.
11. M. Sato, S. Tanaka and K. Kaeriyama, *Makromol. Chem., 188* (1987) 1763.
12. S. Hotta, S. Rughooputh, A. Heeger and F. Wudl, *Macromolecules, 20* (1987) 212.
13. J. Roncali, R. Garreau, A. Yassar, P. Marque, F. Gamier and M. Lemaire, *J. Phys. Chem., 91* (1987) 6706.
14. M. Kobayashi, C. Chung, F. Moraes, A. Heeger and F. Wudl, *Synth. Met., 19* (1984) 77.
15. C. Hotz, P. Covacic and I. Khoury, *J. Polym. Sci., Polym. Chem. Ed., 21* (1983) 2617.
16. G. Koßmehl and G. Chatzitheodorou, *Mol. Cryst. Liq. Cryst., 83* (1982) 291.
17. J. Österhlm, J. Laakso and P. Nyholm, *Polym. Prepr., 30* (1989) 145.
18. E. Ruckenstein and J.S. Park, *Synth. Met., 44* (1991) 293.
19. M. Feldhues, G. Kämpf, H. Litterer, T. Mecklenburg and P. Wegener, *Synth. Met., 28* (1989) C487.
20. A.O. Patil, *Synth. Met., 28* (1989) C495.
21. M. Wakabayashi, Y. Miyazaki, T. Kanbara, K. Osakada and T. Yamamoto, *Synth. Met., 55–57* (1993) 3632.
22. J. Roncali and F. Gamier, *J. Chem. Soc., Chem. Commun.*, (1986) 783.
23. W.-S. Huang and J.M. Park, *J. Chem. Soc., Chem. Commun.*, (1987) 856.
24. S. Yang and E. Ruckenstein, *Synth. Met., 59* (1993) 1.
25. E. Ruckenstein and S. Yang, *Synth. Met., 53* (1993) 283.
26. E. Ruckenstein and S. Yang, *Polymer, 34* (1993) 4655.
27. S. Yang and E. Ruckenstein, *Synth. Met., 60* (1993) 249.
28. Y. Sun and E. Ruckenstein, *Synth. Met., 72* (1995) 261.
29. E. Ruckenstein and L. Hong, *Synth. Met., 66* (1994) 249.
30. E. Ruckenstein and Y. Sun, *Synth. Met., 74* (1995) 107.
31. N.C. Billingham and P.D. Calvert, *Adv. Polym. Sci., 90* (1989) 1.

REFERENCES

1. H. S. O. Chan, S. C. Ng, [illegible] *Synth. Met.*, *69* (1995) 27[illegible]
2. K. Yoshino, [illegible] Kaneto and [illegible] *Jpn. J. Appl. Phys.*, *23* (1984) [illegible]
3. [illegible] V. [illegible] *Synth. Met.* [illegible]
(1985) [illegible]
4. G. Bidan and O. [illegible] *Synth. Met.*, *15* (1986) [illegible]
5. K. Kaneto, K. Yoshino and Y. Inuishi, *Jpn. J. Appl. Phys.*, [illegible]
6. T. Kawai, T. Kuwabara and K. Yoshino, *Synth. Met.*, [illegible]
7. [illegible] *Mol. Cryst. Liq. Cryst.*, [illegible]
8. V. Meyer, *Ber. Dtsch. Chem. Ges.*, *16* (1883) 1465.
9. S. Hotta, T. Hosaka and W. Shimotsuma, *Synth. Met.*, [illegible]
10. M. Sato, S. Tanaka and K. Kaeriyama, *J. Chem. Soc., Chem. Commun.*, (1985) [illegible]
11. M. [illegible] S. Tanaka and K. Kaeriyama, *Makromol. Chem.*, *188* (1987) [illegible]
12. [illegible] *Macromolecules*, *20* (1987) [illegible]
13. [illegible] *J. Chem. Soc., Chem. Commun.*, (1987) 70[illegible]
14. M. Kobryanskii, [illegible] *Synth. Met.*, [illegible]
15. [illegible]
16. [illegible]
17. [illegible] *Synth. Met.*, [illegible]
18. E. Ruckenstein and [illegible] *Synth. Met.*, *44* (1991) [illegible]
19. [illegible] *Synth. Met.*, [illegible]
(1987) [illegible]
20. [illegible]
21. [illegible] *Synth. Met.*, [illegible]
[illegible]
22. [illegible] *J. Chem. Soc., Chem. Commun.*, (1986) [illegible]
23. [illegible] *J. Chem. Soc., Chem. Commun.*, (1987) [illegible]
24. S. Yang and E. Ruckenstein, *Synth. Met.*, *59* (1993) [illegible]
25. E. Ruckenstein and S. Yang, *Synth. Met.*, [illegible]
26. E. Ruckenstein and S. Yang, *Polymer*, *34* (1993) [illegible]
27. S. Yang and E. Ruckenstein, *Synth. Met.*, *60* (1993) [illegible]
28. [illegible] and E. Ruckenstein, *Synth. Met.*, *72* (1995) [illegible]
29. E. Ruckenstein and [illegible] *Synth. Met.*, *66* (1994) [illegible]
30. E. Ruckenstein and Y. Sun, *Synth. Met.*, *74* (1995) [illegible]
31. C. [illegible] *J. Am. Chem. Soc.*, [illegible]

2 Core-Shell Latex Particles

CONTENTS

Core-shell particles are structured composite particles consisting of two different materials, one forming the core, the other forming the shell. Such a special structure may render the particles with new or enhanced properties, which cannot be achieved by simply blending the two materials. Because the core is covered by the shell, unless the shell is porous, surface functionalities and properties of the particle are fully determined by the shell. The synthesis of such two-layer particles can provide opportunities to tailor material properties for diverse applications, such as coatings, controlled release, drug delivery, and polymer modification.

The core-shell particles are often prepared by seeded emulsion polymerization. Seed particles are pre-placed in an emulsion containing monomers, initiator, and surfactant. As the polymerization proceeds, the seed nuclei can grow gradually to give the final latex particles. Seed particles need to be small and colloidally stable in the polymerization medium. If the surface area of the seed particles is large enough to absorb all of the precipitating oligomer radicals and all particles grow at the same time scale with well-maintained colloidal stability, the final latex will be monodispersed. In such a case, the number of the final latex particles is equal to the number of seed particles added, and their sizes are determined by the feeding amount of monomer(s) per seed particle, monomer conversion, and porosity of the shell layer.

Section 2.1 describes the preparation of core-shell particles consisting of a soft core with a very low glass transition temperature (T_g) and a rigid shell with a high T_g. In the first step, poly(dimethyl siloxane) particles are obtained via ring-opening polymerization of octamethylcyclotetrasiloxane (D4). Using the resulting PD4 latex as seed, a seeded emulsion polymerization of styrene-methyl methacrylate-acrylic acid (St-MMA-AA) is conducted to obtain PD4-P(St-MMA-AA) core-shell particles.

In Section 2.2, seeded emulsion polymerization of St-MMA-AA, with the monomers deposited onto monodispersed particles of poly(St-MMA), is conducted. Pore generation of the particles, with volume expansion ratio around 12, is observed. A discussion regarding the mechanism of pore generation and particle expansion is provided.

Section 2.3 presents a pathway, other than seeded emulsion polymerization, to prepare core-shell particles. The hydrophilic polyacrylamide core is prepared by concentrated emulsion polymerization and is transferred to a hexane solution of a hydrophobic vinyl monomer. The use of initiators, which are present only on the hydrophilic surface, induces the polymerization of the hydrophobic monomer to form shells that encapsulate the polyacrylamide latexes. Under the proper conditions, a porous outer shell can be generated; as a result, the inside hydrophilic chains can be accessed by external media. Such particles with hydrophilic core and hydrophobic porous shell can be dispersed in water with well-maintained colloidal stability for a long time period; when they are dispersed in hydrophobic liquids, they remain stable for at least 24 h.

The preparation of particles with hydrophobic core and hydrophilic shell is presented in Section 2.4. Because of the similarity in synthetic approaches, Section 2.4 can be considered parallel to Section 2.3.

Section 2.5 demonstrates the preparation of particles with inorganic core and organic shell by encapsulation of inorganic particles with polymer. This synthetic process can be considered as seeded concentrated emulsion polymerization.

2.1 Core-Shell Latex Particles Consisting of Polysiloxane-Poly(styrene-methyl methacrylate-acrylic acid): Preparation and Pore Generation*

Xiang Zheng Kong and Eli Ruckenstein
Department of Chemical Engineering, State University of New York at Buffalo, Amherst, New York 14260

ABSTRACT Latexes with a poly(dimethyl siloxane) core and a poly(styrene-methyl methacrylate-acrylic acid) [poly(St-MMA-AA)] shell have been prepared in two steps in order to generate particles that have a core with a very low glass transition temperature. In the first step, poly(dimethyl siloxane) particles were obtained via the ring-opening emulsion polymerization of octamethyl tetracyclosiloxane (D_4). The polymerization was carried out using either an anionic or a cationic catalyst. In the first case, sodium hydroxide was used as catalyst and sodium dodecylbenzene sulfonate as surfactant, while in the second, the alkylbenzene sulfonic acid (ABSA) was used both as catalyst and surfactant. Using a PD_4 latex as seed, a seeded emulsion polymerization of St-MMA-AA was conducted to obtain PD_4-P(St-MMA-AA) core-shell particles. Numerous recipes were attempted and the most successful were those in which the seed was prepared with a cationic catalyst (ABSA) at a

* *Journal of Applied Polymer Science*. Vol. 73, p. 2235–2245, 1999.

relatively low temperature (75°C). The core-shell structure of the particles was identified by transmission electron microscopy, but also via wetting angle, water absorption, and T_g measurements. Finally, pores were generated in the core-shell particles via an alkali-acid treatment. Because PD_4 has a very low glass transition temperature, it cannot be easily handled. However, protected by a shell, it could be used as a constituent of composite materials with enhanced impact strength, even at very low temperatures.

2.1.1 INTRODUCTION

A common way to prepare new materials is to combine two or more polymers. This can be achieved by grafting, block polymerization, or simple blending. The greater the difference between the combining polymers, the more likely synergistic properties will result from the combination. Polystyrene (PSt) and poly(methyl methacrylate) (PMMA) are among the most popular polymers. Their homo- and copolymers consist of rigid molecular chains and possess high glass transition temperatures (T_g) and high mechanical moduli. In contrast, the organosilicon polymers consist of soft chains, possess very low T_g's, and have properties, such as polarity, surface energy, temperature performance, and mechanical characteristics, very different from those of PSt and PMMA. Consequently, their combination is of great interest. In fact, studies regarding the combination of vinyl polymers, particularly the various polyacrylates, with the polysiloxanes have been carried out.[1–9] Nakamura[1] blended an acrylic rubber with poly(heptamethyl monovinyl tetracyclosiloxane) using poly(methyldodecyl siloxane) as compatibilizer and obtained a rubber with significantly improved mechanical properties. By changing the compatibilizer or the pending group on siloxane, a number of blends have been prepared.[2,3] The copolymerization of siloxanes with vinyl monomers, in the presence of a vinyl-containing siloxane, by bulk, solution, or emulsion polymerization, was also carried out.[4–9]

On the other hand, research on polymer nanoparticles, particularly regarding their applications in biological and pharmaceutical areas, has increased continuously in the last decade.[10–14] The silicon polymers can be employed for biological applications as well as tougheners. However, the silicon polymer particles have a very low T_g and, therefore, cannot be handled easily. The goal of the present work is to prepare core-shell particles with a polysilicon core (PD_4) protected by a shell of P(St-MMA-AA). To make the polysilicon core accessible to the external medium, pores were generated in the shell via an alkali-acid treatment.[15] The core-shell particles thus obtained could be used as a constituent of composite materials with enhanced impact strength even at very low temperatures.

2.1.2 EXPERIMENTAL

2.1.2.1 Materials

Styrene (St), methyl methacrylate (MMA), and acrylic acid (AA) were purchased from Aldrich, and the inhibitors were removed prior to polymerization by passing the monomers through an inhibitor remover prepacked column (Aldrich). The monomer, octamethyl tetracyclosiloxane (D_4, Aldrich), the coupling agents between the core, and the shell polymers, namely, 2,4,6,8-tetramethyl-2,4,6,8-tetravinyl cyclotetrasiloxane (VD_4, Aldrich) and 3-(trimethoxylsilyl)propyl methacrylate (MATS, 98%, Aldrich), as well as sodium dodecylbenzene sulfonate (SDBS, Aldrich), were used as received. A linear alkylbenzene sulfonic acid (ABSA, assay 98%, Alfa) was employed as catalyst and surfactant in the cationic polymerization of D_4. ABSA ($RC_6H_4SO_3H$) is a mixture of alkylbenzene sulfonic acids with 1 wt % R = C_{10}, 40 wt % R = C_{11}, 28 wt % R = C_{12}, and 31 wt % C_{13} and higher alkyls. A 3*N* solution of sodium hydroxide (99.99%, Aldrich) was prepared and employed as catalyst in the anionic emulsion polymerization of D_4. The initiators employed in radical polymerizations (ammonium persulfate, APS, 99.99%, Aldrich; or azobis-isobutyronitrile, AIBN, Polysciences Inc.) were used as received. Distilled and deionized water with a conductivity of 0.05 μS/cm was employed.

2.1.2.2 Emulsion Polymerization of D_4

The polymerization of D_4 was conducted using a cationic (ABSA) or an anionic (NaOH) catalyst. In the former process, 40 g of water, 12.5 g of D_4, and a selected amount of ABSA and SDBS (the total of the two varied between 1.6–5.0 wt % of the monomer) dissolved in 10 g of water were successively charged to a three-necked, round-bottom flask equipped with a stirrer. The flask was then purged of oxygen by bubbling nitrogen for 30 min at room temperature, with stirring at about 400 rpm. The polymerization started once the flask was introduced into an oil bath of controlled temperature. 1 wt % MATS was added in some experiments. For the anionic polymerization, 50 g of water, 12.5 g of D_4, and 0.625 g of MATS were introduced into the flask reactor, and then various amounts of SDBS (0.06–0.16 g) were added. After the system was purged of oxygen by bubbling nitrogen, the flask was introduced in an oil bath of a controlled temperature of 80 ± 1°C, with stirring at 200 or 400 rpm. Various amounts of a solution of NaOH (3*N*) were then added to start the polymerization.

2.1.2.3 Seeded Copolymerization of St-MMA-AA onto PD_4 Latex Particles

In most of the seeded polymerizations, no additional surfactant was added to the system. While the total amount of monomers, St-MMA-AA, was varied in the experiments, their ratios were kept constant (89.6/4.7/5.7 by weight and 87.2/4.8/8.0 by moles). The amount of initiator was 2.5 wt % of the monomers for APS and 1.8 wt % for AIBN. A selected amount of seed latex of PD_4 and water were first charged to the reactor, which was then purged of oxygen by bubbling nitrogen for 30 min. The reactor was then placed in an oil bath (80 ± 1°C), and the monomer mixture added drop-wise at the low rate of about 10 mL/h. In some experiments, a coupling agent between PD_4 and the shell copolymer P(St-MMA-AA), namely, MATS or VD_4, was added and allowed to react for 6 h with the PD_4 particles prior to the seeded polymerization of St-MMA-AA. The amount of coupling agent was

4–5 wt % of PD_4 in the seed particles. When VD_4 was used, the pH of the media was about 2.0 after adding water and VD_4 to the seed latex. The reaction between the seed particles and VD_4 was carried out at 40°C. When MATS was used, the pH of the media had to be adjusted to about 4.5 by adding a solution of sodium hydroxide (0.1*N*) in order to avoid its homopolymerization. Its reaction with PD_4 particles was conducted at room temperature.

2.1.2.4 Particle Size and Monomer Conversion

The attempts to determine the particle sizes of PD_4 latexes by transmission and scanning electron microscopy (TEM and SEM) have failed because, due to their very low T_g, the particles collapsed and even flowed to form a film on the TEM grid.[16] Even those prepared with 1 wt % MATS (55.8% crosslinked, as tested by Soxhlet extraction with toluene) have been seriously deformed, and no particles could be identified (Figure 2.1.1). The size of the latex particles was determined via light scattering, which provided both the arithmetic average diameter and the deviation from the average. The size and morphology of the latex particles after the seeded polymerization was determined by TEM (Jeol-2010 or Hitachi-H500) and SEM (Hitachi S-4000). An aqueous solution of phosphotungstic acid was used to increase the contrast in the TEM pictures.[17] The conversion of the monomers was determined gravimetrically.

2.1.2.5 Alkali-Acid Treatment

The seeded latex was diluted to a solid content of about 1 wt %, and 100 g of the diluted latex was introduced into a three-necked, round-bottom flask equipped with a stirrer; 0.1–0.2 g of SDBS was then added to ensure the latex stability upon addition of NaOH and HCl. A selected amount of methyl ethyl ketone (MEK) was first added to the flask with stirring, and this was followed by the alkali and, subsequently, by the acid treatment. In the alkali treatment, a selected pH was first achieved by the addition of a solution of sodium hydroxide (0.1*N*). The flask, with the mixture in the flask stirred at 100 rpm, was then placed into an oil bath at 90°C, where it was kept for 3 h, after which it was allowed to cool to room temperature. After the pH was changed to a selected lower value by using a solution of hydrochloric acid (1*N*), the flask was reintroduced into the oil bath at 90°C for another 3 h.

2.1.2.6 Additional Experiments

For comparison purposes, a latex (denoted hereafter as S_1) of P(St-MMA-AA) with the same composition as that for the second stage of seeded polymerization was prepared in the absence of seed particles. Dry polymers of S_1 and PD_4, their blends as well as the polymer of the PD_4/P(St-MMA-AA) core-shell latex were obtained by evaporating the water from the latexes at 40°C in a vacuum oven at 29.5 in. Hg for 40 h. The T_g's of the polymers were determined by using a DuPont

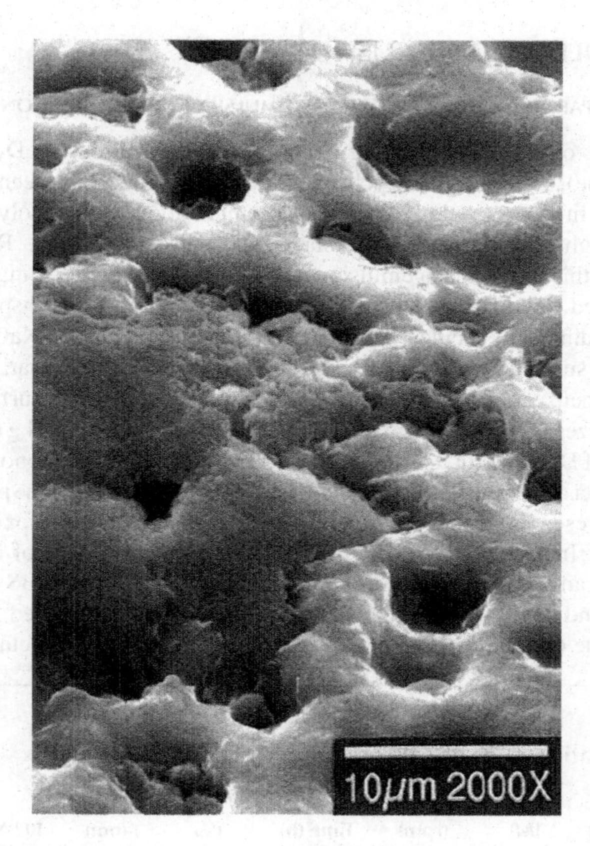

FIGURE 2.1.1 SEM micrograph of PD_4 latex particles prepared with MATS (1 wt % of D_4 monomer) showing that no particles can be identified due to their deformation.

910 DSC instrument and a temperature scanning from −170 to 230°C at 10°C/min. The contact angles of water droplets on various polymers were determined with an NRL C.A. Goniometer manufactured by Rame-Hart Inc. Disc-shaped polymer tablets of about 0.1 g, which were prepared using a Carver Laboratory Press (model C) with an applied load of 4000 psi at room temperature, were employed in the contact angle experiments. Such tablets were also used for determining the water absorption of the polymers: A tablet was immersed in water for 20 h at room temperature, and the water on its surface was wiped off with an absorbent tissue; the water absorption was obtained from the difference of the weights of the tablet before and after immersion. The surface charge density[18] of the latex particles was determined by first diluting 5 mL latex with 145 mL water and then by following the change of the conductivity in the diluted latex with the slow addition of a base (NaOH, 0.1*N*).

2.1.3 RESULTS AND DISCUSSION

2.1.3.1 Preparation of PD_4 Latex by Emulsion Polymerization

The synthesis of PD_4 by the ring-opening polymerization of D_3 (hexamethyl tricyclosiloxane), D_4, or other cyclic siloxane monomers has been described in literature.[19–24] In general, their solution polymerization yields polymers of low-molecular-weight and broad-molecular-weight distribution.[19,24] Regarding the preparation of this polymer via emulsion polymerization, the following contributions should be noted: Hyde and Wehrly[21] obtained a stable aqueous dispersion of PD_4 particles, containing molecules of high molecular weight, using NaOH as catalyst and a cationic surfactant (R_4NCl); Bey et al.[16,22] succeeded in preparing a PD_4 latex using dodecylbenzene sulfonic acid (DBSA) as both catalyst and surfactant; latexes with particle sizes between 50 and 500 nm were obtained with 1.5–4 g of DBSA (or a combination of DBSA with its sodium salt) for 100 g of siloxane monomer. Recently, de Gunzbourg et al.[24] published some preliminary results regarding the polymerization of D_4 in the presence of benzyldimethyl dodecyl ammonium hydroxide as catalyst and surfactant. In the present study, the emulsion polymerization of D_4 was carried out with either an anionic or a cationic catalyst. In the first case, SDBS was employed as surfactant and sodium hydroxide as catalyst. The results are listed in Table 2.1.1, from which one can see that three main factors (stirring rate, surfactant, and NaOH

TABLE 2.1.1
D_4 Polymerization with NaOH as Anionic Catalyst

Runs[a]	SDBS (g)	NaOH [*M*]	Stirring (rpm)	Reaction Time (h)	Conversion (%)	D_p[b] (nm)	$10^{-15}N_p$[c] (L^{-1})	Latex Stability
A41	0.06	0.51	400	45	62	—	—	F[d]
A42	0.06	0.26	400	50	48	—	—	F
A43	0.08	0.26	400	35	51	—	—	F
A44	0.10	0.26	400	75	60	—	—	F
A45	0.10	0.30	400	80	66	—	—	F
A46	0.12	0.26	400	80	70	—	—	F
A47	0.14	0.26	400	70	90	392 ± 45	5.53	S[d]
A48	0.16	0.26	400	70	95.9	143 ± 10	121.87	S
A21	0.08	0.34	200	50	67.3	351 ± 29	—	F
A22	0.08	0.26	200	50	50.1	—	—	F
A23	0.10	0.26	200	90	82.2	260 ± 9	17.26	S
A24	0.12	0.26	200	120	88.4	268 ± 14	11.54	S
A25	0.14	0.26	200	90	96.7	192 ± 12	50.22	S
A26	0.16	0.26	200	160	96.0	180 ± 6	62.47	S

[a] H_2O, 50 g; D_4, 12.5 g; MATS, 0.625 g; polymerization at 80°C.
[b] Diameter of particles determined by light scattering.
[c] Number of particles per liter of latex.
[d] S stands for stable latexes; F implies that the latex flocculated during polymerization or was phase-separated when the stirring was stopped.

concentrations) affect the polymerization process. At a stirring rate of 400 rpm and an amount of SDBS less than 0.12 g (A41 to A46), the latexes became unstable at relatively low conversions; the latexes were, however, stable at higher SDBS contents and high conversions were reached. At a stirring rate of 200 rpm, the latexes became stable for SDBS amounts greater than 0.10 g (A23). In the stable latexes, the particle size decreased with increasing surfactant amount at a given stirring rate (compare A47 and A48, A24, A25, and A26). It should be pointed out that, except the latexes A25 and A26, which remained stable on the shelf at room temperature after 3 months, the other latexes, classified as stable immediately after polymerization, flocculated within two months of shelf standing.

The emulsion polymerization of D_4 by cationic catalysis was carried out either with ABSA alone as catalyst and surfactant, or with a combination of ABSA and SDBS. In contrast to the latexes prepared by anionic catalysis, all the latexes prepared by cationic catalysis were stable, even after 5 months. The experiments summarized in Table 2.1.2 indicate that the particle size is significantly reduced by (1) the presence of the coupling agent MATS (compare C6 to C7, and C8 to C9), probably due to the crosslinking induced by MATS; (2) an increase in the amount of surfactant (ABSA and SDBS; compare C6 to C8, C7 to C9, and C4 to C5). The decrease in particle size with increasing amount of surfactant was also observed in the polymerization of D_4 via the anionic catalysis (see A24, A25, A26, A47, and A48 in Table 2.1.1).

A general observation was that, along with a constant increase in particle size during polymerization, the particle number has constantly decreased. The data obtained for one of the cationic polymerizations of D_4 (C6 in Table 2.1.2) are presented in Table 2.1.3 as a representative example. These results indicate that the particle growth during polymerization is, at least partially, accomplished via coagulation.

TABLE 2.1.2
D_4 Polymerization with ABSA as Cationic Catalyst

Runs	ABSA (g)/ SDBS (g)	Temperature (°C)	Reaction Time (h)	Conversion (%)	D_p[b] (nm)	$10^{-15} N_p$ (L^{-1})
C1	0.20/0.10	60	30	69	290 ± 14	11.03
C2	0.25/0.10	60	30	84	270 ± 8	17.01
C3	0.25/0.10	80	40	88	280 ± 5	15.62
C4	0.63/0.10	80	23	80	290 ± 5	12.78
C5	0.63/0.00	80	23	79	302 ± 6	11.40
C6[a]	0.25/0.00	90	45	78	330 ± 11	8.31
C7	0.25/0.00	90	45	80	390 ± 5	5.21
C8[a]	0.45/0.00	90	29	85	275 ± 3	15.93
C9	0.45/0.00	90	29	84	295 ± 10	12.75
C10[a]	0.25/0.00	75	60	71	240 ± 9	20.60

[a] Polymerization with MATS 1 wt % of D_4 monomer.
[b] Diameter of particles determined by light scattering.

TABLE 2.1.3
Particle Size Evolution in Cationic Polymerization (C6)

Conversion (%)	D_p (nm)	$10^{-15} N_p$(L^{-1})
43.63	190 ± 11	24.74
51.49	214 ± 14	23.25
52.33	225 ± 9	21.02
56.18	252 ± 10	16.01
76.59	310 ± 8	10.00
78.18	330 ± 11	8.31

2.1.3.2 Preparation of PD_4-P(St-MMA-AA) Core-Shell Latex Particles

Among the PD_4 latexes prepared by anionic catalysis, A23, A25, and A48 of Table 2.1.1 were selected for seeded polymerization of St-MMA-AA because of their stability and small particle sizes; and among those prepared by cationic catalysis, C3, C6, and C10 of Table 2.1.2 were selected as seeds because of the low surfactant amounts employed during their preparation, thus reducing the probability of generation of new particles involving only P(St-MMA-AA).

The oil-soluble AIBN was first used as initiator in the seeded polymerization. When no additional SDBS was added to the system, none of the polymerizations was successful due to a complete or partial flocculation of the seed and newly formed P(St-MMA-AA) particles. When additional (0.1 g) SDBS was added, the flocculation could be avoided, and a large number of tiny particles (<60 nm) of P(St-MMA-AA) were produced (SP1 in Table 2.1.4). It is clear that the monomers have not polymerized on the surface of the seed particles but in the monomer droplets produced by stirring. The particles flocculated in the absence of surfactant but remained as a stable dispersion when additional surfactant was introduced.

Further, the water-soluble initiator APS was employed. Experiments were first carried out with PD_4 latexes prepared by anionic catalysis. As a general observation, stable latexes, with limited flocculation during polymerization, were obtained. In the run SP2 (where 0.1 g of SDBS was employed in the seed latex preparation), the latex was partially flocculated (about 20%); in SP3 (where 0.14 g of SDBS was employed in the seed latex preparation), less than 10% flocculated; and little flocculation was found in SP4 (0.16 g SDBS used in the seed latex preparation).

The latexes were subjected to TEM observation, and the results are presented in Table 2.1.4. This table indicates that new particles were generated, particularly when the seed particles were large (SP2 and SP3 in Table 2.1.4), and no core-shell particles were detected (Figure 2.1.2A constitutes a micrograph of SP3). The seeded latex (SP4), prepared under the same conditions but with smaller seed particles (D_p = 143 nm), however, contains core-shell particles (Figure 2.1.2B). The core size (~150 nm) in some of the particles is in agreement with that of the seed particles (A48) of 143 nm. Table 2.1.4 indicates that SP4 also contains numerous small particles formed during the seeded polymerization.

TABLE 2.1.4
Seeded Polymerization of St-MMA-AA with PD$_4$ Latex Prepared with Anionic and Cationic Catalysts

	Seed Latex		Seeded Polymerization			
Runs[a]	Code	D_p (nm)	N_p[b]	M/P[c]	D_f[d] Calcd	D_f[e] Found
SP1	A23	260	1.0×10^{15}	4	444	mostly < 60 nm
SP2	A23	260	2.5×10^{15}	8	522	250, 100
SP3	A25	192	2.5×10^{15}	10	414	180, 60
SP4	A48	143	3.0×10^{15}	8	286	260, 150
SP5	C6	330	1.0×10^{15}	5	600	475, 225
SP6[f]	C3	280	3.0×10^{15}	3.2	450	560, 320
SP7[f]	C10	240	0.73×10^{15}	5.5	448	425, 415

[a] Except SP1, for which AIBN was used as the initiator with 0.1 g SDBS added, all the other experiments were carried out using APS as the initiator without further SDBS addition. For SP1 to SP4, the PD$_4$ seed latexes were prepared by anionic polymerization; for SP5 to SP7, the seed latexes were prepared by cationic polymerization.
[b] Number of particles per liter of latex by assuming that no new particles are formed in the second-stage polymerization.
[c] Weight ratio of (St-MMA-AA)–to PD$_4$.
[d] Calculated particle size (nm) by assuming that no new particle is formed in the second-stage polymerization.
[e] Polydispersed particles; only the largest and the smallest particle sizes are given.
[f] VD$_4$ (5% of the PD$_4$ in the seed) was allowed to react with the seed particles prior to the seeded polymerization of St-MMA-AA.

The subsequent experiments were carried out using seed latexes prepared by cationic catalysis. The results are also presented in Table 2.1.4. The TEM micrographs of SP5, SP6, and SP7 of Table 2.1.4 are given in Figure 2.1.3A–C, which clearly shows that particles with core-shell morphology are present. In Figure 2.1.3A, the sizes of the cores are very different, with the largest being around 450 nm, and the smallest being around 200 nm, while the average size of the seed particles used was 330 nm (SP5 in Table 2.1.4). This could be attributed to the particle size dispersity in the seed latex itself. However, the relatively small size dispersity of the seed particles (Table 2.1.2) suggests that the large dispersity of the core-shell particles is caused by coalescence during the second stage of polymerization for the larger sizes and by a possible contraction of the flexible PD$_4$ particles for the smaller ones. When the seed particle size decreased from 330 to 280 nm (SP6 in Table 2.1.4), the core-shell morphology could be identified in all the particles, but again the core size varied between 200 and 300 nm, a range smaller than that observed for SP5. The different behaviors of SP5 and SP6 are due most likely to the employment of the coupling agent, VD$_4$, before the seeded polymerization. When a seed latex with a smaller particle size of 240 nm and the coupling agent VD$_4$ were employed, the core-shell particles became of more uniform size (SP7 and Figure 2.1.3C).

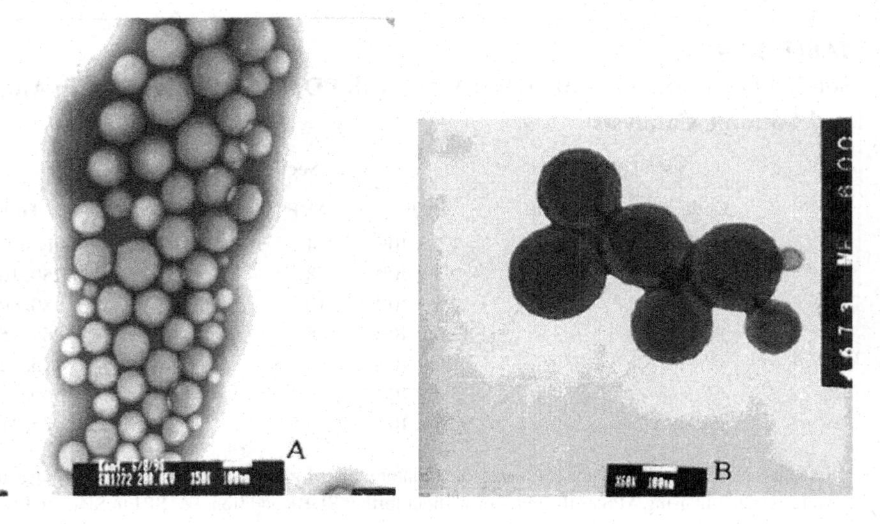

FIGURE 2.1.2 TEM micrographs of latex particles after the seeded polymerization of St-MMA-AA with seed latex of PD_4 prepared by anionic polymerization and APS as initiator. (A) Seeded polymerization based on PD_4 seed particles of 192 nm, showing that numerous new particles were formed. (B) Seeded polymerization based on PD_4 seed particles of 143 nm showing that core-shell particles were generated along with new particles.

To ensure that this core-shell morphology is not an artifact caused by a possible phase separation in P(St-MMA-AA) itself, a latex of P(St-MMA-AA), with the same monomer composition as in the seeded polymerization, was prepared under similar experimental conditions, in the absence of a PD_4 seed latex. The TEM micrograph in Figure 2.1.4 shows that no core-shell morphology was generated.

2.1.3.3 Properties of the Core-Shell Latex and Its Polymer

The experimental results summarized in Tables 2.1.5 and 2.1.6 were carried out to provide supplementary information regarding the core-shell morphology of the PD_4-P(St-MMA-AA) particles. The T_g determinations (Table 2.1.5) revealed that the core-shell polymer, as well as the polymer blends of PD_4 with S_1 exhibit the following two T_g's: one around −106°C, representing that of PD_4, and another at 103°C, representing that of P(St-MMA-AA) copolymer. This provides evidence that polysiloxane is indeed present in the core-shell particles. The contact angles of water droplets and the water absorption of the polymers, listed in Table 2.1.5, provide additional support for the presence of a core-shell morphology in the PD_4-P(St-MMA-AA) latex particles. The table shows that PD_4 is characterized by a very large contact angle and a low water absorption, and S_1 is characterized by a small contact angle and a high water absorption due to its higher hydrophilicity. The blend samples display values between those of PD_4 and S_1 polymers. By comparing the core-shell polymer SP7 with any other polymers of Table 2.1.5, one can see that its contact angle and water absorption are the closest to those of S_1.

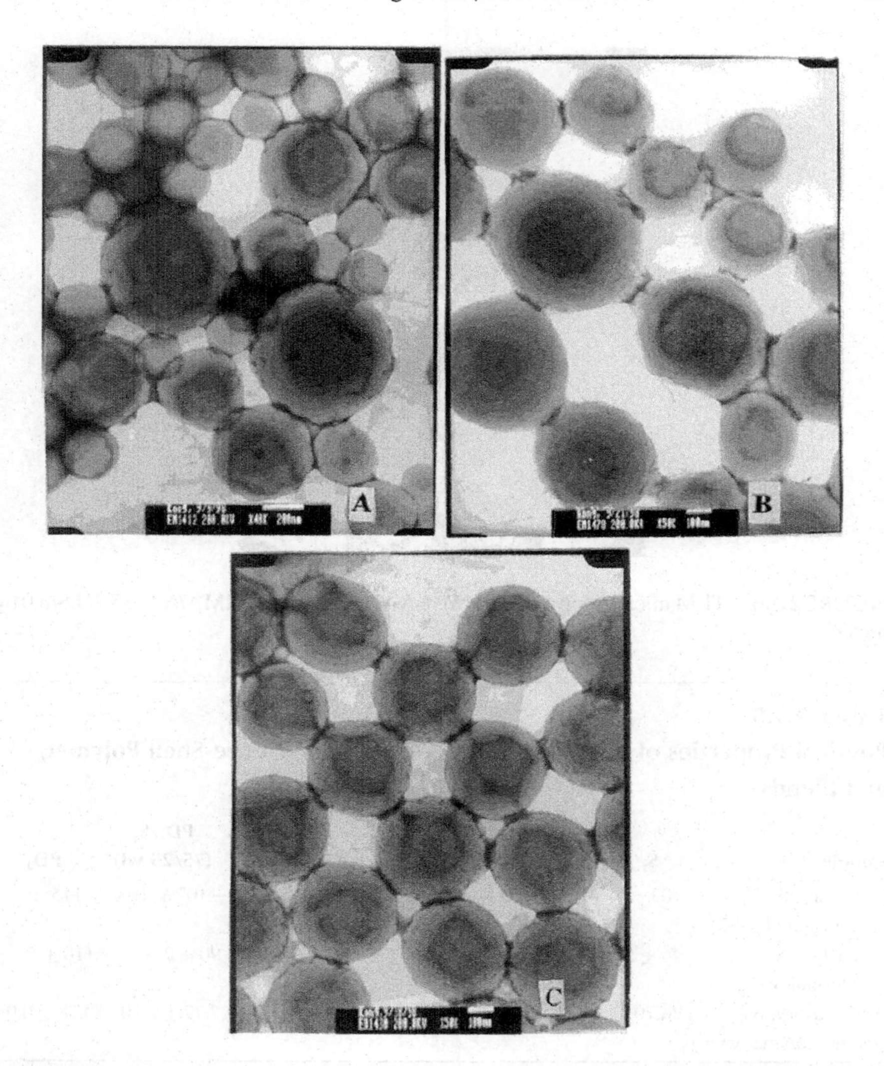

FIGURE 2.1.3 TEM micrographs of latex particles after seeded polymerization of St-MMA-AA using a PD_4 seed latex prepared by cationic polymerization and APS as initiator: (A) seed particle size, D_p = 330 nm (SP5 in Table 2.1.4); (B) seed particle size, D_p = 280 nm (SP6 in Table 2.1.4), with VD_4 coupling before the seeded polymerization; (C) seed particle size, D_p = 240 nm (SP7 in Table 2.1.4), with VD_4 coupling before the seeded polymerization.

This indicates that P(St-MMA-AA) constitutes the shell of the core-shell particles in the SP7 sample. Finally, the surface charge density due to the carboxyl groups was determined for the latexes SP7, S_1, and PD_4 (Table 2.1.6). As expected, PD_4 did not display any detectable carboxyl groups. A value of 1.30×10^{15} COO^-/cm^2, comparable to that of the S_1 latex particles, was obtained for the core-shell particles of SP7.

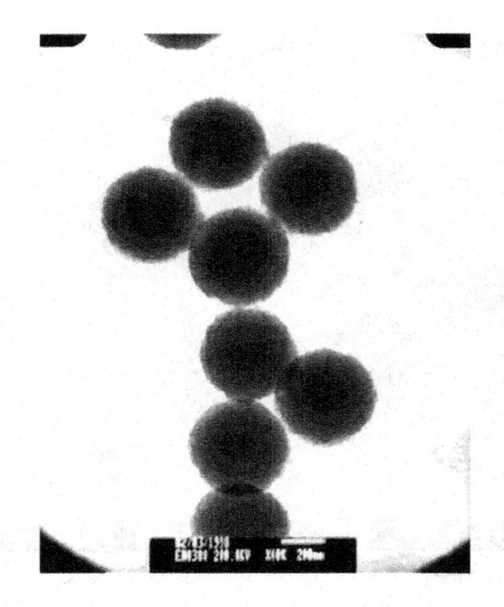

FIGURE 2.1.4 TEM micrograph of P(St-MMA-AA) latex (S_1) (St/MMA/AA: 87.2/4.8/8.0 by moles).

TABLE 2.1.5
Physical Properties of PD_4, P(St-MMA-AA) (S_1), Their Core-Shell Polymer, and Blends

Samples	S_1	SP7	PD_4/S_1 (15/85 wt)	PD_4/S_1 (50/50 wt)	PD_4/S_1 (75/25 wt)	PD_4
Glass transition temperature (T_g/°C)	103	−106 & 103	−105 & 103	−104 & 105	−105 & 103	−115
Contact angles (water-polymer)	46 ± 5°	60 ± 4°	70 ± 2°	92 ± 3°	90 ± 2°	110 ± 7°
Water adsorption (water-polymer wt %)	26.70	18.61 ± 0.09	15.14 ± 0.05	15.05 ± 0.08	7.77 ± 0.10	0.78 ± 0.03

TABLE 2.1.6
Surface Charges on Latex Particles of PD_4, P(St-MMA-AA) (S_1), and of the Core-Shell Polymer

Samples	S_1	SP7	PD_4/S_1[a] (15/85 wt)	PD_4
D_p (nm)	450	420	—	250
Spec surface (10^5 cm²/g)	1.26	1.37	1.44	2.45
OH^--polymer (mmol/g)	0.53 ± 0.02	0.29 ± 0.02	0.45 ± 0.02	0
10^{-15} COO^- (cm^{-2})	2.55 ± 0.10	1.30 ± 0.09	1.90 ± 0.08	0

[a] Calculated based on the composition using the data obtained for S_1 and PD_4 latexes.

2.1.3.4 Pore Generation in the Silicon-Containing Core-Shell Particles

Pores have been generated in these particles by an alkali-acid treatment[17] (Table 2.1.7), and some typical TEM micrographs are presented in Figure 2.1.5.

As noted in a previous article regarding the core-shell latex particles of P(St MMA)-P(St-MMA-AA),[25] a larger volume increase of the particles was obtained for lower pH values during the acid treatment. This observation is confirmed in the present case as well, the volume increase being 31% at a pH of 1.2–1.5 (Figure 2.1.5A) and only 15% at a pH of 2.0–2.2 (Figure 2.1.5B). In the alkali treatment step, at pH 12.0, the volume expansion was larger than that obtained at pH 12.2. Figure 2.1.6 presents micrographs of the particles just after the alkali treatment at pH 12.0 (Figure 2.1.6A) and pH 12.2 (Figure 2.1.6B).

TABLE 2.1.7
Alkali-Acid Treatment of the Core-Shell Particles of PD_4/P(St-MMA-AA)

pH in Alkali-Acid Treatment	D_p (nm) after Alkali Treatment	D_p (nm) after Acid Treatment (D_f)	D_f/D_p[a]	ΔV[b] (%)	Pores
12.0/1.5	475	460	1.095	31.3	Yes
12.0/1.2	475	460	1.095	31.3	Yes
12.2/2.2	440	440	1.048	15.1	Yes
12.2/2.0	440	440	1.048	15.1	Yes

[a] Particle size ratios after (D_f) and before (D_p = 420 nm) the alkali-acid treatment.

[b] Particle volume increase after alkali-acid treatment.

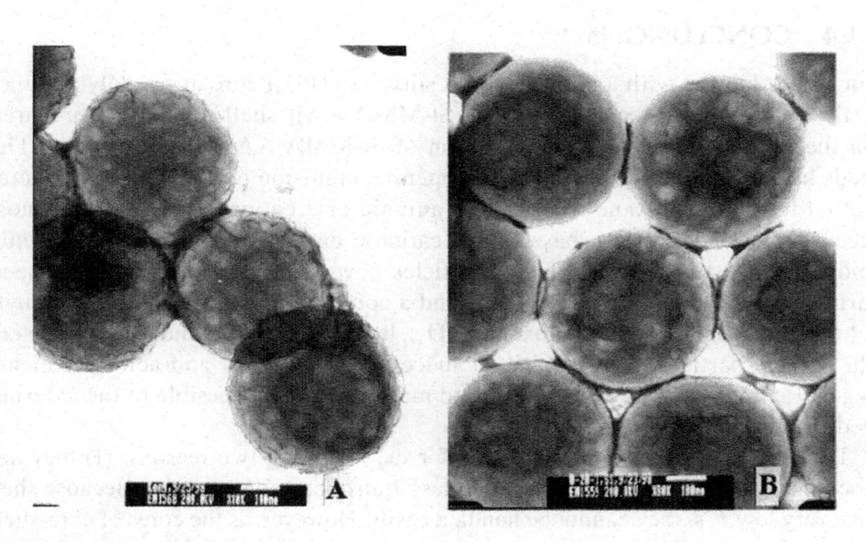

FIGURE 2.1.5 PD_4-P(St-MMA-AA) core-shell latex particles after the alkali-acid treatment for the pH's: (A) 12.0/1.2; (B) 12.2/2.0.

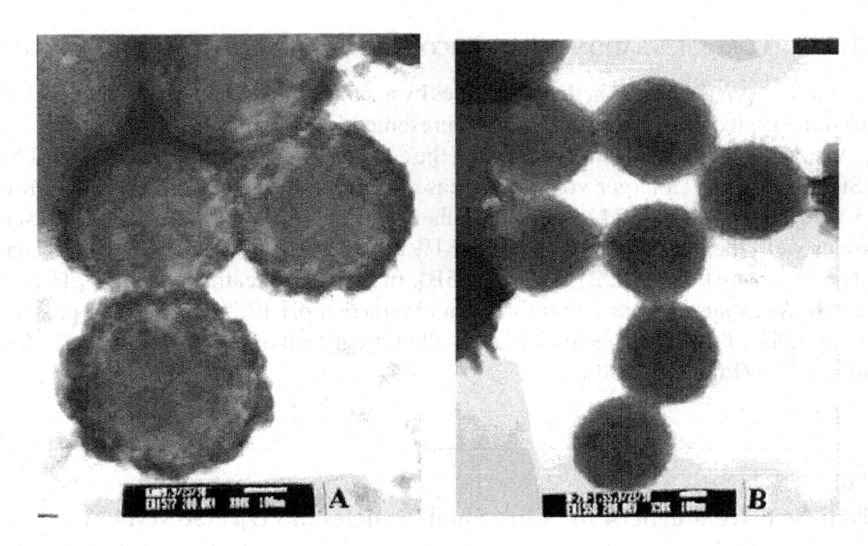

FIGURE 2.1.6 PD_4-P(St-MMA-AA) core-shell latex particles after the alkali treatment for the pHs: (A) 12.0; (B) 12.2.

A further examination of the latex particles of Figure 2.1.5A shows that the following two types of porous latex particles are present: particles with a darker core, similar to those present before the alkali-acid treatment, as well as particles in which the cores disappeared after the alkali-acid treatment. The diffusion of the PD_4 polymer chains away from the particle center during the treatment constitutes a possible explanation.

2.1.4 CONCLUSIONS

Core-shell latexes with a poly(dimethyl siloxane) [PD_4] core and a poly(styrene-methyl methacrylate-acrylic acid) [P(St-MMA-AA)] shell have been prepared via the seeded emulsion polymerization of St-MMA-AA upon PD_4 seeds. The seeds have been obtained via the ring-opening emulsion polymerization of octa-methyl tetracyclosiloxane by using an anionic or a cationic catalyst. The most successful procedure was based on a cationic catalyst (alkylbenzene sulfonic acid). The most uniformly seeded particles have been achieved when the seed particles have been sufficiently small and a coupling agent, 2,4,6,8-tetramethyl-2,4,6,8-tetravinyl cyclotetrasiloxane (VD_4), between the seed and shell was used. The seeded particles were subjected successively to alkali and acid treatments to generate pores in the shell in order to make the core accessible to the external medium.

The silicon polymers are interesting for the following two reasons: (1) they are biocompatible, and (2) they have low glass transition temperatures. Because they have very low T_g's, they cannot be handled easily. However, as the cores of core-shell latexes, they could be useful as constituents of composite materials with enhanced impact strength, even at very low temperatures.

REFERENCES

1. Nakamura, A. UK Pat. Appl. 2,019,417, 1978.
2. Hiroshi, M. Jpn. Pat. 55-007,814, 1980.
3. Yuichi, F. Jpn. Pat. 02-127,458, 1990.
4. James, A. P. Ger. Offen. 2,710,084, 1984.
5. Yasutarou, Y. Jpn. Pat. 58-167,606, 1983.
6. Burzynski, A. US Pat. 3,449,293, 1963.
7. Backderf, R. H. US Pat. 3,706,697, 1972.
8. Akira, Y. Jpn. Pat. 08-41,149, 1996.
9. Bourn, T.; Bufkin, B.; Wildman, G.; Grawe, J. *J Coatings Technol* 1982, 54, 69.
10. Krenter, J., Ed., in *Colloidal Drug Delivery Systems*; Marcel Dekker: New York, 1994; p. 219.
11. Pichot, C.; Delair, T.; Elaissari, A. in *Polymeric Dispersion: Principles and Applications*; Asua, J. M., Ed.; Kluwer Academic Publishers: Dordrecht, the Netherlands, 1997; p. 515.
12. Fritz, H.; Maier, M.; Bayer, E. *J Colloid Polym Sci* 1997, 195, 272.
13. Galisteo-Gonzalez, F.; Rodriguez, A. M.; Hidalgo-Alverez, R. *J Colloid Polym Sci* 1994, 272, 352.
14. Muller, J. J.; Lukowski, G.; Krober, R.; Damuschun, G.; Dittgen, M. *J Colloid Polym Sci* 1994, 272, 755.
15. Okubo, M.; Ichikawa, K. *Colloid Polym Sci* 1994, 272, 933.
16. Weyenberg, D. R.; Findlay, D. E.; Cekata, J., Jr.; Bey, A. E. *J Polym Sci, Part C* 1969, 27, 27.
17. Kong, X. Z.; Ruckenstein, E. *J Appl Polym Sci* 1999, 71, 1455.
18. Kong, X. Z.; Pichot, C.; Guillot, J. *J Colloid Polym Sci* 1987, 265, 791.
19. Patai, S.; Rappoport, Z., Eds. in *The Chemistry of Organic Silicon Compounds*; John Wiley & Sons: New York, 1989; p. 1289.
20. Du, Z.; Chen, J., Ed. in *Organosilicon Chemistry* (*in Chinese*); High Education Publishers: Beijing, China, 1992; p. 232.
21. Hyde, J. F.; Wehrly, J. R. US Pat. 2,891,920, 1959.
22. Bey, A. E.; Weyenberg, D. R.; Seibles, L. *Polym Prepr* 1970, 11, 995.
23. Zhang, X.; Yang, Y.; Liu, X. *Polym Commun* 1982, 4, 310.
24. de Gunzbourg, A.; Favier, J.; Hemery, P. *Polym Int* 1994, 35, 179.
25. Ruckenstein, E.; Kong, X. Z. *J Appl Polym Sci* 1999, 72, 419.

2.2 Control of Pore Generation and Pore Size in Nanoparticles of Poly(styrene-methyl methacrylate-acrylic acid)*

Eli Ruckenstein and Xiang Zheng Kong

Department of Chemical Engineering, State University of New York at Buffalo, Buffalo, New York 14260, USA

ABSTRACT Seeded emulsion polymerization of styrene-methyl methacrylate-acrylic acid onto seed latexes of monodisperse particles of poly(styrene-methyl methacrylate) was conducted with and without divinyl benzene as a crosslinking agent. Experiments revealed that almost no new particles were formed during the second stage of polymerization, and that the seeded latex particles obtained were almost monodisperse. An alkali-acid treatment was then applied to the seeded latex particles swollen in 2-butanone. Experimental results indicated that: (1) for uncrosslinked particles, an optimum volume expansion of >50% is reached for a ratio of the swelling agent, 2-butanone, to polymer (methyl-ethyl-ketone/polymer by weight) between 2.0 and 2.9; the volume expansion is much lower outside the above range. (2) For crosslinked particles, the particle volume expansion follows the same pattern, but with smaller values. (3) pH plays an important role in pore generation and volume expansion. Pore generation is optimized by decreasing pH to a value as low as 1.5 during acid treatment, and by keeping pH in the optimum range between 11.98 and 12.20 during alkali treatment. Based on the above observations, a discussion regarding the mechanism of pore generation and particle expansion is provided.

---

* *Journal of Applied Polymer Science*. Vol. 72. 419–426, (1999).

2.2.1 INTRODUCTION

Polymer colloidal particles have been used in research regarding colloids, as well as dispersable materials in a variety of applications. Numerous studies have been focused on latex preparation, including the control of particle size and surface properties.[1,2] Since the 1970s, the research was directed toward the preparation of latex particles for specific applications and with various characteristics, such as reactive particles,[3,4] core-shell particles,[5] magnetic particles,[6] fluorescent particles,[6] and porous latex particles.[7,8] Porous particles of micron size, particularly poly(styrene-*co*-divinyl benzene) particles, have been extensively studied since the 1960s. They have been prepared *via* suspension polymerization[9] or dynamic swelling polymerization.[10] In the last decade, the emphasis was on nanoparticles, which can be used in drug delivery or as drug targeting systems.[11] Immobilization of biomolecules onto nanoparticles was conducted through either physical adsorption or chemical attachment. An alternative for immobilization of biomolecules is their encapsulation in polymers or their location into porous polymer nanoparticles. However, in contrast to porous particles of micron size, few studies regarding the porous nanoparticles have been reported.[7,8,12] In this article, monodisperse core-shell latex particles containing a carboxylic acid are first prepared and then subjected to the alkali-acid process suggested by Okubo and colleagues,[13] to generate pores. The influence of experimental conditions upon pore generation is studied, and a mechanism of pore generation suggested.

2.2.2 EXPERIMENTAL

2.2.2.1 MATERIALS

Styrene (St; Aldrich Chemical Co., Milwaukee, WI), stabilized with 4-*tert*-butylcatechol, methyl methacrylate (MMA; Aldrich), and acrylic acid (AA; Aldrich), stabilized with 4-methoxyphenol, were passed through an Inhibitor Removal Prepacked Column (Aldrich) to remove the inhibitors. Ammonium persulfate (APS; 99.99%; Aldrich) and sodium dodecylbenzene sulfonate (SDBS; Aldrich) were used as the initiator and the surfactant, respectively. A solution of 2 wt % of phosphotungstic acid (Sigma Chemical Co., St. Louis, MO) was used to increase the contrast of the particles for transmission electron microscopy (TEM) observations. Distilled and deionized water with a conductivity of 0.05 μS cm^{-1} was used. 2-Butanone [methyl ethyl ketone (MEK); Aldrich, >99%], the swelling agent of the polymer particles, and divinyl benzene (DVB; Fluka; 70–85% by gas chromatography) were used as received.

2.2.2.2 POLYMERIZATION, POLYMER, AND PARTICLE CHARACTERIZATIONS

A seed latex was first prepared at 70°C, using 180 g of water, 28.5 g of St, 1.5 g of MMA, 0.15 g of SDBS, and 0.36 g of APS; this seed is denoted as S_1. The subsequent seeded polymerization was conducted as follows: 165 g of water was mixed with 3 g of seed latex in a reactor, and the reactor, stirred at 200 rpm, was located into a bath. After the temperature of 85°C was attained and an aqueous solution of the initiator (0.36 g in 15 mL of H_2O) injected, a slow dropwise addition of the monomer mixture was completed within 5 h. In the second polymerization stage, the amounts of most of the ingredients (including water) introduced were the same as in the seed preparation; however, SDBS was no longer used, but AA (8 mol% of the monomers in the second stage, because, as noted previously,[8] this value provides an optimum pore generation) and DVB (when the polymer was crosslinked) were added. The polymer crosslinking was investigated *via* Soxhlet extraction, using benzene as the solvent; the percentage of crosslinked polymer was obtained by determining the amount of unextracted polymer after 20 h of extraction. The monomer conversion was determined gravimetrically. Particle size (diameter in all cases) was obtained and particle morphology was examined by TEM (JEOL-2100).

2.2.2.3 Alkali-Acid Treatment

The seeded latex was diluted to a solid content of ~1 wt%, and 100 g of the diluted latex was introduced into a three-neck round-bottomed flask equipped with a stirrer; 0.1 g of SDBS was then added to ensure that the latex would remain stable upon the addition of NaOH and HCl. A selected amount of MEK was added to the flask while stirring, and this was followed by alkali and, subsequently, by acid treatment. In the alkali treatment, a selected pH was first achieved by the addition of a solution of sodium hydroxide (1 *N*). The flask, with the mixture in the flask stirred at 100 rpm, was then introduced into an oil bath at 90°C, where it was kept for 3 h. The flask was then taken out from the bath and cooled to room temperature. After the pH was changed to a selected lower value with a solution of hydrochloric acid (1 *N*), the flask was reintroduced into the oil bath at 90°C for another 3 h.

2.2.3 RESULTS AND DISCUSSION

2.2.3.1 Crosslinking of Latex Particles

The diameter of the seed particles, S_1, was ~108 nm, and the seeded emulsion polymerization of St-MMA-AA onto S_1 was conducted using various amounts of DVB. Table 2.2.1, which summarizes the results, shows good agreement between the experimental and calculated values of the particle size and particle number. The calculated values were obtained by assuming that, in the second polymerization stage, all the monomers were polymerized onto the seed particles. The micrographs of Figure 2.2.1 indicate that the particles are almost monodisperse, and suggest that no new particles were formed during the second polymerization stage. They also show that the crosslinked particles ($S_{1\times 2}$), prepared by using 1 wt% DVB (see Table 2.2.1), possess a smoother surface than the uncrosslinked particles (S_{11}).

TABLE 2.2.1
Properties of Seeded Latex Particles of St-MMA-AA Polymerized on Seed Latex of St-MMA and Polymer Crosslinking

Sample	DVB (wt %)[a]	D_n (nm)	N_p (10^{15} L^{-1})	Crosslinking (wt %)[b]
S_{11}	0	450 (459)[c]	2.88 (2.69)	0
$S_{1\times1}$	0.5	451 (468)	3.04 (2.68)	33.5
$S_{1\times2}$	1.0	440 (455)	3.00 (2.68)	67.2

[a] Wt % of all monomers.
[b] Wt % of undissolved polymer by Soxhlet extraction with benzene.
[c] Data in parentheses are calculated values obtained by assuming that all monomers were polymerized onto the seed particles.

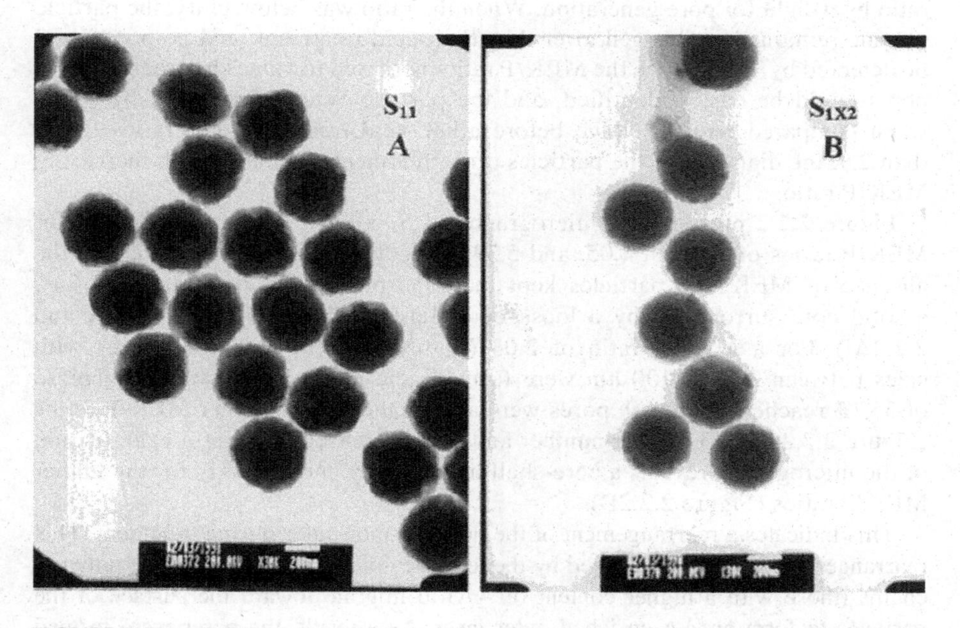

FIGURE 2.2.1 Seeded latex particles of the uncrosslinked (A) S_{11} and the crosslinked (B) $S_{1\times2}$ (1 wt % DVB).

2.2.3.2 Effect of MEK on Pore Generation in Uncrosslinked Seeded Latex

As previously described, 2-butanone (MEK) was used as a swelling agent during the alkali-acid pore generation process. At a fixed amount of latex, S_{11}, the amount of MEK was varied to investigate its effect on pore generation. Results are listed in Table 2.2.2, which shows that there is an optimum MEK/Polymer (MEK/P)

TABLE 2.2.2
Effect of 2-Butanone on Pore Generation and Particle Volume Expansion in the Uncrosslinked Seeded Latex Particles (S_{11})

MEK/P[a]	0	0.58	2.03	2.70	2.90	4.05	5.39
D_f (nm)[b]	450	450	521	517	518	480	485
ΔV (%)[c]	0	0	55.7	51.9	53.0	21.4	25.2
Pores	No	No	Yes	Yes	Yes	Yes	Yes

[a] MEK/polymer ratio by weight.
[b] Number average particle diameter after alkali-acid treatment.
[c] Particle volume increase after alkali-acid treatment.

ratio by weight for pore generation. When the ratio was below unity, the particle volume remained unchanged after the alkali-acid treatment, and no pore could be detected by TEM. Once the MEK/P ratio increased to values between 2 and 3, pores could be easily identified, and the particle volume increased by >50% when compared with its value before alkali-acid treatment. For ratios larger than 2.9, the diameter of the particles after treatment decreased with increasing MEK/P ratio.

Figure 2.2.2 presents the micrographs of S_{11} after alkali-acid treatment for MEK/P ratios of 0, 2.03, 4.05, and 5.39, respectively. One can see that, in the absence of MEK, the particles kept their morphology before treatment (i.e., a solid core surrounded by a loose outer layer [compare Figures 2.2.1A and 2.2.2A]). For a MEK/P ratio of 2.03 (Figure 2.2.2B), numerous pores with sizes between 30 and 100 nm were formed, and a maximum volume increase of 55.7% reached. Although pores were also generated for higher MEK/P ratios (Figure 2.2.2C and D), their number and size decreased. A careful examination of the micrographs reveals a core-shell morphology, more clearly for the higher MEK/P ratios (Figure 2.2.2D).

This indicates a rearrangement of the polymer molecules during treatment. This rearrangement is probably caused by the tendency of the more hydrophilic polymer chains (those with a higher content of AA) to migrate toward the surface of the particles to form an AA-enriched outer layer. As a result, the pores were formed mainly in the outer layer. The smaller volume expansion in Figure 2.2.2C and D than in Figure 2.2.2B might be due to the collapse of the large pores when the MEK/P ratio is high.

2.2.3.3 Effect of MEK on Pore Generation in Crosslinked Latexes

Similar experiments were conducted with crosslinked latex particles prepared with two different amounts of DVB, and the results are listed in Table 2.2.3. From these data, one can conclude that (1) as for the uncrosslinked latex particles, an MEK/P ratio between 2 and 3 favors pore generation, and a maximum particle expansion of ~20% is reached for a ratio of ~2.90. (2) As for the uncrosslinked

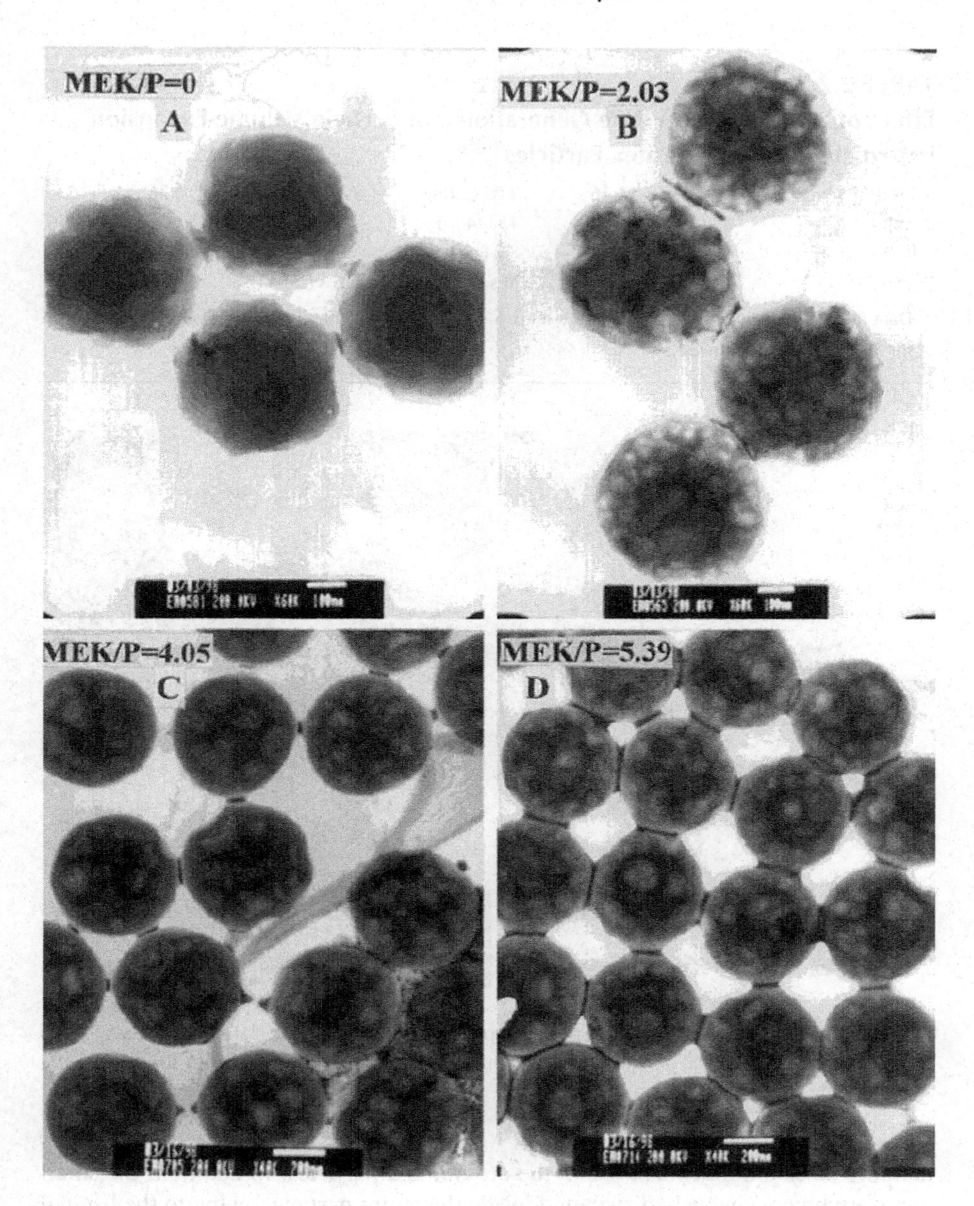

FIGURE 2.2.2 Micrographs of S_{11} seeded latex particles after the alkali-acid treatment for different MEK/P weight ratios (A: 0; B: 2.03; C: 4.05; D: 5.39) at pH 12.20 during alkali treatment and pH 2.20 during acid treatment.

latex particles, a too large amount of swelling agent is not effective in increasing the particle volume expansion for $S_{1\times1}$ and $S_{1\times2}$. For a MEK/P ratio of 5.80, the particle remained the same size as before treatment, and no pores were observed. For comparison purposes, Figure 2.2.3 presents the micrographs of the uncrosslinked particles, S_{11}, and the crosslinked particles, $S_{1\times1}$. They clearly show that, as expected,

TABLE 2.2.3
Effect of 2-Butanone on Pore Generation and Particle Volume Expansion in Crosslinked Seeded Latex Particles

MEK/P (by wt)	0.58	1.16	2.03 (2.03)[a]	2.90 (2.90)	4.36	5.80
D_f (nm)	450	454	452 (440)	460 (480)	—	440
ΔV (%)	0	10	8.2 (≈ 0)	14.3 (20.6)	—	≈ 0
Pore	No	Yes	Yes (No)	Yes (Yes)	No	Yes

[a] Data in parentheses were obtained for the latex $S_{1\times1}$ (33.5% crosslinked); all other data were obtained for $S_{1\times2}$ (67.2% crosslinked).

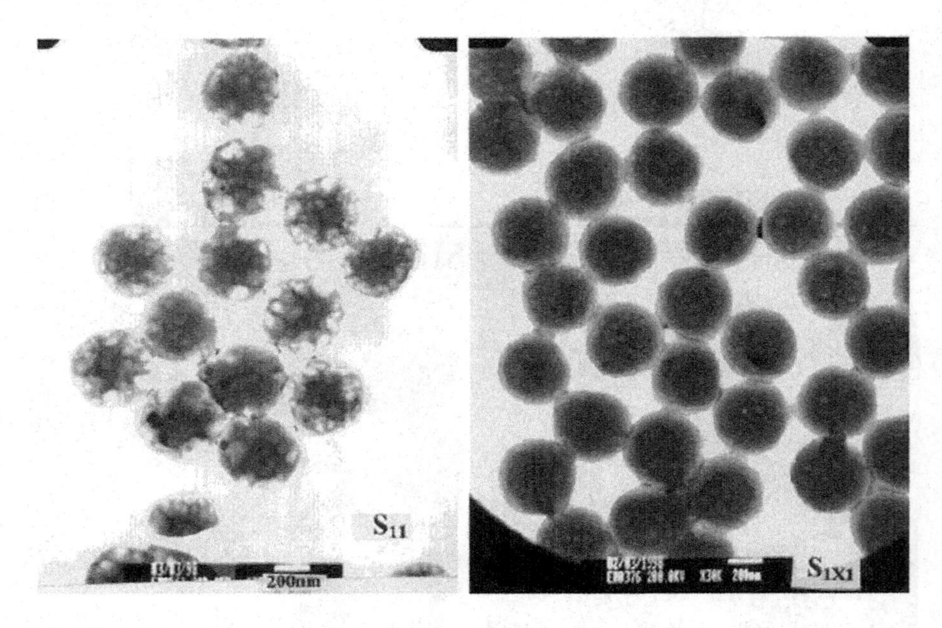

FIGURE 2.2.3 Comparison of the micrographs of the uncrosslinked latex particles S_{11} (left) and the crosslinked latex particles $S_{1\times1}$ (right) (MEK/P = 2.9, pH 12.20 in alkali and pH 2.20 in acid treatment).

the pore sizes are much smaller in the crosslinked particles. In addition, the small pores are homogeneously distributed inside the entire particle, owing to the limited migration of the polymer segments containing carboxylic groups. In contrast, in the uncrosslinked particles, the pores are mostly located near the surface of the particles.

2.2.3.4 Influence of pH in the Alkali-Acid Treatment on Pore Generation

Using the uncrosslinked latex, S_{11}, the pH was varied during alkali treatment at a constant MEK/P ratio of 2.90, followed by acid treatment at pH 2.20. In another group of experiments, the pH during acid treatment was varied, while that during the

TABLE 2.2.4
pH Effect on Pore Generation and Particle Volume Expansion in Alkali-Acid Treatment of the Uncrosslinked Seeded Late Particles (S_{11})

	Acid Treatment[a]			Alkali Treatment[b]					
pH	1.50	2.20	2.90	10.50	11.50	11.98	12.10	12.20	12.35
D_f (nm)	550	518	485	458	455	521	523	519	486
ΔV (%)	82.6	53.1	25.2	5.6	3.7	55.3	56.7	53.1	25.6
Pores	Yes	Yes	Yes	Yes	Yes	Yes	Yes	Yes	Yes

[a] pH was kept at 12.20 in the preceding alkali treatment.
[b] pH was kept at 2.20 in the subsequent acid treatment.

preceding alkali treatment was kept at 12.20. Results are presented in Table 2.2.4, which shows that the pH in the alkali-acid treatment has an important effect on the pore generation and particle volume expansion. Volume expansions of >50% are reached by changing the pH during the alkali treatment between 11.98 and 12.20. The expansion sharply decreases for a pH larger than 12.20 or smaller than 11.98. The effect of pH during the acid treatment on the particle volume expansion is even more pronounced. An expansion of the particle volume of 53.1% is observed at pH 2.20, and a much larger expansion of 82.6% is noted for pH 1.50. However, by increasing pH from 2.2 to 2.9, the particle expansion is reduced to 25.2%. Figure 2.2.4 presents the micrographs of the corresponding latex particles. The number of pores in the particles treated at pH 1.50, as well as their size, are obviously larger. A core-shell structure with a more porous shell can be noted in Figure 2.2.4A (particles treated at pH 2.90); this is probably due to the lower acidity in this case than in those presented in Figure 2.2.4B and C.

2.2.3.5 Mechanism of Pore Generation and Particle Volume Expansion

Our experimental results have clearly indicated that swelling of the particles plays an essential role in pore generation and particle volume expansion during alkali-acid treatment. Indeed, in the absence of MEK, no pores could be generated, and no volume expansion was observed. This swelling agent is very soluble in water[14] and is also a solvent for polystyrene.[15] As a result, it can swell the copolymer and even entrain some water during swelling. The swelling allows the penetration of the sodium hydroxide molecules inside the particles, where they react with the carboxylic groups, amplifying their ionization. Because of electrostatic repulsion, the charges thus generated lead to more extended configurations of the copolymer chains. Because of the particle swelling and the more extended configurations of the polymer chains, the volume of the particles is expanded. The swelling also stimulates the rearrangement of the polymer chains, which have a tendency to expose as much as possible their hydrophilic segments to the water solution. When the acid (HCl) is added, the degree of

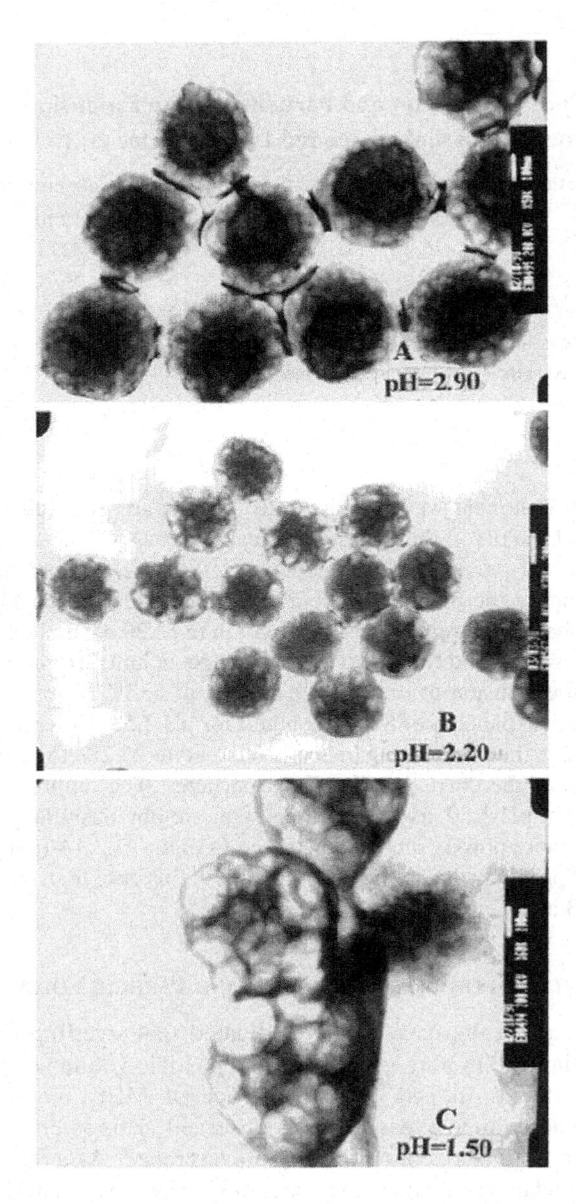

FIGURE 2.2.4 Particle morphologies of latex S_{11} alkali treated at pH 12.20 and acid-treated at different pH values. (A, pH 2.90; B, pH 2.20; and C, pH 1.50).

ionization of the carboxylic groups is decreased and the electrostatic repulsion diminishes. As a result, the chains become more flexible. Because the molecules have a tendency to expose their more hydrophilic segments to the swelling solution, a rearrangement, stimulated by their higher flexibility, occurs, which generates the pores.

Experiment indicates the existence of an optimum pH in the alkali treatment as well as of an optimum amount of swelling agent. The optimum pH probably occurs because a too low concentration of NaOH does not sufficiently ionize the carboxylic groups, whereas a too high pH generates a too high ionic strength, which decreases the electrostatic repulsion among the charges present on the polymer chains. The optimum amount of swelling agent is probably due to a too low dielectric constant of the swelling solution (MEK and water) when the MEK/P ratio is above 2.90, and this decreases the degree of dissociation of the carboxylic groups.

2.2.4 CONCLUSIONS

Seeded emulsion polymerization of St-MMA-AA onto seed latexes of poly(St-MMA) was conducted, and the latex particles obtained were subjected to an alkali-acid treatment to generate pores. The experiments lead to the following conclusions:

1. A swelling agent (2-butanone, MEK) is needed to generate pores, and there is an optimum amount of MEK for which the particle volume expansion is maximum. In the present case, the optimum volume expansion in the particles was reached when the weight ratio of MEK to the polymer was between 2.03 and 2.90.
2. Optimum pore generation and particle expansion are obtained for pH values between 11.98 and 12.20 during alkali treatment, followed by an acid treatment at pH 2.20. The volume expansion is even higher when the pH during acid treatment is lower.
3. A mechanism of pore generation is suggested, on the basis of which the above experimental results are explained.

REFERENCES

1. Fitch, R. M. Ed. *Polymer Colloids; Plenum Press*, New York, 1971.
2. Daniel, E. S.; Sudol, E. D.; El-Aasser, M. S. Eds. *Polymer Latexes: Preparation, Characterization & Applications*; ACS: Washington, DC, 1992.
3. Delair, T.; Pichot, C.; Mandranal, B. *Colloid Polym Sci* 1994, 272, 72.
4. Ganachaud, F.; Mouterde, G.; Delair, T. *Polym Adv Technol* 1994, 6, 480.
5. Sarobe, J.; Forcada, J. *Colloid Polym Sci* 1996, 274, 8.
6. Arshady, R. *Biomaterials* 1993, 14, 5.
7. Okubo, M.; Nakamura, M.; Ito, A. *J Appl Polym Sci* 1997, 64, 1947 [and references therein].
8. Kong, X. Z.; Ruckenstein, E. *J Appl Polym Sci* 1999, 71, 1455.
9. Lewandowski, K.; Svec, F.; Frechet, J. M. J. *J Appl Polym Sci* 1998, 67, 597.
10. Okubo, M.; Minami, H. *Colloid Polym Sci* 1997, 275, 992.
11. Pichot, C.; Delair, T.; Elaissari, A. In *Polymeric Dispersions: Principles and Applications*; Asua, J. M., Ed.; Kluwer Academic Publishers: Dordrecht, the Netherlands, 1997; p. 515.
12. Vanderhoff, J. W.; Park, J. M.; El-Aasser, M. S. In *Polymer Latexes: Preparation, Characterization & Applications*; Daniel, E. S.; Sudol, E. D.; El-Aasser, M. S., Eds.; ACS: Washington, DC, 1992; p. 272.

13. Okubo, M.; Ichikawa, K.; Fujimura, M. *Colloid Polym Sci* 1991, 269, 1257.
14. Weast, R. C., Ed. *Handbook of Chemistry and Physics*, 51st ed; The Chemical Rubber Co.: Cleveland, Ohm 1970–1971; p. C222.
15. Brandrup, J.; Immergut, E. H., Eds. *Polymer Handbook*, 3rd ed.; John Wiley & Sons: New York, 1989; p. VII386.

2.3 Amphiphilic Particles with Hydrophilic Core/Hydrophobic Shell Prepared via Inverted Emulsions*

Hangquan Li and Eli Ruckenstein

Department of Chemical Engineering, State University of New York at Buffalo, Amherst, New York 14260

ABSTRACT An aqueous solution of acrylamide, its crosslinker (*N, N'*-methylenebisacrylamide), and an oxidant (ammonium persulfate) was first used to prepare an inverted concentrated emulsion in hexane. Span 80, which is soluble in hexane, was employed as a dispersant. The polymerization of acrylamide in the concentrated emulsion was greatly accelerated by introducing an aqueous solution of a reductant (sodium metabisulfite); it started at room temperature and was completed in a few seconds, resulting in a pastelike product. The system thus obtained was subsequently diluted with hexane containing a hydrophobic monomer. When styrene was used as the hydrophobic monomer, cumene hydroperoxide (which, together with sodium metabisulfite present in the dispersed phase, constitutes the initiator for the polymerization of styrene) was dissolved in the continuous phase. When vinylidene chloride was employed as the hydrophobic monomer, no additional initiator besides sodium metabisulfite and ammonium persulfate already present in the hydrophilic phase had to be employed. The use of initiators which are present only in the hydrophilic phase, and hence also at the interface between this phase and hexane, ensured the polymerization of the hydrophobic monomer as shells that encapsulate the polyacrylamide latexes. Under the proper conditions, a porous outer shell can be generated, which makes the hydrophilic chains present inside accessible. Such hydrophilic core/hydrophobic porous shell particles can be dispersed in water, where

* *Journal of Applied Polymer Science*. Vol. 61, 2129–2136, (1996).

they remain stable for a long time, and in hydrophobic liquids, where they remain stable for at least 24 h. For this reason, we call these kinds of particles amphiphilic particles.

2.3.1 INTRODUCTION

New materials can be produced by combining two different polymers. The greater the difference between the two polymers, the more likely synergistic properties will result from the combination. For this reason, the combination of a hydrophilic and a hydrophobic polymer has received attention in recent years. The combination of hydrophilic and hydrophobic polymers can be achieved via block, graft, or random copolymerization[1–3]; interpenetrating or semi-interpenetrating networks[4,5]; and hydrophilic-hydrophobic composites based on concentrated emulsions[6] or colloidal pathways.[7] Since both hydrophilic and hydrophobic chains or domains are associated in such materials, amphiphilicity can be achieved, which can find wide applications in membrane separation,[8] controlled release,[9] drug delivery,[10,11] etc. It was reported[12] that a remarkable amphiphilicity can be achieved by preparing hydrophilic core/hydrophobic shell particles. A hydrophilic core of poly(triethylvinyl benzylammonium chloride) (PEVAC) was encapsulated with a porous hydrophobic shell of polydivinyl benzene. The particles thus prepared have been used for the immobilization of catalysts and employed in some catalytic processes.[13] Hydrophilic core/hydrophobic shell particles have also been prepared by seeded emulsion polymerization in water, in which copolymers of methyl methacrylate and methacrylic acid were used as the hydrophilic core and polystyrene as the hydrophobic shell.[14,15] The goal was to achieve a complete and uniform encapsulation with a hydrophobic shell of a hydrophilic core. As a result, the hydrophilic core was unaccessible.

In the present article, a novel, simple procedure to prepare hydrophilic core/hydrophobic shell particles via inverted emulsions is proposed. An aqueous solution of a hydrophilic monomer (acrylamide) and its crosslinker (*N, N′*-methylenebisacrylamide) containing ammonium persulfate as initiator was first used to prepare a concentrated emulsion in hexane. The polymerization of acrylamide in the concentrated emulsion generated a pastelike product. After it was diluted with hexane containing a hydrophobic monomer, the system became a suspension of water-swollen polymeric particles. Depending on the nature of the hydrophobic monomer, either only the initiators used for the polymerization of acrylamide and present in the system could be employed, or another one had to be introduced in the hydrophobic phase. Since either all the initiators or some of them were present in the swollen particles (and hence also at the interface between the two phases), the polymerization of the hydrophobic monomer occurred on the surface of the polyacrylamide (PAAM) latexes. By selecting proper coating conditions, a porous shell structure could be achieved, which made the hydrophilic chains present inside accessible and thus ensured the amphiphilicity of the particles.

2.3.2 EXPERIMENTAL

2.3.2.1 MATERIALS

Styrene (St, 99%), vinylidene chloride (VDC, 99%), and divinyl benzene (DVB, tech, 80%) were filtered through an inhibitor removal column before use. Acrylamide (AAM, 99%), *N, N′*-methylenebisacrylamide (MBA, 99%), hexane (95%), ammonium persulfate (APS, >98%), sodium metabisulfite (SMBS, >97%), cumene hydroperoxide (CHPO, tech, 80%), and sorbitan monooleate (Span 80) were used as received. All chemicals were purchased from Aldrich, except Span 80, which was purchased from Fisher. Water was distilled and deionized.

2.3.2.2 PREPARATION PROCEDURE

A flask containing a magnetic stirring bar was sealed with a rubber septum and kept in an ice bath below 5°C. The air inside was replaced with nitrogen. Hexane (containing 0.1 g/g Span 80), which constitutes the continuous phase of a concentrated emulsion, was introduced into the flask through the rubber with a syringe (when amounts smaller than 0.1 g/g of Span 80 were used, the emulsion became unstable). The dispersed phase, an aqueous solution of AAM, MBA, and an oxidant (APS), was subsequently introduced dropwise with stirring, until its volume fraction became 0.8. After the concentrated emulsion was formed, an aqueous solution of a reductant (SMBS) was injected into the flask. After the reductant solution was uniformly dispersed, the flask was removed from the ice bath. The polymerization of AAM started at room temperature and was completed in a few seconds. The paste-like product obtained was washed with methanol and dried in a vacuum oven until constant weight. We found that the weight ratio of the polyacrylamide thus obtained to the acrylamide monomer was always higher than 0.95; hence the polymerization of acrylamide in the concentrated emulsion was almost complete. After the paste was diluted with hexane, either VDC or a mixture of St, its crosslinker DVB, and another oxidant (CHPO) was added. The flask was then introduced into a thermostated water bath at 30°C for VDC or 60°C for St, for various lengths of time (4, 6, and 8 h), to allow the polymerization of the hydrophobic monomer to occur on the surface of the PAAM latexes. The product thus obtained was washed with methanol and dried in a vacuum oven.

2.3.2.3 SCANNING ELECTRON MICROSCOPY (SEM)

A small amount of the powder sample was glued on a SEM holder and then coated with a thin film of gold. The surface morphology was examined by SEM (Hitachi S-800).

2.3.2.4 Elemental Analysis

Elemental analysis was performed by Quantitative Technologies Inc. (Whitehouse, NJ).

2.3.3 RESULTS AND DISCUSSION

Various preparation conditions were employed; they are listed in Table 2.3.1. In the present article, PAAM latexes were prepared mainly via the concentrated emulsion method.[6,16] A concentrated emulsion differs from the conventional emulsion in that the volume fraction ϕ of the dispersed phase is greater than 0.74 (which represents the volume fraction of the most compact arrangements of monosize spheres). The volume fraction $\phi = 0.8$ was employed. Inverted emulsions with a lower ϕ were also tried. We found that the greater the volume fraction of the continuous phase (hexane), the slower the polymerization rate. When ϕ was between 0.5 and 0.7, more than 1 h was necessary to complete the polymerization of AAM. When ϕ was lower than 0.4, the polymerization lasted as long as 48 h. When the volume fraction of the dispersed phase was in the concentrated emulsion range, the polymerization took place almost instantaneously.

In the range of concentrations employed, the monomer concentration in the dispersed phase (aqueous solution of AAM, MBA, and oxidant) had no detectable

TABLE 2.3.1
Preparation Conditions of the Samples

	Hydrophilic Phase[a]			Coating Monomer			
Figure Number	AAM (g)	MBA (g)	Hexane (g)	Styrene (g)	DVB (g)	VDC (g)	Coating Time (h)
1	2	0	3	0.5	0.5	—	8
2(a)	2	0.1	No subsequent coating applied				
2(b)	2	0.2	No subsequent coating applied				
3	2	0.1	3	—	—	3	8
4	2	0.1	0	1.0	1.0	—	8
5	2	0.1	3	1.0	1.0	—	8
6	2	0.1	6	1 0	1.0	—	8
7(a)	2	0.1	3	1.5	1.5	—	8
7(b)	2	0.1	3	0.5	0.5	—	8
8	2	0.1	3	1.0	1.0	—	4
9	2	0.1	3	—	—	1	8

[a] The hydrophilic phase contains also 2 g water and 0.06 g ammonium persulfate. After the concentrated emulsion was prepared, and aqueous solution of 0.06 g sodium metabisulfite in 0.5 g water was introduced.

effect on the polymerization rate. Different monomer/water weight ratios—1/1, 1/1.5, 1/2, and 1/4—were employed, and almost the same polymerization rate was observed. To remove the water easily after the reaction was completed, a ratio of 1/1 was selected. When the monomer/water weight ratio was greater than 1/1, the subsequent dilution of the polymerized concentrated emulsion with hexane and hydrophobic monomer became difficult because of its high viscosity.

The initiator had an important effect on the polymerization rate. AAM could be polymerized using only APS as an initiator. However, the addition of a reductant (SMBS) greatly accelerated the polymerization. In our experiments, a concentrated emulsion based on APS alone completed the polymerization of AAM in 8 h at room temperature. If an aqueous solution of SMBS was added to the concentrated emulsion, the polymerization was finished almost instantaneously. However, the reductant could not be introduced together with the oxidant before the concentrated emulsion was prepared, because the monomer would have polymerized immediately.

To obtain individual latexes, a suitable crosslinking of the PAAM was necessary. Indeed, if not properly crosslinked, the latexes formed lumps, and, as shown in Figure 2.3.1, the hydrophobic polymer could coat in the subsequent stage only the surface of the lumps and not those of the latexes. The particle morphologies of Figure 2.3.2 are based on AAM/MBA weight ratios of 100/5 and 100/10, respectively (see Table 2.3.1 for other details) and show that individual particles were obtained in

FIGURE 2.3.1 Particle morphology of a sample based on uncrosslinked polyacrylamide (see Table 2.3.1).

FIGURE 2.3.2 Particle morphologies of polyacrylamide latexes. AAM/MBS weight ratio: (a) 100/5; (b) 100/10 (see Table 2.3.1).

both cases. The particles based on an AAM/MBA weight ratio of 100/10 are larger and possess a smoother surface than those based on 100/5. Most of the samples in the present study were prepared with an AAM/MBA weight ratio of 100/5.

With a 100/5 AAM/MBA weight ratio, individual particles were obtained only via the concentrated emulsion pathway. In diluted emulsions, the particles would still stick to one another to some extent. This occurs because in a dilute emulsion, the stirring stimulates the collisions among the polymerizing particles. However, in the case of a concentrated emulsion, the thin films between the particles are not strongly affected by the stirring and the particles maintain their individuality.

After the prepared PAAM latexes were diluted with hexane, a hydrophobic monomer was added to the system. Two monomers were employed—namely, VDC or St (the latter containing a crosslinker, DVB). For the systems based on VDC, no additional initiator besides those soluble only in water and used for the polymerization of AAM (APS and SMBS) were needed. The VDC polymerization is initiated on the surface of the PAAM particles because the initiator is present in the hydrophilic phase. For the systems involving St, the aforementioned redox pair had to be strengthened with an additional oxidant, CHPO, which (being hydrophobic) was dissolved in the oil phase. Since the polymerization of styrene is initiated by CHPO and SMBS together, but by neither of them individually, the polymerization is initiated on the latex surface.

After the polymerization of the monomer was initiated on the surface of the PAAM particles, the polymer molecules could either grow there *in situ* or in the oil phase, near the place where they were initiated. Since both the monomer and the continuous medium are hydrophobic, the latter is more likely to occur.

The growing polymer molecules will remain in the oil phase until the latter can no longer dissolve or swell them because of their increasing molecular weight, degree of crosslinking, or crystallinity. Then the hydrophobic polymer will precipitate on the surface of the PAAM latexes as tiny particles, and thus shells with holes will be generated. A poly(vinylidene chloride) (PVDC) coated sample is presented in Figure 2.3.3. One can see that the sample is coated with a clear, netlike PVDC structure. Because the VDC homopolymer is crystalline and insoluble in both its monomer and in hexane, the PVDC chains are readily precipitated as small crystallites.

For the samples based on styrene, the solubility of polystyrene (PS) in the oil phase plays an even more important role than for PVDC, because both styrene and DVB dissolve or swell the polymer. In a system diluted only with styrene and DVB, the crosslinked PS, dispersed in the hydrophobic phase, aggregated to form individual particles which did not deposit on the surface of the PAAM latexes (Figure 2.3.4). The addition of a large proportion of hexane (which is a solvent for styrene and DVB monomers, but a nonsolvent for PS) in the oil phase reduces the solubility of PS in the latter, and for this reason tiny PS particles are deposited on

FIGURE 2.3.3 Particle morphology of a sample possessing a netlike PVDC coating (see Table 2.3.1).

FIGURE 2.3.4 Particle morphology of polyacrylamide/polystyrene in the absence of hexane (see Table 2.3.1).

the surface of PAAM latexes. In a system in which the PAAM/hexane/St/DVB weight ratios were 2/3/1/1, the coating was composed of a large number of tiny particles loosely packed and having holes, as shown in Figure 2.3.5. When an even larger amount of hexane was employed (i.e., for PAAM/hexane/St/DVB weight ratios of 2/6/1/1; Figure 2.3.6), the coating consisted of even smaller particles, more uniformly distributed.

Consequently, there are conditions under which a netlike coating can be achieved, which allows the hydrophilic core to be accessible to a hydrophilic medium. The particles of Figure 2.3.5 could be dispersed easily both in water, where they remained stable for a long time, and in hydrophobic media, where they remained stable for at least 24 h; for this reason they can be called amphiphilic particles.

When the amphiphilic particles with PAAM core/PS shell were dispersed in water, the PAAM cores had swollen and became semitransparent. The unswollen hydrophobic shells had the appearance of a large number of white spots scattered in the suspension. When the particles were dispersed in toluene, the thin, swollen PS shell could not ensure a stability as high as that in water. However, when the outer

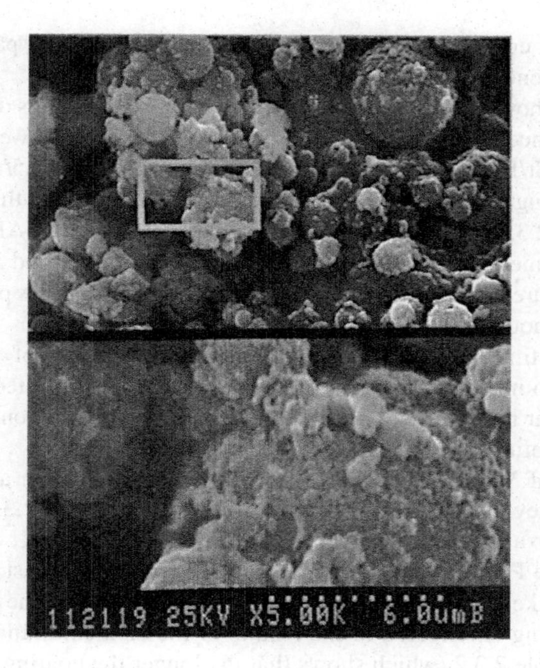

FIGURE 2.3.5 Particle morphology of a sample possessing a netlike PS coating (see Table 2.3.1).

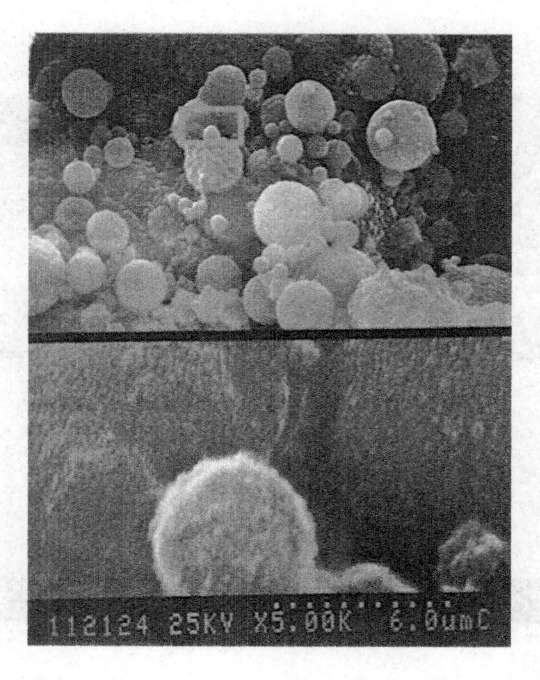

FIGURE 2.3.6 Particle morphology of a sample obtained with a large amount of hexane (see Table 2.3.1).

shell completely encapsulated the core, as in Figure 2.3.6, the particles could be dispersed in toluene but not in water.

Figure 2.3.7 shows that the thickness of the coating layer varies dramatically with the monomer concentration. The samples in Figure 2.3.7a and b were prepared with PAAM/hexane/St/DVB weight ratios of 2/3/0.5/0.5 and 2/3/1.5/1.5, respectively. The latexes in Figure 2.3.7a are only partially covered with a thin PS layer, and those in Figure 2.3.7b are completely covered with a thick layer. Although the same amount of monomer was used for the samples of Figures 2.3.5 and 2.3.6, the coating is thinner in Figure 2.3.6, because the monomer in the continuous phase was diluted with a double amount of hexane.

The coating time (the time used for the polymerization of the hydrophobic monomer) is also important. The sample in Figure 2.3.8 was coated under the same conditions as that in Figure 2.3.5, but for a shorter polymerization time (4 vs. 8 h); therefore, the coating layer has only a moderate thickness.

In the case of VDC, the effects of monomer concentration and coating time were as in the previous case. The samples in Figures 2.3.3 and 2.3.9 were prepared with PAAM/hexane/VDC weight ratios of 2/3/3 and 2/3/1, respectively. In Figure 2.3.3, the PVDC shell has a netlike structure; in contrast, in Figure 2.3.9, only a few netlike domains can be identified. The effect of the coating time on the PVDC coating is reflected in the results of the elemental analysis, which are presented in Table 2.3.2 (which shows that the longer the coating time, the higher the Cl wt % in the samples).

(a)

(b)

FIGURE 2.3.7 Particle morphologies of samples with different monomer concentrations. PAAM/hexane/St/DVB weight ratio: (a) 2/3/1.5/1.5; (b) 2/3/0.5/0.5 (see Table 2.3.1).

FIGURE 2.3.8 Particle morphology of a sample obtained after a short coating time (see Table 2.3.1).

TABLE 2.3.2
Elemental Analysis of Some Samples

Samples	AAM (g)	MBA (g)	Hexane (g)	VDC (g)	Coating Time (h)	N (Wt %)	Cl (Wt %)
a	2	0.1	3	1	8	15.73	1.92
b	2	0.1	3	1	6	15.71	1.33
c	2	0.1	3	1	4	15.90	1.12

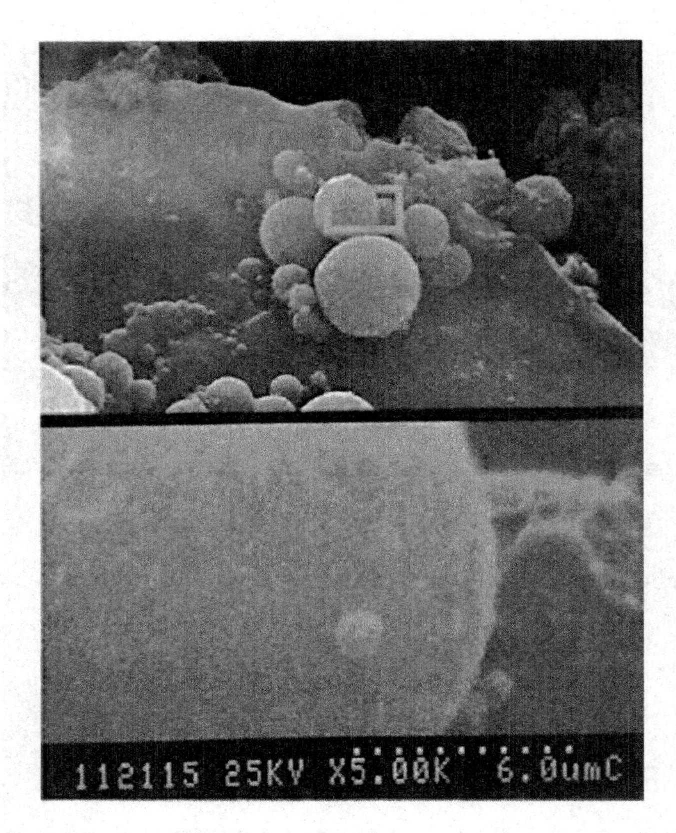

FIGURE 2.3.9 Particle morphology of a sample obtained with a low concentration of VDC (see Table 2.3.1).

2.3.4 CONCLUSION

Hydrophilic core/hydrophobic shell particles were prepared via an inverted emulsion method. Polyacrylamide latexes were first obtained, using a concentrated emulsion of acrylamide in hexane as precursor. Subsequently, hexane and a hydrophobic monomer were introduced into the polymerized concentrated emulsion, and the polyacrylamide latexes were coated, through surface-initiated polymerization, with a hydrophobic polymer. In proper conditions, tiny particles of hydrophobic polymer are present on the surface of polyacrylamide latexes as a netlike shell. The netlike structure of the outer shell makes the hydrophilic chains present inside accessible to the outside medium. Because such particles can be compatible with both water and oil phases, we call them amphiphilic particles.

REFERENCES

1. H. Mori, A. Hirao, S. Nakahama, and Z. Osawa, *Macromolecules,* **27**, 4093 (1994).
2. T. Kimura, K. Tanji, and M. Minabe, *Polymer J.,* **24**, 1311 (1992).
3. M. Akashi, D. Chao, and N. Yashima, *J. Appl. Polym. Sci.,* **39**, 2027 (1990).

4. F. O. Eschbach and S. J. Huang, *J. Bioact. Compat. Polym.*, **9**, 29 (1994).
5. S. Murayama, S. Kuroda, and Z. Osawa, *Polymer*, **34**, 3893 (1994).
6. E. Ruckenstein and J. S. Park, *J. Polym. Sci. Chem. Lett. Ed.*, **26**, 529 (1988).
7. E. Ruckenstein and J. S. Park, *Chem. Mater.*, **1**, 343 (1989).
8. M. Pulat, *React. Polym.*, **24**, 59 (1995).
9. H. Alhmoud, M. Efentakis, and N. H. Choulis, *Int. J. Pharm.*, **68**, R1 (1991).
10. R. Yoshida, K. Sakai, T. Okano, and Y. Sakurai, *Polym. J.*, **23**, 1111 (1991).
11. C. Sung, J. E. Raeder, and E. W. Merrill, *J. Pharm. Sci.*, **79**, 829 (1990).
12. E. Ruckenstein and L. Hong, *Chem. Mater.*, **4**, 1032 (1992).
13. L. Hong and E. Ruckenstein, *J. Molec. Catal.*, **A101**, 115 (1995).
14. J. W. Vanderhoff, J. M. Park, and M. S. El-Aaser, *ACS Symp. Ser.*, **492**, 272 (1992).
15. S. Lee and A. Rudin, *ACS Symp. Ser.*, **492**, 234 (1992).
16. E. Ruckenstein and K. J. Kim, *J. Appl. Polym. Sci.*, **36**, 907 (1988).

2.4 Hydrophobic Core/Hydrophilic Shell Amphiphilic Particles*

Yang Yun[†], *Hangquan Li*[†], *and Eli Ruckenstein*[‡]

[†]School of Materials Science and Engineering, Beijing University of Chemical Technology, Beijing 100029, China

[‡]Department of Chemical Engineering, State University of New York at Buffalo, Amherst, New York 14260

ABSTRACT Amphiphilic colloidal particles with hydrophobic cores and hydrophilic shells were prepared via a two-step method. First, polystyrene cores were obtained through the concentrated emulsion polymerization. A mixture of styrene, ethyl benzene, divinyl benzene, azobisisobutyronitrile, and cumene hydroperoxide (CHPO) was partially polymerized at 80°C for 40 min and subsequently used as the dispersed phase of a concentrated emulsion in water. The concentrated emulsion was subjected to complete polymerization at 60°C for 12 h; colloidal particles of crosslinked polystyrene were thus obtained. In the second step, the polystyrene particles were dispersed in water, after which acrylamide, *N,N′*-methylenebisacrylamide, and ferrous sulfate (FS) were added. The system was heated (typically at 30°C) to conduct the polymerization of the hydrophilic monomers. The CHPO present on the surface of the polystyrene particles and the FS present in the aqueous phase (both together constitute a redox initiator) ensured that the initiation occurred mostly on the surface of the particles and that the hydrophilic polymer obtained formed a shell encapsulating the particles. Under proper conditions, a porous outer shell could be generated, making the hydrophobic core accessible to the outside medium.

2.4.1 INTRODUCTION

Traditionally, emulsion polymerization was employed to prepare colloidal particles (1–10). These particles were surface treated or functionalized and found wide applications (11–22) in coating, in controlled drug release, as catalyst supports, etc. The core/shell particles have particularly attracted a great deal of attention (23–27).

* *Journal of Colloid and Interface Science*. Vol. 238, p. 414–419, 2001.

In previous papers (28,29), a novel, simple procedure for preparing hydrophilic core/hydrophobic shell particles starting from concentrated emulsions was proposed. In a concentrated emulsion the volume fraction of the dispersed phase is larger than 0.74, which represents the volume fraction of the most compact arrangement of spheres of equal radius and can be as large as 0.99 (30). It has a paste-like appearance and behavior. When the volume fraction of the continuous phase is sufficiently small, the dispersed phase is formed of polyhedral cells, separated by thin films of continuous phase. In Refs. 28 and 29, an inverted concentrated emulsion in hexane was first prepared by using an aqueous solution containing acrylamide, a crosslinker (*N, N′*-methylenebisacrylamide), and an initiator (ammonium persulfate) as the dispersed phase. The polymerization of acrylamide and its crosslinker generated hydrophilic particles. The system was subsequently diluted with a hexane solution containing hydrophobic monomers and a hydrophobic initiator and subjected to polymerization, thus generating a porous shell on the surface of the particles. There were no chemical links between the core and shell, the latter simply encapsulating the former. The porous structure made the hydrophilic chains present in the core accessible to the external fluid.

In the present paper, particles possessing a hydrophobic core and a hydrophilic shell were prepared. To obtain such particles, the core was first generated from a concentrated emulsion in water and encapsulated in a hydrophilic shell. The preparation conditions are described in detail in what follows.

2.4.2 EXPERIMENTAL

2.4.2.1 Materials

Styrene (St, AR, Tianjin reagent manufacturer No. 6) and divinyl benzene (DVB, 45%, Kunshan Additive Manufacturer) were passed through an inhibitor removal column, and azobisisobutyronitrile (AIBN, CP) was recrystallized from methanol. The following chemicals were purchased from Beijing Chemical Reagent Co. and used as received: acrylamide (AA, AR), *N,N′*-methylenebisacrylamide (MBA, 98%), hexane (99%), ferrous sulfate (FS, AR), ethyl benzene, cumene hydroperoxide (CHPO, 70%), dodecyl sulfate sodium salt (SDS), and diesel oil ($0^{\#}$). The water was distilled and deionized.

2.4.2.2 Preparation of the Hydrophobic Core

A mixture of St (4.5 g), ethyl benzene (5.0 g), DVB (0.5 g), AIBN (0.07 g), and CHPO (typically 0.2 g) was partially polymerized at 80°C for 40 min and subsequently used as the dispersed phase of a concentrated emulsion. The continuous phase, 3.0 g of aqueous solution of SDS (0.03 g/g water), was first introduced into a 100-mL flask equipped with a magnetic stirrer. The dispersed phase was then added dropwise with stirring using a syringe for about 10–15 min. A concentrated emulsion with a volume fraction of dispersed phase of 0.8 was thus generated and transferred to a test tube; the latter was sealed with a rubber septum and the air inside replaced with nitrogen. The test tube was subsequently introduced into a water bath at 60°C for 12 h for complete polymerization. The product was washed twice with methanol, and colloidal particles of polystyrene were finally obtained as a white powder.

2.4.2.3 Determination of the Stability of the Concentrated Emulsion

To select suitable conditions for the preparation of polystyrene particles, the concentrated emulsion was subjected to polymerization at various temperatures and surfactant concentrations for 1 h. Some liquid separated from the concentrated emulsion forming a new phase. The separated phase was removed with a syringe and weighed. The weight ratio of the separated phase to the whole system was considered a measure of the stability of the concentrated emulsion.

2.4.2.4 Encapsulation with a Hydrophilic Shell

An aqueous solution containing water, AA, MBA, and FS was first placed in a 100-mL flask provided with a magnetic stirrer. The proportion of various constituents varied in different experiments, a typical one being water, 4.0 g; AA, 0.2 g; MBA, 0.1 g; and FS, 0.02 g. Polystyrene particles (2.0 g) were dispersed in the aqueous solution, the flask was sealed with a rubber septum, and the air inside was replaced with nitrogen. The system was heated at a selected temperature (typically 30°C) for varying lengths of time (typically 8 h). The hydrophilic monomers polymerized, encapsulating the hydrophobic cores. The amphiphilic particles thus obtained were washed three times with water and finally dried in a vacuum oven at room temperature.

2.4.2.5 Scanning Electron Microscopy

A small amount of amphiphilic particles was glued to a conductive paper and coated with gold; their morphology was examined by scanning electron microscopy (Cambridge Steroscan 250MK3, UK).

2.4.2.6 Determination of the Absorption Capacity for Water or Diesel Oil

A small amount of powder of the hydrophobic core/hydrophilic shell particles (0.5 g) was wrapped in a filter paper and located at the bottom of a funnel with its outlet blocked. Water or diesel oil (10 mL) was introduced to immerse the wrapped sample. After 1 h the outlet of the funnel was opened and the water or diesel oil was allowed to flow out. After no drops of water or diesel oil were observed, the wet particles were weighed and the water or diesel absorption capacity calculated using the following expression: water (diesel) absorption capacity (g/g) = [(weight of the wet sample-weight of the dry sample)/(weight of the dry sample)]. Each point on the obtained curves represents the average of at least three experiments with a standard error less than 10%.

2.4.2.7 Stability of the Hydrophilic Shells

A small amount of amphiphilic particles (0.5 g) and 10 mL water or diesel oil were introduced into a 100-mL flask, and the system was subjected to vigorous stirring for 2 h. The particles were retrieved and examined by scanning electron microscopy.

2.4.3 RESULTS AND DISCUSSION

2.4.3.1 Stability of the Concentrated Emulsion

The fraction of the phase separated after 1 h of polymerization was used as a measure of the stability of the concentrated emulsion. The effects of the amount of surfactant and the polymerization temperature on stability are presented in Figure 2.4.1. As expected, Figure 2.4.1a indicates that the larger the amount of surfactant, the more stable the concentrated emulsion. For a surfactant content larger than 0.03 g/g of water, only a small amount of separated phase was observed. For this reason, in the preparation of the hydrophobic core, a surfactant concentration of 0.03 g/g of water was selected. Figure 2.4.1b shows that the higher the temperature, the less stable the concentrated emulsion. To maintain the stability of the concentrated emulsion, one should use a temperature as low as possible. However, a low temperature provides a low polymerization rate. We observed that for temperatures lower than 60°C the fraction of separated phase increased moderately and for temperatures higher than 60°C it increased more rapidly. To ensure that the polymerization will occur at a sufficiently high rate, 60°C was selected as the polymerization temperature. The stability of the concentrated emulsion had an important effect on the generation of particles. At the selected surfactant concentration (0.03 g/g of water) and polymerization temperature (60°C), the concentrated emulsion was sufficiently stable and provided independent particles (Figure 2.4.2a). When the separated phase became greater than 2.5%, the polystyrene particles were glued together, as shown in Figure 2.4.2b, and could not be subsequently coated individually with hydrophilic polymers.

2.4.3.2 Encapsulation by a Hydrophilic Shell

Because the hydrophobic cores contained an oxidant (cumene hydroperoxide, CHPO) and the continuous phase contained a reductant (ferrous sulfate, FS), which together constitute a redox initiator, the polymerization of the hydrophilic

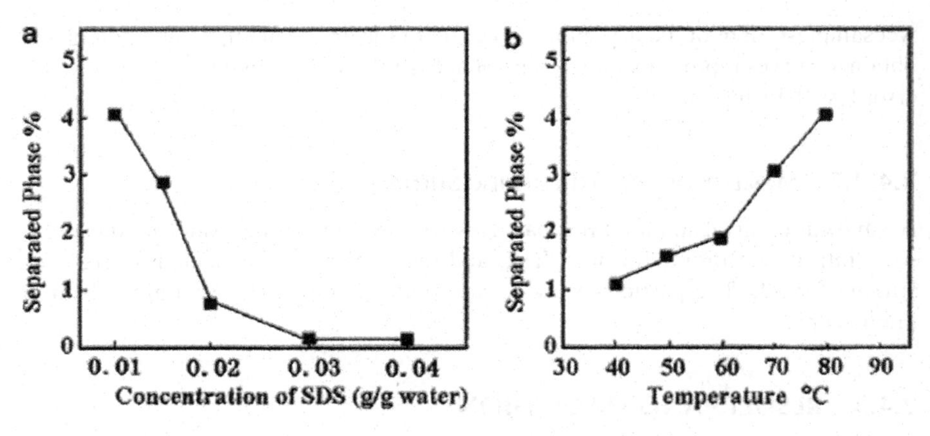

FIGURE 2.4.1 Effects of the amount of surfactant and temperature on the stability of concentrated emulsions. Dispersed phase: St, 4.5 g; ethyl benzene, 5.0 g; DVB, 0.5 g; AIBN, 0.07 g; and CHPO, 0.2 g; partially polymerized at 80°C for 40 min. (a) Continuous phase: 3.0 g of aqueous solution of SDS (various concentrations); polymerization temperature, 40°C. (b) Continuous phase: 3.0 g of aqueous solution of SDS (0.03 g/g of water); various polymerization temperatures.

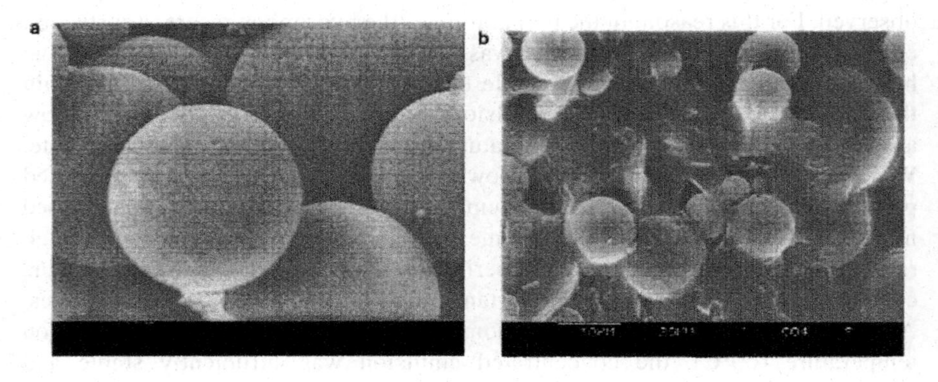

FIGURE 2.4.2 SEM micrographs of polystyrene particles. Dispersed phase: St, 4.5 g; ethyl benzene, 5.0 g; DVB, 0.5 g; AIBN, 0.07 g; and CHPO, 0.2 g; partially polymerized at 80°C for 40 min. (a) Continuous phase: 3.0 g of aqueous solution of SDS (0.03 g/g of water); polymerization temperature, 60°C. (b) Continuous phase: 3.0 g of aqueous solution of SDS (0.01 g/g of water); polymerization temperature, 80°C.

monomers was mostly initiated on the surface of the hydrophobic cores. While ferrous sulfate can initiate alone the polymerization of acrylamide, the rate of polymerization is low. The linear polyacrylamide is water-soluble; however, once crosslinked by MBA, the polyacrylamide will become insoluble and hence will remain deposited on the surface of the hydrophobic particles. Because the accumulation of the hydrophilic polymer affects the water and oil absorptions, the process of encapsulation can be characterized by their changes with the coating time (Figure 2.4.3). This figure shows that, with increasing reaction time, the absorption capacity for water increased and that of the diesel oil decreased. However, the increase and decrease were not linear, but they underwent a number of fast and slow stages of deposition. In the first stage, all hydrophilic polymers formed were linear or weakly crosslinked, hence readily dissolved or swollen in water, and the rate of accumulation on the core was low. In the second stage, the extent of crosslinking of the hydrophilic polymer became sufficiently high, leading to enhanced encapsulation rate. The morphology of the particles at the end of this stage is shown in Figure 2.4.4a. As the coating process proceeded, parts of the core surface became covered with the hydrophilic polymer and the oxidant present on the surface of the core became largely consumed. Because the diffusion of fresh oxidant from the inside of the core to the surface was slow, the rate of deposition also became slow (stage 3). At the end of this stage, a thin shell with a porous structure was deposited on the core, as shown in Figure 2.4.4b. As already mentioned, the ferrous sulfate alone can slowly initiate the polymerization of acrylamide in the aqueous phase; perhaps a part of the crosslinked polyacrylamide formed in the aqueous solution has deposited on the surface of the core, resulting in the fourth stage of fast accumulation (Figure 2.4.4c). In the last stage, the hydrophobic cores were almost fully coated with the hydrophilic polymer, and this slowed down the accumulation. The morphology of a completely coated particle is shown in Figure 2.4.4d.

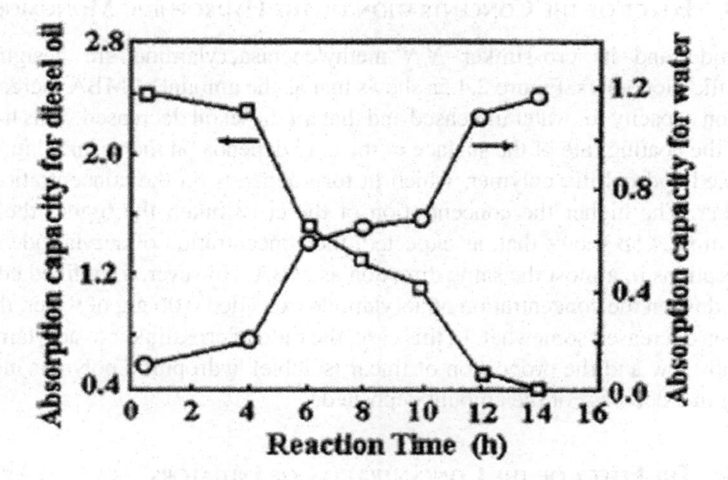

FIGURE 2.4.3 Effect of coating time on the water and diesel oil absorption capacities of the particles. The recipe of the core is as in Figure 2.4.1b. Aqueous solution for coating: water, 4.0 g; AA, 0.2 g; MBA, 0.1 g; FS, 0.02 g; and polystyrene particles, 2.0 g; coating temperature, 30°C.

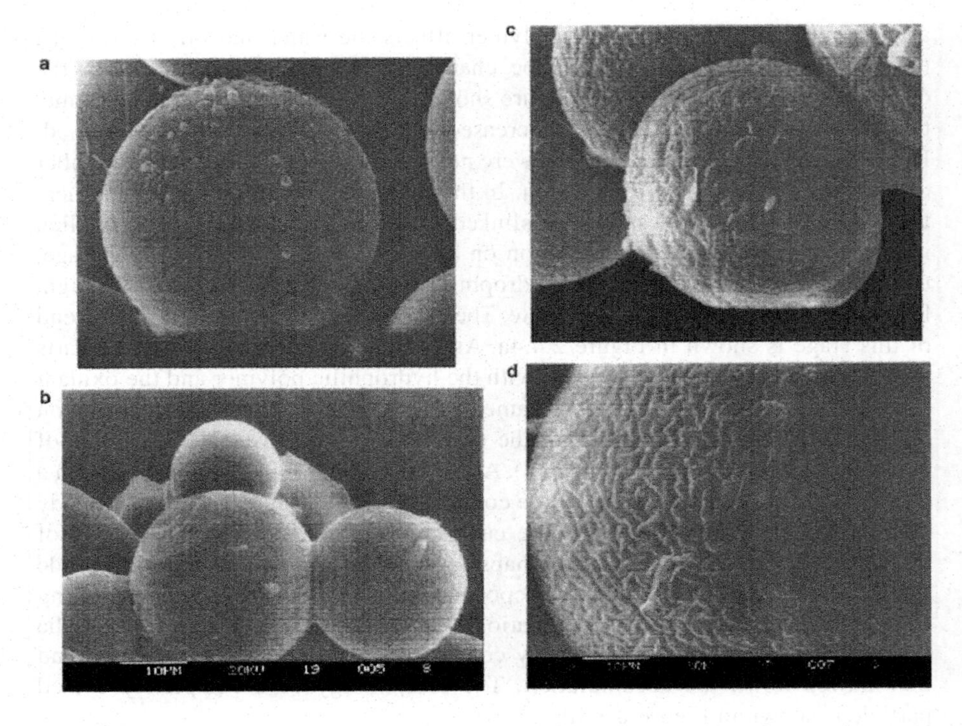

FIGURE 2.4.4 Morphology of the particles as a function of coating time, recipe as in Figure 2.4.3: (a) 6 h, (b) 10 h, (c) 12 h, and (d) 14 h.

2.4.3.3 Effect of the Concentration of the Hydrophilic Monomers

Acrylamide and its crosslinker *N,N'*-methylenebisacrylamide are designated as hydrophilic monomers. Figure 2.4.5a shows that as the amount of MBA increased, the absorption capacity for water increased and that for diesel oil decreased. This happened because the coating rate of the surface of the core depends on the accumulation of the crosslinked hydrophilic polymer, which in turn depends on the concentration of the crosslinker. The higher the concentration of the crosslinker, the higher the coating rate. Figure 2.4.5b shows that, as expected, the concentration of acrylamide affected the absorptions in almost the same direction as MBA. However, for a fixed content of MBA and when the concentration of acrylamide exceeded 0.06 g/g of water, the water absorption decreased somewhat. In this case, the ratio of crosslinker to acrylamide was sufficiently low and the proportion of linear (soluble) hydrophilic polymer increased, resulting in a decrease of the amount deposited.

2.4.3.4 The Effect of the Concentration of Initiators

The polymerization of hydrophilic monomers was initiated by a redox system consisting of an oxidant CHPO, which was present in the hydrophobic phase, and a reductant FS, which was present in the hydrophilic phase. For this reason, most

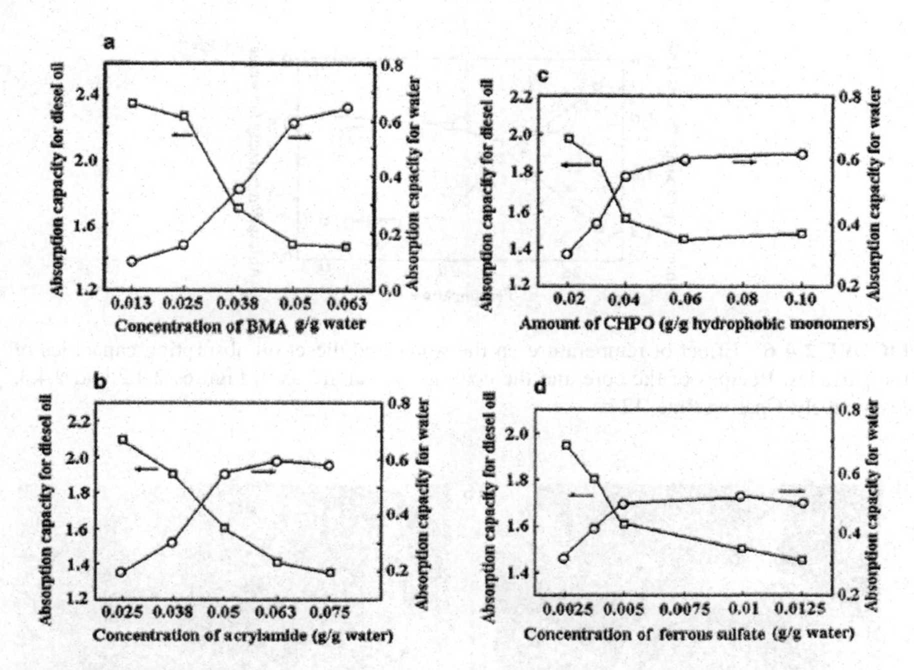

FIGURE 2.4.5 Effect of the ingredients on the water and diesel oil absorption capacities of the particles. Recipes of the core and the coating system are as in Figures 2.4.2 and 2.4.3, respectively, except (a) various contents of MBA; (b) various contents of AA; (c) various contents of CHPO; and (d) various contents of FS. Coating temperature: 30°C. Coating time: 12 h.

of the initiation occurred on the surface of the hydrophobic core. It is expected that the greater the amounts of initiators, the greater the amount of hydrophilic coating. Of course, for amounts of initiators that are too large, saturation is expected to occur. Indeed, Figure 2.4.5c and d show such a general tendency. When the amount of FS became too high, the amount of deposited polymer decreased. This happened because FS initiated the formation of linear soluble polymers, which did not deposit on the surface of the core. Furthermore, the amount of CHPO must be kept under a certain value. When the concentration of CHPO exceeded 0.1 g/g of hydrophobic monomers and the content of FS was 0.005 g/g of water, the polymerization of the hydrophilic monomers became very rapid (even explosive, being finished in a few seconds), the temperature increased tremendously, and little coating occurred. This probably happened because some molecules of CHPO dissolved in water and the increased temperature due to the reaction increased the solubility of CHPO; as a result, most of the initiation took place in water.

2.4.3.5 Effect of Temperature on Encapsulation

Generally, high temperatures ensured high rates of encapsulation. As shown in Figure 2.4.6, below 28°C the coating rate increased rapidly, while above 30°C, the increase in the coating rate was moderate. For these reasons 30°C was selected as a suitable temperature for encapsulation.

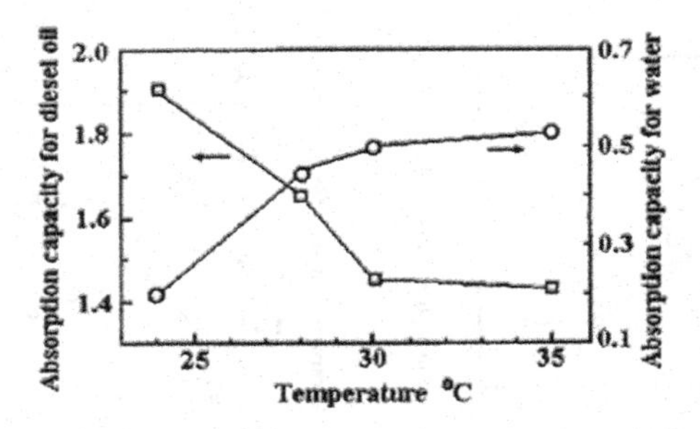

FIGURE 2.4.6 Effect of temperature on the water and diesel oil absorption capacities of the particles. Recipes of the core and the coating system are as in Figures 2.4.2 and 2.4.3, respectively. Coating time, 12 h.

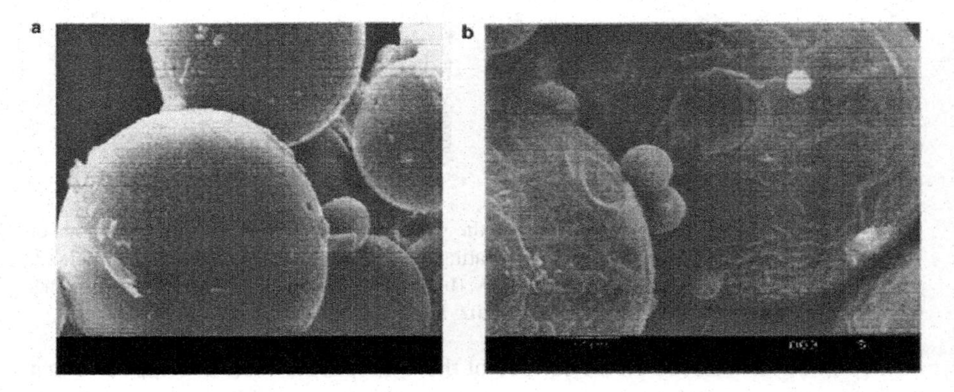

FIGURE 2.4.7 Morphology of particles with porous shells after being stirred in diesel oil or water: (a) thin porous shells, stirred in diesel oil; and (b) thick porous shells, stirred in water. Recipes are as in Figure 2.4.3, with 10 h of coating for (a) and 14 h of coating for (b).

2.4.3.6 The Stability of the Shell

A small sample of amphiphilic particles (0.5 g) was dispersed in 10 mL of water or diesel oil and stirred for 2 h. Subsequently the particles were dried and examined under scanning electron microscope. Comparing Figure 2.4.7a and b with Figure 2.4.4b and d, respectively, one can notice that the hydrophilic coating remained the same after it was swollen and stirred. Although there are no chemical bonds between the hydrophobic cores and the hydrophilic shells, the polymer chains in the shells are crosslinked and entangled and thus the stability to detachment is ensured.

2.4.4 CONCLUSION

Amphiphilic particles with hydrophobic cores and porous hydrophilic shells were prepared through a two-step procedure. Polystyrene cores were first prepared via the concentrated emulsion polymerization, and the particles obtained were subsequently encapsulated with polyacrylamide. The concentrated emulsion should be sufficiently stable to ensure the formation of independent particles. To ensure the deposition of polyacrylamide on the surface of the core, a redox initiator, consisting of an oxidant (cumene hydroperoxide) in the hydrophobic core and a reductant (ferrous sulfate) in the aqueous solution, was employed. For this reason, most of the initiation took place on the surface of the core. To obtain a uniform hydrophilic porous coating, a sufficiently large amount of hydrophilic crosslinker (*N,N'*-methylenebisacrylamide) was necessary. A content of crosslinker that was too low provided a large amount of linear polyacrylamide, which, being soluble, did not completely deposit onto the surface of the core.

ACKNOWLEDGMENT

This work was supported by the Ph.D. Program Foundation, Ministry of Education, China, and by National Science Foundation, U.S.A.

REFERENCES

1. Pichot, C., Delair, T., and Elaissari, A., in *Polymeric Dispersion: Principles and Applications* (Asua, J. M., Ed.), p. 515. Kluwer Academic Publishers, Dordrecht, the Netherlands, 1997.
2. Vanderhoff, J. W., Ark, J. M., and El-Aasser, M. S., in *Polymer Latexes: Preparation, Characterization & Applications* (Daniel, E. S., Sudol, E. D., and El-Aasser, M. S., Eds.), p. 272, ACS, Washington DC, 1992.
3. Delair, T., Pichot, C., and Mandranal, B., *Colloid Polym. Sci.* **272**, 72 (1994).
4. Ganachaud, F., Mouterde, G., and Delair, T., *Polym. Adv. Technol.* **6**, 480 (1994).
5. Lewandowski, K., Svec, F., and Frechet, J. M. J., *J. Appl. Polym. Sci.* **67**, 597 (1998).
6. Coutinho, F. M. B., Teixeira, V. G., and Barbosa, C. C. R., *J. Appl. Polym. Sci.* **67**, 781 (1998).
7. Okubo, M., Ito, A., and Hashiba, A., *Colloid Polym. Sci.* **274**, 428 (1996).
8. Fritz, H., Maier, M., and Bayer, E., *J. Colloid. Polym. Sci.* **195**, 272 (1997).
9. Galisteo-Gonzalez, F., Rodriguez, A. M., and Hidalgo-Alverez, R., *J. Colloid. Polym. Sci.* **272**, 755 (1994).
10. Weyenberg, D. R., Findlay, D. E., Cekata, J., Jr., and Bey, A. E., *J. Polym. Sci. Part C* **27**, 27 (1969).
11. Okubo, M., Yamada, A., and Matsumoto, T., *J. Polym. Sci., Polym. Chem. Ed.* **16**, 3219 (1980).
12. Okubo, M., Katsuta, Y., and Matsumoto, T., *J. Polym. Sci., Polym. Lett. Ed.* **20**, 45 (1982).
13. Okubo, M., Ito, A., and Nakamura, M., *Colloid Polym. Sci.* **275**, 82 (1997).
14. Okubo, M., Yamashita, T., Suzuki, T., and Shimizu, T., *Colloid Polym. Sci.* **275**, 288 (1997).
15. Weiss, A., Hartenstein, M., Dingenouts, N., and Ballauff, M., *ColloidPolym. Sci.* **276**, 794 (1998).
16. Voigt, I., Simon, F., Komber, H., Jacobasch, H. J., and Spange, S., *Colloid Polym. Sci.* **278**, 48 (2000).
17. Marciano, V., Minore, A., and Turco, L. V., *Colloid Polym. Sci.* **278**, 250 (2000).
18. Pochard, I., Foissy, A., and Couchot, P, *Colloid Polym. Sci.* **277**, 818 (1999).
19. Okubo, M., and Ichikawa, K., *Colloid Polym. Sci.* **272**, 933 (1994).

20. Okubo, M., Ichikawa, K., and Fujimura, M., *Colloid Polym. Sci.* **269**, 1257 (1991).
21. Okubo, M., and Minami, H., *Colloid Polym. Sci.* **275**, 992 (1997).
22. Okubo, M., Takekoh, R., Izumi, J., and Yamashita, T., *Colloid Polym. Sci.* **277**, 972 (1999).
23. Ruckenstein, E., and Hong, L., *J. Catalysis* **136**, 378 (1992).
24. Ruckenstein, E., and Hong, L., *Chem. Mater.* **4**, 1032 (1992).
25. Ruckenstein, E., and Sun, F., *J. Appl. Polym. Sci.* **46**, 1271 (1992).
26. Kong, X. Z., and Ruckenstein, E., *J. Appl. Polym. Sci.* **71**, 1455 (1999).
27. Kong, X. Z., and Ruckenstein, E., *J. Appl. Polym. Sci.* **73**, 2235 (1999).
28. Li, H. Q., and Ruckenstein, E., *J. Appl. Polym. Sci.* **61**, 2129 (1996).
29. Li, H. Q., Zhao, J., and Ruckenstein, E., *Colloids Surf.* **161**, 489 (2000).
30. Ruckenstein, E., *Adv. Polym. Sci.* **127**, 1 (1997).

2.5 Encapsulation of Solid Particles by the Concentrated Emulsion Polymerization Method*

Jun Seo Park and Eli Ruckenstein

Department of Chemical Engineering, State University of New York at Buffalo, Buffalo, New York 14260, USA

ABSTRACT The encapsulation of submicron sizes of alumina and fumed silica particles in crosslinked polyacrylamide is described. In a first step, a colloidal dispersion was prepared by dispersing the solid particles in an aqueous monomer (acrylamide) solution containing a water-soluble dispersant, a crosslinking agent, and a suitable initiator. In the second step, a concentrated emulsion was prepared at room temperature by dispersing the above colloidal system in decane containing a suitable surfactant, the volume fraction of the continuous phase being about 0.1. Upon heating at 40°C, polymerization of the dispersed phase of the gel-like emulsion obtained took place, thus encapsulating the solid particles in capsules with a polyhedral shape. Electron microscopy revealed that the solid particles were covered by polymer, and that the sizes of the capsules were around 4–5 μm for alumina and 1.0–1.5 μm for silica.

2.5.1 INTRODUCTION

Encapsulation is a well-known process in which tiny particles or droplets are covered by a coating or a membrane.[1–6] The role of encapsulation is either to isolate the active ingredient or to control the rate by which it leaves the capsule.

* *Polymer*. Vol. 31, 75–179, (1990).

As examples for the first case, one can mention the isolation of vitamins from oxygen or of a reactive core from chemical attack, and for the second case, the control of the rate of release of drugs or pesticides. A great many encapsulation techniques have been suggested and new ones are continually being developed. In encapsulation by coacervation,[3] which is a phase separation phenomenon that occurs in colloidal systems, the coacervate layer is deposited uniformly around the individual particles of the active core material, which are uniformly dispersed in the medium. The capsules are provided with rigid walls by crosslinking the precipitated coacervates. Interfacial polymerization can also be used for encapsulation.[4,5] In this case, the active ingredient is dissolved or dispersed in an organic phase containing diacid chloride. The system thus obtained is dispersed in water containing a suitable surfactant. The instantaneous polymerization at the interface of each oil droplet by addition of diamine leads to the formation of a thin film enclosing the droplet containing the active ingredient. The encapsulation of inorganic powder by the soapless emulsion polymerization of methyl methacrylate in water in the presence of powders was also attempted.[7,8] Until now, most of the encapsulations of the active solid materials have been carried out for solids larger than 1 μm.

In the present paper, the concentrated emulsion polymerization method[9,10] is employed to encapsulate submicron inorganic powders. In the conventional emulsion polymerization, the volume fraction of the continuous phase is large, and polymerization occurs in micelles and the monomer molecules migrate from the monomer droplets to the micelles.[11] In contrast, in concentrated (gel-like) emulsion polymerization the volume fraction of the continuous phase is small (as small as 0.05), and polymerization occurs in the monomer cells of the dispersed phase.[9] The stability of this gel is ensured by the adsorption of a surfactant, which is dissolved in the continuous phase, on the interface of the droplets. In a concentrated emulsion, the dispersed phase consists of polyhedral cells separated by a network of thin layers of the continuous phase.[10] In a first step, a stable colloidal dispersion of the powder in an aqueous solution of a monomer containing an appropriate dispersant and a suitable initiator was prepared. This colloidal dispersion was subsequently employed as the dispersed phase of a concentrated emulsion whose continuous phase, decane, contained a surfactant. The role of the surfactant is to stabilize the gel-like concentrated emulsion. Upon heating at 40°C, polymerization took place and the solid particles were encapsulated in the polymer.

2.5.2 EXPERIMENTAL

2.5.2.1 MATERIALS

Acrylamide (Polysciences) was purified by recrystallization in methanol. Potassium persulphate (Aldrich) was recrystallized from water. *N,N′*-Methylenebisacrylamide (Kodak), sorbitan monooleate (Fluka), Triton X-45 (Rohm and Haas), decane (Fluka), and cyclohexane (Aldrich) were used as received. α-Alumina (M-300, 0.05 μm) and fumed silica (HS-5, ~10^{-3} μm) were obtained from Metlade Corp. and Carbot Corp., respectively. Water was deionized and distilled.

2.5.2.2 Preparation of the Capsules

A small amount of decane containing sorbitan monooleate was placed in a 250 mL three-neck flask equipped with a mechanical stirrer, an addition funnel, and a nitrogen inlet. In addition, a colloidal dispersion was prepared by dispersing the solid powder in an aqueous solution of acrylamide, *N,N′*-methylenebisacrylamide (crosslinking agent), and Triton X-45 (dispersant) under stirring. Subsequently, potassium persulphate (initiator) was added to the system. The preparation of the concentrated emulsion was carried out at room temperature by dropwise addition of the colloidal dispersion to the continuous phase within about 10 min under a nitrogen atmosphere. Polymerization was carried out in a water bath at 40°C under a nitrogen stream for 6 h.

2.5.2.3 Electron Microscopy

Scanning electron microscopy (SEM, Amray 100A) and transmission electron microscopy (TEM, Hitachi HS-8) were employed to examine the state of encapsulation of the powder. The polymerized gel was dispersed in cyclohexane to produce a latex solution. The specimens were prepared by placing a drop of the latex solution on carbon films coated copper grids for TEM and on a clean cover glass for SEM. The grids were allowed to dry before observation. In the case of SEM, a thin layer of gold was deposited prior to observation.

2.5.3 RESULTS AND DISCUSSION

The effect of a water-soluble dispersant on the stability of the colloidal dispersion is illustrated in Figure 2.5.1, which presents scanning electron micrographs of alumina powders dispersed in water in the absence and presence of dispersant. They show that the agglomeration of the solid particles decreases tremendously in the presence of the dispersant. The agglomerates are larger than 10 μm in the absence of dispersant and smaller by more than one order of magnitude in its presence. Figure 2.5.2 presents scanning electron micrographs of crosslinked polyacrylamide latexes, free of solid particles, prepared by the concentrated emulsion polymerization method. The amounts of the components used in their preparation are listed in Table 2.5.1 under PL1. The latexes have spherical shape and range in size from 1.0 to 4.0 μm. Figure 2.5.3 shows scanning electron micrographs of latexes containing a water-soluble dispersant. The shape of the latexes is polygonal and their sizes are in the range of 0.5–3.0 μm. The amounts of various components involved are listed in Table 2.5.1 under PL2. Scanning electron micrographs of latexes which contain a water-soluble dispersant and a smaller amount of polyacrylamide (PL3 in Table 2.5.1) are presented in Figure 2.5.4. The sizes of the latexes are in the range of 0.5–1.5 μm. By including a dispersant and decreasing the amount of polyacrylamide, the polymer latex particles became polygonal and smaller in size.

TABLE 2.5.1
Representative Composition in the Preparation of Polyacrylamide Latexes Free of Solid Particles

	PL1	PL2	PL3
Dispersed phase			
Acrylamide (g)	5	5	2.5
Crosslinking agent (*N,N′*-methylenebisacrylamide) (g)	0.5	0.5	0.25
Initiator (potassium persulphate) (g/g acrylamide)	0.01	0.01	0.01
Dispersant (Triton X-45) (g)	—	2	2
Water (g)	20	20	20
Continuous phase			
Decane (mL)	3	3	3
Surfactant (sorbitan monooleate) (mL)	1.5	1.5	1.5

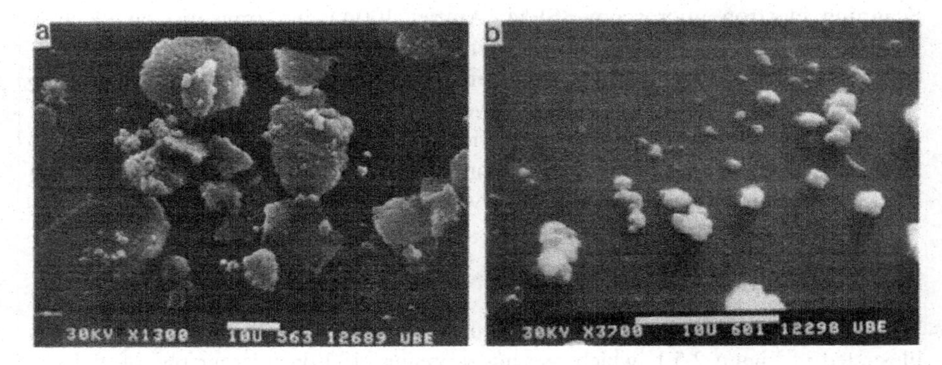

FIGURE 2.5.1 Scanning electron micrographs of alumina particles dispersed in water (a) in the absence of dispersant and (b) in the presence of dispersant.

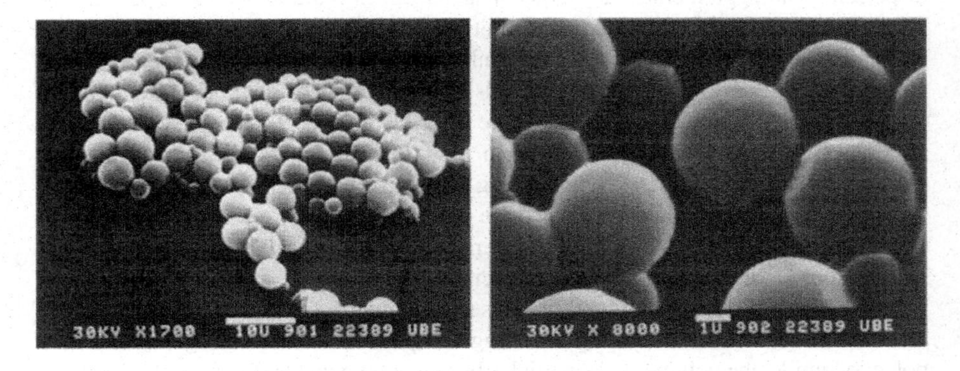

FIGURE 2.5.2 Scanning electron micrographs at two magnifications of crosslinked polyacrylamide latex particles with the composition PL1 (Table 2.5.1).

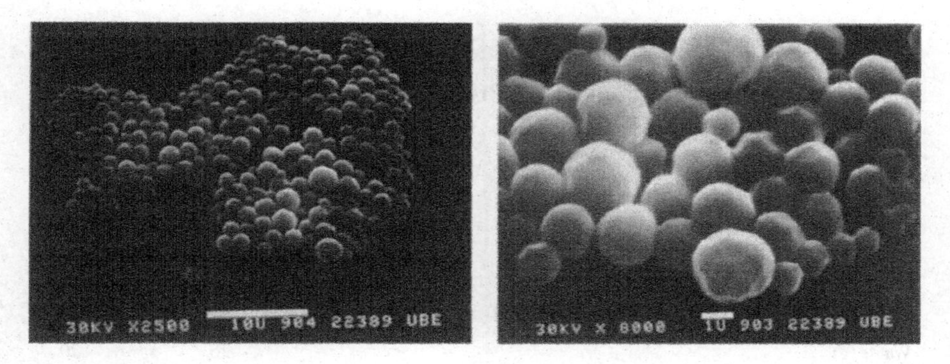

FIGURE 2.5.3 Scanning electron micrographs at two magnifications of crosslinked polyacrylamide latex particles with the composition PL2 (Table 2.5.1).

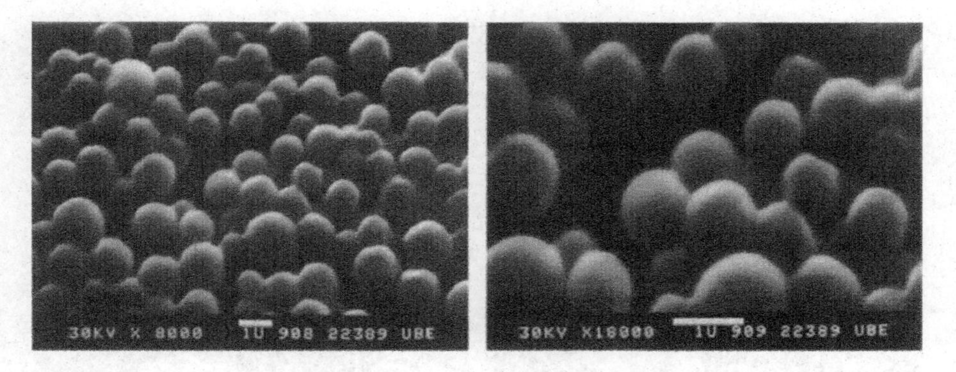

FIGURE 2.5.4 Scanning electron micrographs at two magnifications of crosslinked polyacrylamide latex particles with the composition PL3 (Table 2.5.1).

Figure 2.5.5 presents a scanning electron micrograph of a capsule, in which alumina particles are encapsulated in crosslinked polyacrylamide. Table 2.5.2 lists under PLA1 the amounts of the components involved in the preparation of these capsules. The capsules have a polyhedral shape and their sizes are larger (around 5 μm) and more uniform than the polymer latexes free of solid particles. As reported before,[10] the stability of the gel depends on the polymerization temperature and the amount of acrylamide in the dispersed phase. Some of the cells of the gel coalesce during polymerization, forming bulk phases. As a result, some unencapsulated solid particles are observed. Figure 2.5.6a and b present transmission electron micrographs of alumina particles and capsules which contain alumina particles, respectively. The amounts of the components involved in the preparation of capsules are listed in Table 2.5.2 under PLA1. Figure 2.5.7 presents scanning electron micrographs of PLA2 capsules (Table 2.5.2), in which the amount of polymer is smaller than for PLA1. The size of the capsules is slightly smaller, and their shape is changed (perhaps because of the loss of water?).

Very fine solid particles, namely, fumed silica, were also encapsulated by the concentrated emulsion polymerization method. Figure 2.5.8 shows scanning electron

TABLE 2.5.2
Representative Compositions in the Preparation of Alumina Capsules

	PLA1	PLA2
Dispersed phase		
Acrylamide (g)	5	2.5
N,N'-Methylenebisacrylamide (g)	0.5	0.25
γ-Alumina (g)	1.5	1.5
Dispersant (Triton X-45) (g)	2	2
Initiator (sodium persulphate) (g/g acrylamide)	0.01	0.01
Water (g)	20	20
Continuous phase		
Decane (mL)	3	3
Surfactant (sorbitan monooleate) (mL)	1.5	1.5

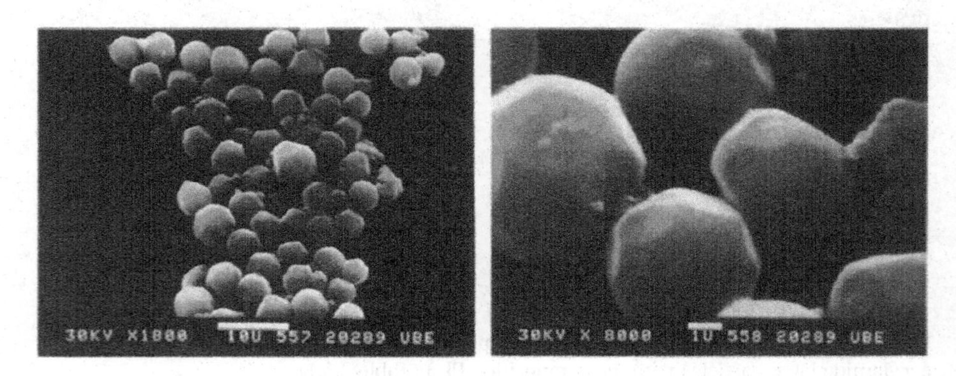

FIGURE 2.5.5 Scanning electron micrographs at two magnifications of capsules of alumina particles encapsulated in crosslinked polyacrylamide with the composition PLA1 (Table 2.5.2).

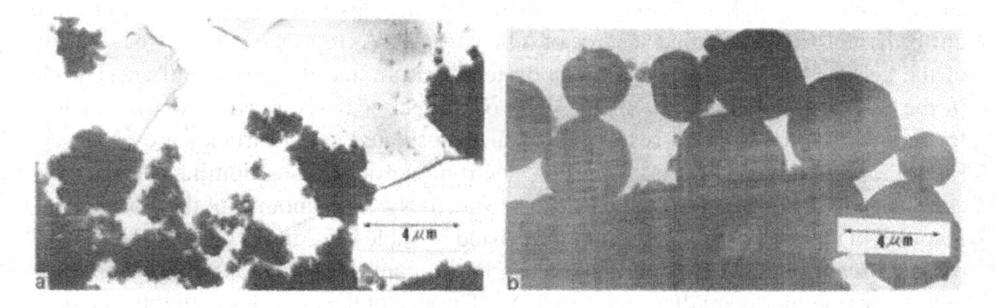

FIGURE 2.5.6 Transmission electron micrographs of (a) alumina and (b) alumina encapsulated in crosslinked polyacrylamide with the composition PLA1 (Table 2.5.2).

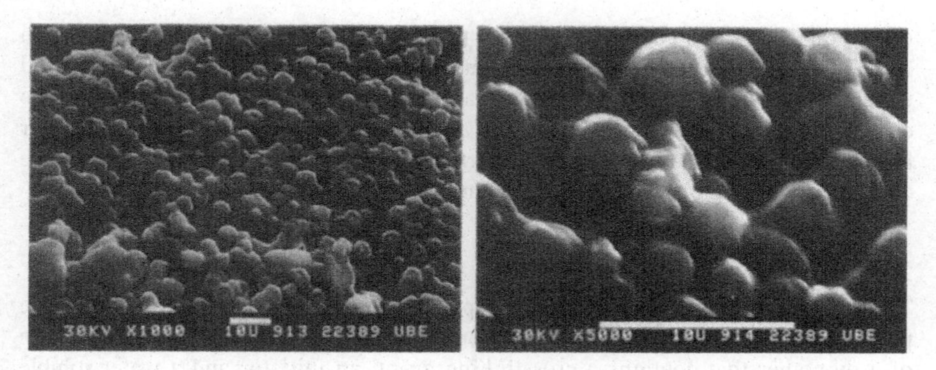

FIGURE 2.5.7 Scanning electron micrographs at two magnifications of alumina encapsulated in crosslinked polyacrylamide with the composition PLA2 (Table 2.5.2).

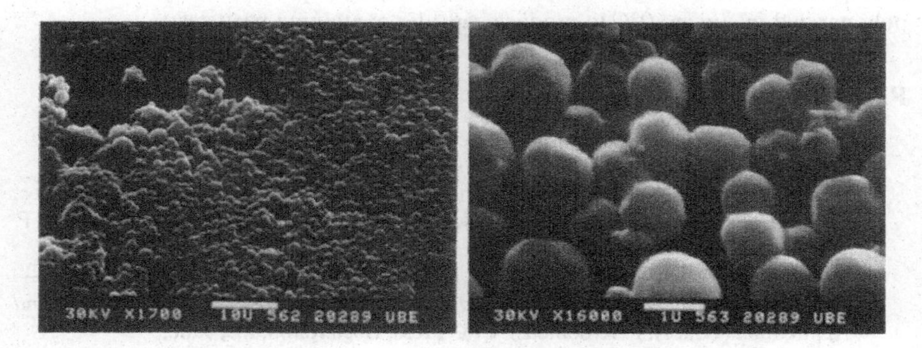

FIGURE 2.5.8 Scanning electron micrographs at two magnifications of fumed silica encapsulated in crosslinked polyacrylamide with the composition PLS1 (Table 2.5.3).

TABLE 2.5.3
Representative Compositions in the Preparation of Silica Capsules

	PLS1
Dispersed phase	
Acrylamide (g)	5
N,N′-Methylenebisacrylamide (g)	0.5
Fumed silica (g)	0.75
Dispersant (Triton X-45) (g)	2
Initiator (sodium persulphate) (g/g acrylamide)	0.01
Water (g)	20
Continuous phase	
Decane (mL)	3
Surfactant (sorbitan monooleate) (mL)	1.5

micrographs of silica capsules, in which fumed silica particles are encapsulated in crosslinked polyacrylamide. The amounts of the components involved are listed under PLS1 in Table 2.5.3. The PLS1 capsules range from 1.0 to 1.5 μm. These capsules also have polyhedral shape and are slightly smaller than the PL3 polymer latex particles. Some unencapsulated particles have been observed because of the coalescence during polymerization of some cells of the gel.

2.5.4 CONCLUSIONS

Three steps are employed in the encapsulation process. The first involves the preparation of a stable colloidal dispersion of solid particles in an aqueous solution of a monomer that contains a crosslinking agent, an initiator and a water-soluble dispersant. In the second step, the colloidal dispersion is used as the dispersed phase in the preparation at room temperature of a concentrated, gel-like emulsion having decane containing a surfactant as the continuous phase. Finally, the system is polymerized *in situ* at 40°C to produce capsules containing particles.

REFERENCES

1. Nixon, J. R. *Microencapsulation*, Marcel Dekker, New York, 1976.
2. Das, K. G. *Controlled Release Technology*, Wiley, Interscience, New York, 1983.
3. Bakan, J. A. Microencapsulation using coacervation/phase separation techniques. In *Controlled Release Technologies: Method, Theory, and Applications* (Ed. A. F. Kydonieus), CRC Press, Boca Raton, FL, 1980.
4. Koestler, R. C. Microencapsulation by interfacial polymerization techniques—Agricultural applications. In *Controlled Release Technologies: Methods, Theory, and Applications* (Ed. A. F. Kydonieus), CRC Press, Boca Raton, FL, 1980.
5. Brynko, C. U.S. Patent 2969 330, 1961.
6. Goldenhersh, K. K., Fluang, W., Manson, N. S. and Sparks, R. E. *Kidney Int.* 1976, **10**, 251.
7. Hasegawa, M., Arai, K. and Saito, S. *J. Polym. Sci., Polym. Chem. Edn.* 1987, **25**, 3117.
8. Hasegawa, M., Arai, K. and Saito, S. *J. Appl. Polym. Sci.* 1987, **33**, 411.
9. Ruckenstein, E. and Kim, K. J. *J. Appl. Polym. Sci.* 1988, **36**, 907; Kim, K. J. and Ruckenstein, E. *Macromol. Chem., Rapid Commun.* 1988, **9**, 285.
10. Ruckenstein, E. and Park, J. S. *J. Polym. Sci., Polym. Lett. Edn.* 1988, **26**, 529.
11. Bovey, F. A., Kolthoff, I. M., Medalia, A. I. and Meehan, E. E. *Emulsion Polymerization*, Wiley-Interscience, New York, 1955.

3 Enzyme/Catalyst/ Herbicide Carriers

CONTENTS

Immobilized enzyme systems are innovative catalysts with distinct merits. Compared with chemical catalysts, these enzyme-based catalysts can function at mild, more environmentally friendly conditions. They can be readily recovered after reaction. Moreover, they can accelerate reactions without changing the reaction equilibrium. The techniques used to immobilize enzymes onto carriers include reversible physical adsorption and ionic interactions, as well as irreversible conjugation with the formation of stable covalent bonds.

The characteristics of the carrier are crucial in determining the effectiveness of the immobilized enzyme systems. Important carrier characteristics include hydrophobicity, inertness towards enzymes, biocompatibility, resistance to microbial attack, and resistance to compression. In addition, the preparation of such a carrier should be inexpensive to provide commercial feasibility. The most commonly used carriers are polymers (such as cellulose, ion exchange resins, and several other types of polymers), active charcoal, silica, and ceramic. Porous carriers are generally preferred, because their large surface area permits a higher enzyme loading and the immobilized enzyme can also receive better protection from the environment.

As aforementioned, the volume fraction of dispersed phase is very high and can be as high as 0.99 in a concentrated emulsion. When polymerizable hydrophobic components (e.g., styrene and corresponding crosslinker) constitute the continuous phase and are subjected to polymerization, a carrier material with a high porosity can be obtained. Therefore, concentrated emulsion polymerization offers an appropriate pathway to prepare the porous polymeric carriers that are hydrophobic and inert to enzymes, and that also have other preferred properties for enzyme immobilization application. Sections 3.1–3.2 described studies of immobilization of different enzymes on such carriers, and their catalysis performance in various reactions.

In Section 3.1, *Phanerochaete chrysosporium* (white rot fungi) is immobilized on the porous polystyrene carrier and used for successive batch production of lignin peroxidase in shake cultures based on a C-limited medium. The best results are achieved when a spore inoculum is used for immobilization. The immobilized *Phanerochaete chrysosporium* is also used as a biocatalyst for the degradation of 2-chlorophenol.

In Section 3.2, lipase from *Candida rugosa* is immobilized on the internal surface of porous polystyrene supports and used as a biocatalyst for the hydrolysis of triacylglycerols. The effects of the hydrophilicity of the carriers on the adsorption, activity, and stability of the immobilized lipase are investigated. It is concluded that the porous polystyrene constitutes an excellent substrate for lipase because the amount of lipase adsorbed is large and the immobilized enzyme also leads to higher catalytic efficacy than free enzyme.

In a similar way, concentrated emulsion polymerization is employed to prepare the carriers for chemical catalysts rather than enzymes. In Section 3.3, four kinds of porous molecular substrates are prepared via concentrated emulsion, including (1) crosslinked polystyrene particles, (2) crosslinked polyacrylamide particles, (3) core-shell particles with poly(vinylbenzyl chloride) cores and porous crosslinked polystyrene shells, and (4) core-shell particles with poly(ethylene oxide) core and crosslinked polyacrylamide shell. These particles can be readily loaded or functionalized to make supported catalyst systems.

Section 3.4 provides an example in which copolymer particles of styrene-vinylbenzyl chloride are prepared via concentrated emulsion polymerization, yielding core-shell particles with crosslinked poly(vinylbenzyl chloride) cores and non-uniform polystyrene shells. Such particles are then quaternized to generate a polymer substrate with bound quaternary onium chloride. The resulting polymer-supported onium salts are used as phase transfer catalysts in the alkylation of isopropylidene malonates.

Section 3.5 describes the preparation of a colloidal carrier, not for enzyme/catalyst, but for a model hydrophobic herbicide, 2-(2,4-dichlorophenoxy)propionic acid. A mixture of styrene, herbicide, and an initiator is dispersed into a small amount of water, forming a concentrated emulsion. Polymerization at 40°C leads to the formation of latex containing the herbicide. It is found that the model herbicide dispersed in the polymer latex is released to water over a period of several weeks.

3.1 Production of Lignin Peroxidase by *Phanerochaete chrysosporium* Immobilized on Porous Poly(styrene-divinylbenzene) Carrier and Its Application to the Degrading of 2-Chlorophenol*

Eli Ruckenstein and Xiao-Bai Wang
Department of Chemical Engineering,
State University of New York at Buffalo,
Buffalo, New York 14260

ABSTRACT Porous poly(styrene-divinylbenzene) carriers for the immobilization of white rot fungus *Phanerochaete chrysosporium* have been prepared by the concentrated emulsion polymerization method. The concentrated emulsion consists of a mixture of styrene and divinylbenzene containing a suitable surfactant and an initiator as the continuous phase, and water as the dispersed phase. The polymerization of the monomers of the continuous phase

* *Biotechnology and Bioengineering*, Vol. 44, 79–86, (1994).

generated the polymer carrier with a porous structure. The white rot fungus *Phanerochaete chrysosporium* has been immobilized on porous poly(styrene-divinylbenzene) carriers and used for the batch production and the repeated batch production of lignin peroxidase in shake cultures based on a carbon-limited medium containing veratryl alcohol. The best results were achieved when a spore inoculum was used for immobilization instead of 1-day-old mycelial pellets, for both the batch production and the repeated batch production. The porous poly(styrene-divinylbenzene) immobilized *Phanerochaete chrysosporium* and freely suspended mycelial pellets were used as biocatalysts for the degradation of 2-chlorophenol in a 2-L bioreactor. The porous poly(styrene-divinylbenzene) particle (diameter $\cong$ 0.2 cm) immobilized spores exhibited a much higher activity in the degradation of 2-chlorophenol than the freely suspended mycelial pellets. © 1994 John Wiley & Sons.

3.1.1 INTRODUCTION

Phanerochaete chrysosporium belongs to a family of wood-rotting fungi. They excrete a highly effective extracellular oxidative enzyme (ligninase) capable of degrading not only lignin, but, being nonspecific, also a wide variety of compounds. Several articles[1–6] have described the degradation of a wide spectrum of chlorocarbons and polycyclic aromatics, such as 1,1,1, trichloro-2,2 -bis(4-chlorophenynl) ethane (DDT) polychlorinated biphenyls, polychlorinated dibenzo-*p*-dioxins, lindane, and chlorinated alkanes.[2–4]

The ligninase was first identified as an extracellular oxygenase[7,8] and later as multiple peroxidases.[9–12] The production of lignin peroxidases in agitated shake cultures is enhanced by veratryl alcohol.[13] This observation was followed by a report that immobilized *Phanerochaete chrysosporium* can be efficient in producing extracellular lignin peroxidase.[14] Numerous carriers have been tried for the immobilization of *Phanerochaete chrysosporium,* such as nylon web,[15–27] polyurethane foam,[17,22,28–34] silicon tubing,[35,36] sintered glass,[37] porous ceramics,[38] polypropylene,[39] stainless steel,[10,39] agarose,[14,22] and agar gel beads.[14,22] Good results have been obtained with nylon web, polyurethane foam, sintered glass, or silicon tubing as carriers. In the present study, a new carrier, namely porous poly(styrene-divinylbenzene) prepared by the concentrated emulsion polymerization method,[40,41] is investigated.

3.1.2 MATERIALS AND METHODS

3.1.2.1 Chemicals

Styrene (ST; 99%; Aldrich) and divinylbenzene (DVB; 55%; Aldrich) were distilled to remove the polymerization inhibitor (4-*tert*-butylcatechol), azobisisobutyronitrile (AIBN; Alfa) was recrystallized from methanol, and veratryl alcohol (96%, Aldrich) was purified by distillation. The dispersant Span-80 (Fluka), potassium dihydrogen phosphate (99%; Aldrich), magnesium sulfate heptahydrate (98+%; Aldrich), calcium chloride dihydrate (98+%; Aldrich), ammonium chloride (ACS reagent; Aldrich), thiamine (Sigma), 2-chlorophenol (99+%; Aldrich), hydrogen peroxide (30 wt% solution in water; ACS reagent; Aldrich), sodium tartrate (99.8%; Sigma), Tween-80 (enzyme grade; Fisher), and glucose (≥99.8%; Fisher) were used as received. Water was deionized and distilled.

3.1.2.2 Preparation of Porous Poly(styrene-divinylbenzene) Carrier by the Concentrated Emulsion Polymerization Method

For the preparation of the carrier, a mixture of styrene (8.32 g) and divinylbenzene (2.6 g), in which AIBN (0.022 g) and Span-80 (2.18 g) were dissolved (by heating for a short time at 40°C), was placed in a 100-mL single-neck flask equipped with a mechanical stirrer. Water (36 mL) was added dropwise to the stirred mixture with a syringe, at room temperature. A gel-like emulsion was thus generated, in which water constitutes the dispersed phase. The concentrated emulsion thus prepared was transferred to 20-mL test tubes, which were then sealed with rubber septa. Prior to polymerization, oxygen was removed from the test tubes by purging with nitrogen for 20 min. The polymerization was carried out in a water bath in two steps: at 40°C for 24 h, and at 60°C for additional 20 h. After polymerization, the polymer carrier was washed with methanol in a Soxhlet for 24 h to remove the unreacted monomers, and finally subjected to drying.

3.1.2.3 Structural Characteristics of Porous Poly(styrene-divinylbenzene) Particles

The morphology of poly(styrene-divinylbenzene) was examined with a Hitachi S-800 scanning electron microscope.

The specific surface area of the porous poly(styrene-divinylbenzene) carrier was determined by the Brunauer–Emmett–Teller (BET) method[42] using a Micromeritics Accusorb 2100D physical adsorption analyzer, with N_2 as adsorbate.

3.1.2.4 Culture of Microorganism

Phanerochaete chrysosporium ATCC 24725 was maintained on 3% malt agar slants. The organism was grown aerobically on a carbon-limited medium containing 2 g/L glucose, 0.5 g/L $MgSO_4{\cdot}7H_2O$, 0.1 g/L $CaCl_2{\cdot}2H_2O$, 0.12 g/L NH_4Cl, 2 g/L KH_2PO_4, and 0.001 g/L thiamine. The medium was buffered to pH 4.5 with 20 mM sodium tartrate solution, which is less costly than the commonly employed sodium 2.2-dimethyl succinate buffer and has been previously shown to be almost as efficient in the lignin peroxidase production.[24] For pellet (agglomerate of fungus cells) formation, spores (1 mL; ca. 1.5×10^7 spores/mL) were inoculated into 75 mL of the medium placed in 250-mL Erlenmeyer flasks and shaken at 100 rpm at 39°C for 1 day when they were further used for immobilization, or 2 days when they were used to obtain freely suspended pellets.

3.1.2.5 Immobilization

Approximately 0.2-cm-diameter round particles or 0.4-cm cubes of porous poly(styrene-divinylbenzene) (2.5 g) were placed in 250-mL Erlenmeyer flasks containing either freely suspended spores or 1-day-old pellets of fungus cells in 75 mL of medium and shaken for 2 days at 100 rpm at 39°C.

3.1.2.6 Production of Lignin Peroxidase

When immobilized *Phanerochaete chrysosporium* or freely suspended pellets of fungus cells were employed, the volume of the medium was decreased after a 2-day growth period to 40 mL and 25 mL, respectively, as suggested by Linko et al.[22] by decanting. In the immobilized case, the volume was reduced less because of practical experimental reasons. Veratryl alcohol was added at the final concentration of 1.5 mM to enhance the lignin peroxidase production,[13,17–19,21,24,43,44] and the enzyme production was carried out at 39°C under a pure oxygen atmosphere and under shaking at 30 rpm. Because we compare, later in the article, the activities for ligninase generation in the immobilized and free cases when the volume of the medium was reduced to 40 and 25 mL, respectively, a few experiments were performed by reducing the volume of the medium containing the free pellets to 40 mL. The mean activity for ligninase remained, however, almost the same as for 25 mL. All experiments were carried out in triplicate.

3.1.2.7 Analytical Assay of Lignin Peroxidase Activity

Lignin peroxidase activity was determined spectrophotometrically at the wavelength of 310 nm from the initial rate of the oxidation at room temperature of veratryl alcohol to veratraldehyde according to the method of Tien and Kirk,[45] using a Beckman DU-70 UV spectrophotometer. To 0.5 mL of supernatant containing 1.5 mM veratryl alcohol, 0.47 mL sodium tartrate buffer of pH 3.0 and 0.03 mL of 30 wt% hydrogen peroxide were added to achieve concentrations of 0.1 M and 0.3 mM, respectively. One unit of activity is defined as 1 μmol of veratryl alcohol oxidized in 1 min, and the activities are reported in units (U) per liter. Peroxidase activities were calculated based on activities in 40 mL for immobilized *Phanerochaete chrysosporium* and in 25 mL for freely suspended pellets, respectively.

3.1.2.8 Repeated Batch Production of Lignin Peroxidase with Immobilized Biocatalyst

The spores immobilized on the 2.5 g of porous poly(styrene-divinylbenzene) carrier were first grown at 39°C in 75 mL of carbon-limited medium (containing 1 g/L glucose, the other components having the concentrations mentioned in the section Culture of Microorganism) in 250-mL Erlenmeyer flasks with shaking at 100 rpm until the entire carbon was consumed in order to obtain a biocatalyst containing active growing mycelia. The liquid was reduced to 40 mL by decanting, veratryl alcohol was added to achieve a final concentration of 2.0 mM, and the enzyme production was carried out at 39°C under a pure oxygen atmosphere with shaking at 30 rpm.

The time interval for each batch was between about 48 and 96 h. At the end of each batch the culture supernatant was decanted and 40 mL of fresh carbon-limited medium (with the composition given at the beginning of this section) containing 2.0 mM veratryl alcohol was added to the Erlenmeyer flask. Whenever the flasks were opened, either for sampling or for changing the medium, they were flushed with pure oxygen for 10 min before closing. Each batch of enzyme production was continued until the activity of lignin peroxidase in a given batch has reached approximately its maximum value attainable, which was evaluated from the single batch experiments. The biocatalyst showed no sign of contamination even when it had been briefly exposed to nonsterile air, and the culture supernatant remained clear throughout the experiments.

3.1.2.9 Degradation of 2-Chlorophenol with Freely Suspended Mycelial Pellets

For freely suspended pellet formation, spores (1 mL; ca. 1.5×10^7 spores/mL) were inoculated into 75 mL of the carbon-limited medium (whose composition is given in the section Culture of Microorganism) placed in a 250-mL Erlenmeyer flask and shaken at 100 rpm at 39°C for 2 days. The culture was then used to inoculate the bioreactor. In the bioreactor, a 2-L New Brunswick Scientific Microferm MF114 batch fermentar, the freely suspended pellets were grown at 39°C in 1350 mL of the medium with stirring at 100 rpm, at an aeration rate of 0.3 L/min. The enzyme production was enhanced by adding, after the glucose was consumed (about 2 days), veratryl alcohol to achieve a concentration of 1.0 mM. After 5 days, 0.675 g (500 ppm) of 2-chlorophenol was introduced into the fermentar along with the surfactant Tween-80 at a concentration of 0.5 g/L. All bioreactor experiments were performed in triplicate.

3.1.2.10 Degradation of 2-Chlorophenol with the Immobilized System

Mycelial inocula were also immobilized. For immobilization, freely suspended 2-day-old pellets were introduced into a 2-L New Brunswick Scientific Microferm MF 114 batch fermentar containing 2.5 g of approximately 0.2 cm of round particles or 0.4-cm cubes of porous poly(styrene-divinylbenzene) carriers and 1350 mL of the medium. The agitation speed was 100 rpm, and the aeration rate was 0.3 L/min. The immobilization and the growth of the cells were carried out at 39°C. The enzyme production was enhanced by adding veratryl alcohol to a concentration of 1.0 m*M* after the glucose was consumed (after about 2 days). After 5 days, 0.675 g (500 ppm) of 2-chlorophenol were introduced into the fermentor along with Tween-80 at a concentration of 0.5 g/L.

In other experiments, 2.5 g of approximately 0.2-cm round particles or 0.4-cm cubes of porous poly(styrene-divinylbenzene) carriers were placed in a 250-mL Erlenmeyer flask containing freely suspended spores in 75 mL of medium and shaken at 100 rpm at 39°C for 2 days. Further, the immobilized system was introduced into a 2-L New Brunswick Scientific Microferm MF 114 batch fermentor and the process continued under conditions identical to those described above.

3.1.2.11 Analytical Method for Determining the Degrading of 2-Chlorophenol

The analysis of 2-chlorophenol was performed with a HPLC (ISCO-2350 pump/ISCO V_4 detector) possessing a 150 × 4.6 mm Hypersil Green ENV column; a solution of 1 vol% acetic acid in methanol was employed as the mobile phase and a wavelength of 254 nm as the UV detector wavelength. The system was calibrated using solutions of known concentrations.

The stripping rate by air of 2-chlorophenol during aeration was determined using dead fungus in a batch fermentor. This was accomplished by growing the freely suspended fungus or immobilized fungus in a 2-L batch fermentor as described above, and then heat sterilizing the fermentor and its content in an autoclave at 121°C for 20 min. After cooling, 0.675 g of 2-chlorophenol solution was added to the fermentor. The fermentor was sparged with filtered air at an aeration rate of 0.3 L/min. The off-gas was passed through three flasks in series, each containing 500 mL of water in order to absorb the 2-chlorophenol. Every 12 h, the water in the absorption flasks was changed, and samples were analyzed from the fermentor and gas traps. A mass balance was then performed to calculate the amount of 2-chlorophenol, which was biodegraded.

3.1.3 RESULTS AND DISCUSSION

3.1.3.1 Structural Characteristics of Porous Poly(styrene-divinylbenzene)

In the present experiments, a concentrated emulsion of water in oil constitutes the precursor of the porous polymer. A concentrated emulsion has a large volume fraction (between 0.74 and 0.99) of dispersed phase. It is prepared by the dropwise addition of the dispersed phase (water in the present case) to a small amount of continuous phase (here, styrene and divinylbenzene) containing a surfactant as dispersant and an initiator for polymerization. The adsorption of the dispersant molecules from the continuous phase on the surface of the droplets of the dispersed phase ensures the stability of the emulsion. Electrostatic repulsion and/or steric repulsion generated by the adsorbed molecules are responsible for the kinetic stability of this emulsion.[46] The polymerization of the monomers of the continuous phase generates the polymer carrier with a porous structure.

A scanning electron micrograph of the porous poly(styrene-divinylbenzene) is presented in Figure 3.1.1. The specific surface areas of the two porous poly(styrene-divinylbenzene) carriers employed are 246 m^2/g for 0.2-cm round particles, and 224 m^2/g for 0.4-cm cubes, respectively.

3.1.3.2 Batch Production of Lignin Peroxidase

The effect of porous poly(styrene-divinylbenzene) carrier on lignin peroxidase production by *Phanerochaete chrysosporium* in shake cultures is presented in

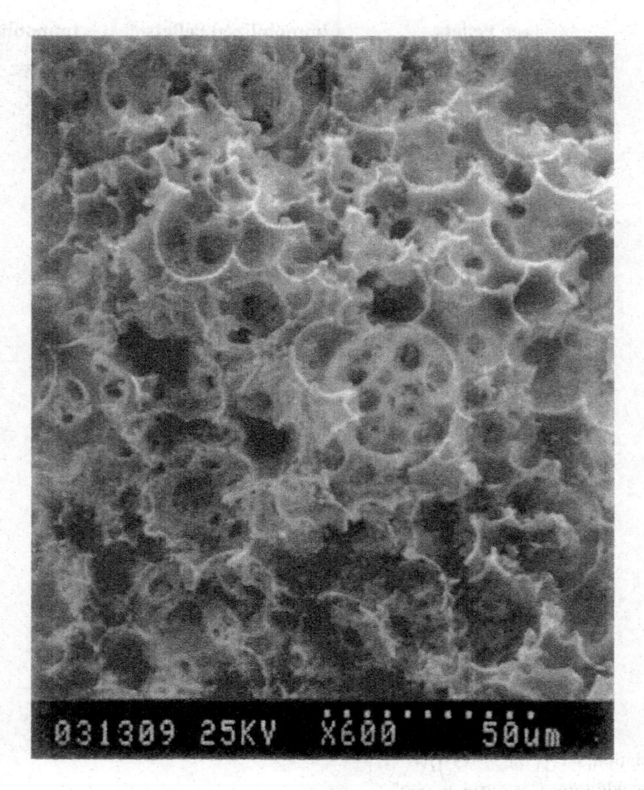

FIGURE 3.1.1 Scanning electron micrograph of porous poly(styrene-divinylbenzene) carrier.

Table 3.1.1. For comparison purposes, results obtained with freely suspended mycelial pellets are also included. The results have been obtained by reducing the volume of the medium to 25 mL. To be sure that the greater reduction of the volume for free pellets does not affect the comparison with the immobilized pellets and spores, a few experiments have been carried out by reducing the volume to 40 mL as for the immobilized systems. The values for 40 mL (in the parentheses in Table 3.1.1) show that the change in the volume affects, in a minor way, the lignin peroxidase production. The superiority of the porous poly(styrene-divinylbenzene) carrier can

TABLE 3.1.1
Lignin Peroxidase Mean Activity (U/L) and Standard Deviation[a] as a Function of Time with Free Pellets, Immobilized Pellets, and Immobilized Spores on 0.4-cm Cubes

	Free Pellets		Immobilized Pellets		Immobilized Spores	
Time[b] (days)	Mean Activity (U/L)	Standard Deviation (s)	Mean Activity (U/L)	Standard Deviation (s)	Mean Activity (U/L)	Standard Deviation (s)
2	104 (108)[c]	7.0 (7.5)	127	7.0	322	13.1
3	135 (141)	7.8 (4.1)	178	10.4	409	13.1
4	153 (155)	6.2 (4.2)	224	12.8	489	12.3
5	170 (177)	10.6 (4.1)	274	10.4	550	15.9
6	186	11.8	321	11.4	602	13.2
7	209	13.4	339	13.5	561	16.6
8	232	11.4	358	13.9	532	13.5
9	248	12.3	373	11.5	508	12.5
10	264	16.1	385	13.5	492	10.4
11	276	14.0	401	13.8	497	11.4
12	285	9.5	413	11.3	491	7.9
13	299	16.1	422	8.2	490	7.6
14	307	15.9	418	12.5	486	13.2
16	292	9.9	402	12.1	472	13.5
18	266	15.7	388	12.2	447	15.1
20	243	18.4	375	9.6	407	15.6
25	204	15.1	347	12.8	384	8.2
30	176	11.4	330	11.4	373	12.1
35	152	8.7	309	13.9	369	9.9
40	134	7.9	280	10.6	358	11.5

[a] Standard deviation, $s = \left[\left(\Sigma_N (x_i - \bar{x})^2\right)/(N-1)\right]^{1/2}$
[b] Time after the addition of veratryl alcohol.
[c] The values in parentheses are for a volume of the medium reduced to 40 mL.

TABLE 3.1.2
The Effect of Immobilization on Lignin Peroxidase Productivity

Phanerochaete chrysosporium	Maximum Lignin Peroxidase Mean Activity (U/L)	Time (days)	Productivity (U/L · h)
Free pellets	307	14	0.91
Immobilized spores	602	6	4.18

be clearly seen. An enzyme activity of 322 U/L is reached with immobilized spores 2 days after the addition of veratryl alcohol, and an activity of 602 U/L is achieved in about 6 days. With the freely suspended mycelial pellets, a maximum activity of 307 U/L is reached after 14 days. Higher activities are obtained with the immobilization of spores than with the immobilization of 1-day-old pellets.

Table 3.1.2 presents the productivities calculated on the basis of the maximum activities obtained, and further stresses the suitability of porous poly(styrene-divinylbenzene) as a carrier, its lignin peroxidase productivity being as high as 4.18 (U/L · h). A number of other carriers for the lignin peroxidase production with immobilized *Phanerochaete chrysosporium* spores were investigated by Linko and Zhong.[22] For nylon, which is the best carrier employed by them, a maximum of 370 U/L was obtained on day 7, and the corresponding productivity was 2.27 (U/L · h).

The activity of immobilized fungus is higher than that of the freely suspended one because immobilization stimulates the growth of the fungus, which consumes glucose at a faster rate, thus enhancing the ability to produce ligninase.

3.1.3.3 Repeated Batch Production of Lignin Peroxidase with Immobilized Biocatalyst

In order to examine the possibility of reusing the immobilized biocatalysts for lignin peroxidase production, repeated batch shake cultures with porous poly(styrene-divinylbenzene) immobilized spores were carried out as described above. The results are listed in Table 3.1.3. The time interval for each batch varied from about 48 to 96 h. When the estimated maximum activity level was reached, the culture supernatant was decanted and 40 mL of fresh carbon-limited medium (containing 2.0 mM veratryl alcohol) was added to the Erlenmeyer flasks. The biocatalyst was used for about 6 weeks, after which it still possessed a significant lignin peroxidase production activity. The highest activity of 737 U/L was reached in the fourth batch after about 8.3 days from the beginning of the repeated batch series; in all batches, the activities have been higher than 500 U/L. The lignin peroxidase production activity decreases gradually after the highest activity is reached in the fourth batch. At batch 14, 0.2 mL of trace element solution[22] [containing 3.0 mg/L $ZnSO_4 \cdot 7H_2O$ 1.5 mg/L Na-nitrilotriacetate, 1.0 mg/L $MnSO_4 \cdot H_2O$, 1.0 mg/L $CoCl_2 \cdot 6H_2O$ 0.01 mg/L

TABLE 3.1.3
Repeated Batch Production of Lignin Peroxidase with Porous 0.2 cm Particles Poly(styrene-divinylbenzene)-Immobilized *Phanerochaete chrysosporium*

Batch No.	Time (h)	Mean Activity (U/L)	Standard Deviation (*s*)[a]
1	24	407	16.7
	48	509	16.6
2	27	436	18.7
	51	594	16.8
3	24	451	15.8
	48	666	25.4
4	24	508	9.8
	48	737	12.1
5	24	479	18.5
	48	686	19.4
6	24	422	14.2
	48	577	20.7
7	24	410	11.1
	48	553	16.5
8	24	388	19.1
	48	537	22.6
9	24	418	16.8
	58	554	17.8
10	24	370	12.8
	48	481	12.1
	66	564	26.9
11	24	360	12.3
	48	449	19.5
	72	517	23.5
12	24	333	13.5
	48	429	18.4
	92	552	25.0
13	24	320	14.0
	48	406	10.8
	96	508	17.6
14[b]	24	446	15.6
	48	614	18.0
	60	675	13.5
15	24	465	9.5
	48	596	25.2
	72	651	22.7
16	24	426	23.5
	48	537	17.7
	72	575	26.9
	96	594	6.9

[a] $s = \left[\left(\Sigma_N (x_i - \bar{x})^2\right)/(N-1)\right]^{1/2}$

[b] At batch 14, 0.2 mL trace element solution and 0.2 mL vitamin solution were added to the Erlenmeyer flask with 40 mL fresh carbon-limited medium containing 2.0 mM veratryl alcohol.

$CuSO_4{\cdot}5H_2O$, 0.01 mg/L $A1K(SO_4)_2{\cdot}12H_2O$, 0.01 mg/L H_3BO_3, and 0.01 mg/L Na_2MoO_4] and 0.2 mL vitamin solution[22] (containing 100 mg/L thiamine-HCl, 20 mg/L pyridoxine-HCl, 10 mg/L riboflavin, 10 mg/L 4-aminobenzoic acid, 10 mg/L thioctic acid, 4 mg/L biotin, 4 mg/L folic acid, and 0.2 mg/L cyanocobalamin) were added to the Erlenmeyer flask with 40 mL fresh carbon-limited medium containing 2 mM veratryl alcohol. As a result, the lignin peroxidase activity increased markedly (Table 3.1.3).

3.1.3.4 Degradation of 2-Chlorophenol by *Phanerochaete chrysosporium*

Figure 3.1.2 presents the degradation results for free pellets for a feed concentration of 500 ppm 2-chlorophenol, and contains curves for the total removal of 2-cholorophenol from the batch fermentor and the removal rate by stripping alone. It shows that the biodegradation rate obtained with free pellets is low. In Figure 3.1.3, curves 2 and 3 present the biodegradation of 2-chlorophenol with ~0.4-cm poly(styrene-divinylbenzene) cube immobilized pellets and ~0.2-cm poly(styrene-divinylbenzene) round particle immobilized pellets, respectively. One can see that, for the same amounts of free pellets and 2-chlorophenol concentrations, the immobilized systems biodegrade 2-chlorophenol more effectively than the nonimmobilized ones. The somewhat lower biodegradation rate obtained with the ~0.4-cm poly(styrene-divinylbenzene) cube immobilized pellets is due to the effect of the size on the rate of diffusion of oxygen and 2-chlorophenol in the pellets.

Curves 4 and 5 of Figure 3.1.3 present the biodegradation of 2-chlorophenol with ~0.4-cm poly(styrene-divinylbenzene) cube immobilized spores and ~0.2-cm poly(styrene-divinylbenzene) round particle immobilized spores, respectively, and exhibit much higher activities than with the immobilized pellets. The ~0.2-cm poly(styrene-divinylbenzene) particle immobilized spores show the highest activity

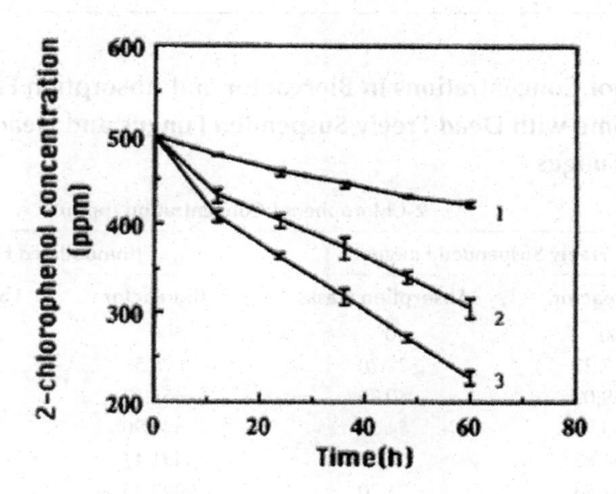

FIGURE 3.1.2 Degradation of 2-chlorophenol by free pellets; 1: stripping only; 2: biodegradation only; 3: total removal. The concentration of 2-chlorophenol is expressed in (g/g) × 10^6.

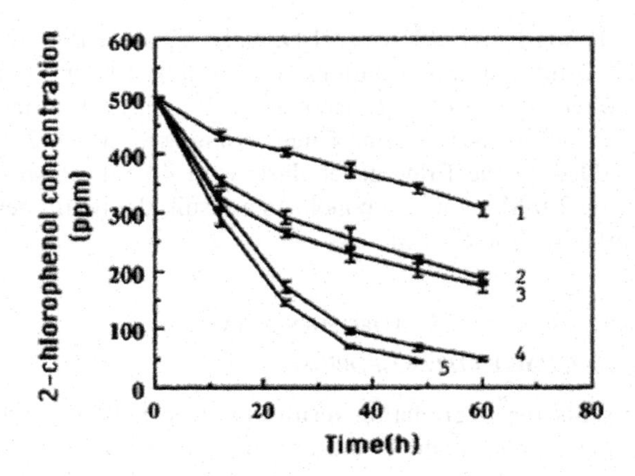

FIGURE 3.1.3 Biodegradation of 2-chlorophenol by various forms of fungi: 1: free pellets; 2: ~0.4 cm poly(styrene-divinylbenzene) cube immobilized pellets; 3: ~0.2 cm poly(styrene-divinylbenzene) particle immobilized pellets; 4: ~0.4 cm poly(styrene-divinylbenzene) cube immobilized spores; 5: ~0.2 cm poly(styrene-divinylbenzene) particle immobilized spores. The concentration of 2-chlorophenol is expressed in (g/g) × 10^6.

in the biodegradation of 2-chlorophenol. These results agree with those obtained with both the immobilized pellets and the immobilized spores in the batch lignin peroxidase production.

The possibility that the increased biodegradation of 2-chlorophenol in the immobilized systems is a result of the adsorption of 2-chlorophenol on the poly(styrene-divinylbenzene) carrier was also investigated, by comparing the results obtained in bioreactor experiments with dead, freely suspended fungus and with dead, immobilized fungus. Table 3.1.4 shows that there is no significant difference

TABLE 3.1.4

2-Chlorophenol Concentrations in Bioreactor and Absorption Flasks as a Function of Time with Dead Freely Suspended Fungus and Dead Immobilized Fungus

	2-Chlorophenol Concentration (ppm)			
	Freely Suspended Fungus		Immobilized Fungus	
Time (h)	Bioreactor	Absorption Flasks	Bioreactor	Absorption Flasks
0	500	0	500	0
12	477.27	22.70	476.58	22.96
24	459.09	40.90	458.57	41.12
36	445.45	54.50	444.96	54.71
48	431.82	68.15	431.47	68.19
60	422.73	77.20	422.35	77.28
72	418.18	81.80	417.81	81.85

between the two kinds of experiments. This indicates that the 2-chlorophenol disappearance cannot be attributed to the adsorption on the pure carrier. In addition, when the immobilized fungus is used to biodegrade 2-chlorophenol, the carrier is covered with the growing fungus. Because around the immobilized fungus the concentration of excreted enzyme is high, any 2-chlorophenol molecule adsorbed on the carrier (if present) will be quickly degraded. Consequently, the possibility of increased 2-chlorophenol disappearance because of its adsorption on the pure polymer carrier should be disregarded.

Lewandowski et al.[47] investigated the degradation of 2-chlorophenol with suspended cultures of *Phanerochaete chrysosporium* and obtained a degradation activity of about 35 mg/L · day. For comparison purposes, one should note that for suspended cultures we obtain 68 mg/L · day (curve 1 in Figure 3.1.3), whereas for immobilized spores curve 5 in Figure 3.1.3 shows a degradation activity of 355 mg/L · day.

The laboratory conditions have not allowed us to scale-up the results obtained at the small scale. The present experiments are, however, useful because they compare the degradation activity of the immobilized fungus with that of the free fungus. It is clear that, in optimizing the conditions at a pilot plant level, the degradation activity can be substantially increased.

3.1.4 CONCLUSIONS

1. Porous poly(styrene-divinylbenzene) carriers have been prepared by the concentrated emulsion polymerization method[40,41] and used for the immobilization of the white-rot fungus *Phanerochaete chrysosporium.*
2. Porous poly(styrene-divinylbenzene)-immobilized *P. chrysosporium* has a very high lignin peroxidase productivity in both batch and repeated batch shake cultures. The porous poly(styrene-divinylbenzene)-immobilized biocatalyst could become an interesting alternative in the production of fungal enzymes such as lignin peroxidase.
3. Spores should be used for immobilization instead of small mycelial pellets.
4. Porous poly(styrene-divinylbenzene) particle immobilized spores exhibit a higher activity in the biodegradation of 2-chlorophenol than the free and immobilized pellets. The immobilized systems can biodegrade hazardous wastes more effectively than the nonimmobilized systems.

REFERENCES

1. Arjmand, M., Sandermann, H. 1985. Mineralization of chloroaniline/lignin conjugates and of free chloroanilines by the white rot fungus *Phanerochaete chrysosporium. Agric. Food Chem.* **33**: 1055–1060.
2. Bumpus, J. A., Aust, S. D. 1987. Biodegradation of DDT [1,1,1, trichloro-2,2-bis(4-chlorophenynl) ethane] by the white rot fungus *Phanerochaete chrysosporium. Appl. Environ. Microbiol.* **53**: 2001–2008.
3. Bumpus, J. A., Tien, M., Wright, D., Aust, S. D. 1985. Oxidation of persistent environmental pollutants by a white rot fungus. *Science* **228**: 1434–1436.
4. Eaton, D. C. 1985. Mineralization of polychlorinated biphenyls by *Phanerochaete chrysosporium:* A ligninolytic fungus. *Enzyme Microb. Technol.* **7**: 194–196.

5. Huynh, V. B., Chang, H., Joyce, T. W., Kirk, T. K. 1985. Dechlorination of chloro-organics by white-rot fungus. *TAPPI J.* **68**: 98–101.
6. Sanglard, D., Leisola, M., Fiechter, A. 1986. Role of extracellular ligninases in biodegradation of benzo(a)-pyrene by *Phanerochaete chrysosporium. Enzyme Microb. Technol.* **8**: 209–212.
7. Glenn, J. K., Morgan, M. A., Mayfield, M. B., Kuwahara, M., Gold, M. H. 1983. An extracellular H_2O_2-requiring enzyme preparation involved in lignin biodegradation by the white-rot fungi basidiomycete *Phanerochaete chrysosporium. Biochem. Biophys. Res.* **114**: 1077–1083.
8. Tien, M., Kirk, T. K. 1983. Lignin-degrading enzyme from the hymenomycete *Phanerochaete chrysosporium* burds. *Science.* **221**: 661–663.
9. Kirk, T. K., Croan, S., Tien, M., Murtaght, K. E., Farrell, R. L. 1986. Production of multiple ligninases by *Phanerochaete chrysosporium:* Effect of selected growth conditions and use of a mutant strain. *Enzyme Microb. Technol.* **8**: 27–32.
10. Leisola, M., Troller, J., Fiechter, A., Linko, Y. Y. 1986. Production and characterization of ligninolytic enzymes of *Phanerochaete chrysosporium. Proceedings of the Biotechnology in the Pulp and Paper Industry Third International Conference,* Stockholm, Sweden, June 16–19, pp. 46–48.
11. Paszczynski, A., Huynh, V.-B. Crawford, R. 1986. Comparison of ligninase-I and peroxidase-I from the white-rot fungus *Phanerochaete chrysosporium. Arch. Biochem. Biophys.* **244**: 750–765.
12. Renganathan, V., Miki, K., Gold, M. H. 1986. Role of molecular oxygen in lignin peroxidase reactions. *Arch. Biochem. Biophys.* **246**: 155–161.
13. Leisola, M. S. A. Fiechter, A. 1985. Ligninase production in agitated conditions by *Phanerochaete chrysosporium. FEMS Microbial. Lett.* **29**: 33–36.
14. Linko, Y.-Y., Leisola, M., Lindholm, N. Trailer, J. Linko, P., Fiechter, A. 1986. Continuous production of lignin peroxidase by *Phanerochaete chrysosporium. J. Biotechnol.* **4**: 283–291.
15. Glumoff, T., Winterhalter, K.H., Smit, J.D.G. 1989. Monoclinic crystals of lignin peroxidase. *FEBS Lett.* **257**: 59–62.
16. Kling, S. H., Neto, J. S. A. 1991. Oxidation of methylene blue by crude lignin peroxidase from *Phanerochaete chrysosporium. J. Biotechnol.* **21**: 295–300.
17. Linko, S. 1988. Continuous production of lignin peroxidase by immobilized *Phanerochaete chrysosporium* in a pilot scale bioreactor. *J. Biotechnol.* **8**: 163–170.
18. Linko, S. 1988. Production and characterization of extracellular lignin peroxidase from immobilized *Phanerochaete chrysosporium* in a 10-ℓ bioreactor. *Enzyme Microb. Technol.* **10**: 410–417.
19. Linko, S. 1988. Production of lignin peroxidase by immobilized *Phanerochaete chrysosporium* in an agitated bioreactor. *Ann. N Y Acad. Sci.* **542**: 195–203.
20. Linko, S., Hujanen, M. 1990. Enzyme production by an immobilized fungus bioreactor. *Ann. N Y Acad. Sci.* **613**: 764–770.
21. Linko, S., Zhong, L.-C. 1991. Central composite experimental design in the optimization of lignin peroxidase production in shake cultures by free and immobilized *Phanerochaete chrysosporium. Bioproc. Eng.* **6**: 43–48.
22. Linko, S., Zhong, L.-C. 1987. Comparison of different carrier methods of immobilization for lignin peroxidase production by *Phanerochaete chrysosporium. Biotechnol. Technol.* **1**: 251–256.
23. Linko, S., Zhong, L.-C., Leisola, M., Fiechter, A., Linko, P. 1987. Optimization of lignin peroxidase production by *Phanerochaete chrysosporium* in shake cultures using response surface methodology, In: O. M. Neijssel, R. R. van der Meer, and K. C. A. M. Luyben (eds.), *Proceedings of the 4th European Congress on Biotechnology.* Elsevier, Amsterdam, the Netherlands. pp. 121–124.

24. Linko, S., Zhong, L.-C., Leisola, M., Linko, Y.-Y., Fiechter, A., Linko, P. 1987. Lignin peroxidase production by immobilized *Phanerochaete chrysosporium* in repeated batch shake cultures, In: E. Odier (ed.), *Lignin Enzymic and Microbial Degradation*. INRA Publications, Versailles, France. pp. 209–213.
25. Polvinen, K. Lehtonen, P., Leisola, M., Visuri, K. 1991. Pilot-scale production and properties of lignin peroxidase. *ACS Symp. Ser.* **460**: 225–235.
26. Schmidt, B., Heimgartner, U., Kozulic, B., Leisola, M. S. A. 1990. Lignin peroxidases are oligomannose type glycoproteins. *J. Biotechnol.* **13**: 223–228.
27. Seiskari, P., Linko, Y.-Y., Linko, P. 1985. Continuous production of gluconic acid by immobilized *Gluconobacter oxydans* cell bioreactor. *Appl. Microbiol. Biotechnol.* **21**: 356–360.
28. Bonnarme, P., Delattre, M., Corrieu, G., Asther, M. 1991. Peroxidase secretion by pellets or immobilized cells of *Phanerochaete chrysosporium* BKM-F-1767 and INA-12 in relation to organelle content. *Enzyme Microb. Technol.* **13**: 727–733.
29. Bonnarme, P., Jeffries, T. W. 1990. Selective production of extracellular peroxidases from *Phanerochaete chrysosporium* in an air-lift bioreactor. *J. Ferment. Bioeng.* **70**: 158–163.
30. Capdevila, C., Corrieu, G., Asther, M. 1989. A feed-harvest culturing method to improve lignin peroxidase production by *Phanerochaete chrysosporium* INA-12 immobilized on polyurethane foam. *J. Ferment. Bioeng.* **68**: 60–63.
31. Chen, A. H. C., Dosoretz, C. G., Grethlein, H. E. 1991. Ligninase production by immobilized cultures of *Phanerochaete chrysosporium* grown under nitrogen-sufficient conditions. *Enzyme Microb. Technol.* **13**: 404–407.
32. Kirkpatrick, N., Palmer, J. M. 1987. Semi-continuous ligninase production using foam-immobilized *Phanerochaete chrysosporium. Appl. Microbiol. Biotechnol.* **27**: 129–133.
33. Kirkpatrick, N., Palmer, J. M. 1987. Semi-continuous ligninase production using foam-immobilized *Phanerochaete chrysosporium*. In: E. Odier (ed.), *Lignin Enzymic and Microbial Degradation*. INRA Publications, Versailles, France. pp. 191–195.
34. Ma, D., Hattori, T., Shimada, M., Higuchi, T. 1990. Improvement of lignin peroxidase production by *Phanerochaete chrysosporium* in shaking culture in the presence of polyurethane foam cubes. *Wood Res.* **77**: 35–41.
35. Willershausen, H., Graf, H., Jäger, A. 1987. Production of ligninase in stirred tank fermentors, In: E. Odier (ed.), *Lignin Enzymic and Microbial Degradation*. INRA Publications, Versailles, France. pp. 203–207.
36. Willershausen, H., Jäger, Graf, H. 1987. Ligninase production of *Phanerochaete chrysosporium* by immobilization in bioreactors. *J. Biotechnol.* **6**: 239–243.
37. Jager, A. G., Wandrey, C. 1990. Immobilization of the basidiomycete *Phanerochaete chrysosporium* on sintered glass: Production of lignin peroxidase. In: J. A. M. de Bont, J. Vtssert, B. Mattiasson, and J. Tramper (eds.), *Physiology of Immobilized Cells*. Elsevier, Amsterdam, the Netherlands. pp. 433–438.
38. Cornwell, K. L., Tinland-Butez, M. F., Tardone, P.J., Cabasso, I., Hammel, K. E. 1990. Lignin degradation and lignin peroxidase production in cultures of *Phanerochaete chrysosporium* immobilized on porous ceramic supports. *Enzyme Microb. Technol.* **12**: 916–920.
39. Asther, M., Bellon-Fontaine, M. N., Capdevila, C., Corrieu, G. A. 1990. A thermodynamic model to predict *Phanerochaete chrysosporium* INA-12 adhesion to various solid carriers in relation to lignin peroxidase production. *Biotechnol. Bioeng.* **35**: 477–482.
40. Ruckenstein, E., Kim, K. 1988. Polymerization in gel-like emulsion. *J. Appl. Polymer Sci.* **36**: 907–923.
41. Ruckenstein, E., Park, J. S. 1988. Hydrophilic-hydrophobic polymer composites. *J. Polym. Sci. Polym. Lett.* (*Part C*) **26**: 529–536.

42. Brunauer, S., Emmett, P. H., Teller, E. 1938. The adsorption of gases in multimolecular layers. *J. Am. Chem. Soc.* **60**: 309–319.
43. Faison, B. D. Kirk, T. K. 1985. Factors involved in the regulation of a ligninase activity in *Phanerochaete chrysosporium. Appl. Environ. Microbiol.* **49**: 299–304.
44. Faison, B. D., Kirk, T. K., Farrell, R. L. 1986. Role of veratryl alcohol in regulating ligninase activity in *Phanerochaete chrysosporium. Appl. Environ. Microbiol.* **52**: 251–254.
45. Tien, M., Kirk, T. K. 1984. Lignin-degrading enzyme from *Phanerochaete chrysosporium:* Purification, characterization, and catalytic properties of a unique H_2O_2-requiring oxygenase. *Proc. Natl. Acad. Sci. USA*. **81**: 2280–2284.
46. Hunter, J. R. 1987. *Foundation of Colloid Science*, vol. I., Clarendon, Oxford.
47. Lewandowski, G. A., Armenante, P. M., Pak, D. 1990. Reactor design for hazardous waste treatment using a white rot fungus. *Water Res.* **24**: 75–82.

3.2 Lipase Immobilized on Hydrophobic Porous Polymer Supports Prepared by Concentrated Emulsion Polymerization and Their Activity in the Hydrolysis of Triacylglycerides*

Eli Ruckenstein and Xiao-Bai Wang

Department of Chemical Engineering,
State University of New York at Buffalo,
Buffalo, New York 14260-4200

ABSTRACT Microporous polymer supports for the immobilization of lipase have been prepared by the polymerization of a concentrated emulsion precursor. The concentrated emulsion consists of a mixture of styrene and divinylbenzene containing a suitable surfactant and an initiator as the continuous phase and water as the dispersed phase. The volume fraction of the latter phase was greater than 0.74, which is the volume fraction of the dispersed phase for the most compact arrangement of spheres of equal radius. The lipase from Candida rugosa has been immobilized on the internal surface of the hydrophobic microporous poly(styrene-divinylbenzene) supports and used as biocatalysts for the hydrolysis of triacylglycerides. The effects of the amount of surfactant, of the molar ratio of divinylbenzene/styrene in the continuous phase, and of the aquaphilicity of the supports on the adsorption, activity, and stability of the

* *Biotechnology and Bioengineering*, Vol. 42, 821–828, (1993).

immobilized lipase have been investigated. The microporous poly(styrene-divinylbenzene) adsorbents constitute excellent supports for lipase because both the amount adsorbed is large and the rate of enzymatic reaction per molecule of lipase is higher for the immobilized enzyme than for the free one.

3.2.1 INTRODUCTION

Lipases constitute a group of enzymes whose biological function is to catalyze the hydrolysis and synthesis of triacylglycerides. Many lipases from microbial sources have been investigated and found to be promising catalysts for the hydrolysis and synthesis of fats and oils.[1] Among them, the enzyme from *Candida rugosa* is an interesting lipase because of its high activity in hydrolytic[25] as well as synthesis[6–8] reactions.

With immobilized enzymes, improved stability, reuse, continuous operation, possibility of better control of reactions, high purity and product yields, and hence more favorable economic factors can be expected.[9] Many supports for the immobilization of lipase have been investigated.[2–5]

Brady et al.[2] searched for adsorbents suitable as supports for lipase. Adsorbents, such as celite, cellulose, ethyl cellulose, silica gel, kieselguhr, clay, alumina, CPG-100, carbon, Accurel, Celgard 2500, Profax polypropene (PP), Microthene high-density polyethylene (HDPE), etc., were screened as possible immobilization supports. Most of these supports were found to decrease tremendously the lipase activity upon immobilization. The hydrophobic microporous materials such as Accurel and Celgard 2500 were found to provide better performances. Otero et al.[5] investigated the time evolution of the hydrolysis of tributyrin for several immobilized lipases to demonstrate its dependence on the aquaphilicity (which is a measure of the hydrophilicity) of the support. Lie and Molin[4] studied the hydrolysis and esterification with lipase immobilized on hydrophobic and hydrophilic zeolites. The hydrophilic zeolite blocked the hydrolysis but mediated the esterification. The esterification with the hydrophobic zeolite as support was, however, poor. Kimura et al.[3] immobilized lipase on different inorganic and organic supports and found that the hydrophobic matrices exhibited a higher activity in the hydrolysis of olive oil.

In the present work novel, hydrophobic porous supports are prepared. The concentrated emulsion polymerization method[10] is employed to obtain a porous structure of styrene copolymerized and crosslinked with divinylbenzene, which is used as adsorbent for lipase. The adsorption capacity for lipase and the activity of the immobilized lipase in the hydrolysis of triacylglycerides are investigated.

3.2.2 EXPERIMENTAL SECTION

3.2.2.1 CHEMICALS

Styrene (ST, 99%, Aldrich) and divinylbenzene (DVB, 55%, Aldrich) were distilled to remove the inhibitor, and azobisisobutyronitrile (AIBN, Alfa) was recrystallized from methanol. The dispersant Span-80 (Fluka), disodium hydrogen phosphate (98%+, Aldrich), potassium dihydrogen phosphate (99%, Aldrich), sodium hydroxide (99.9%, Aldrich), sodium carbonate (ACS reagent grade, Aldrich), sodium tartrate (99.8%, Sigma), cupric sulfate pentahydrate (ACS reagent grade, Sigma), Folin–Ciocalteu phenol reagent (2.0 *N*, Sigma), hydrochloric acid (1.005–0.995 *N*, Fisher), and lipase (EC 3.1.1.3, type VII, from *Candida rugosa*, 900 units/mg solid, 4865 units/mg protein, Sigma) were used as received. Water was deionized and distilled.

3.2.2.2 Preparation of Polymer Supports by Concentrated Emulsion Polymerization Method

A concentrated emulsion has a large volume fraction of dispersed phase, which can be as large as 0.99. It is prepared by the dropwise addition of the dispersed phase (water in the present case) to a small amount of continuous phase (styrene and vinylbenzene in the present case) containing a surfactant as dispersant and an initiator for polymerization. The adsorption of the surfactant on the surface of the droplets of the dispersed phase ensures the stability of the system. For sufficiently large volume fractions, the globules are deformed into polyhedral cells separated by a network of thin films of the continuous phase.

For one of the supports, a mixture of styrene (8.32 g) and divinylbenzene (2.6 g) in which AIBN (0.022 g) and Span-80 (2.18 g) were dissolved (by heating for a short time at 40°C) was placed in a 100-mL single-neck flask equipped with a mechanical stirrer. Water (36 mL) was added dropwise to the stirred mixture with a syringe at room temperature. A gel-like emulsion was thus generated, in which water constituted the dispersed phase. The concentrated emulsions thus prepared were transferred to 20-mL test tubes, which were then sealed with rubber septa. Prior to polymerization, oxygen was removed from the test tubes by purging with nitrogen for 20 min. The polymerization was carried out in a water bath in two steps: at 40°C for 24 h and at 60°C for an additional 20 h. After polymerization, the polymer supports were extracted with methanol in a Soxhlet for 24 h to remove the unreacted monomers and the excess of surfactant. To examine the effects of the ratio of vinylbenzene/styrene, the amount of surfactant, and the volume fraction of the dispersed phase on the properties of the polymer supports, numerous supports (listed in Table 3.2.1) have been prepared.

3.2.2.3 Measurement of Aquaphilicity of Polymer Support

The aquaphilicity of the polymer support, which is a measure of its hydrophilicity, was determined as follows.[6] To 50 mg of a fine powder (0.2–0.5 mm diameter) of dry support, 250 mg of water-saturated diisopropyl ether was added. After 16 h, the water content of the solvent (μL H_2O in the solvent) was measured by gas chromatography using a Porapak Q column (3.2 mm × 1.8 m) provided with a thermal conductivity detector. Helium was used as the gas carrier at a flow rate of 50 mL/min (NTP). After 2 min at 130°C to determine the water content, the

TABLE 3.2.1
Series of Styrene/Divinylbenzene Copolymers with Different Monomer Ratios, Phase Ratios, and Surfactant Concentrations

		Continuous Phase[b]							Disperse Phase	
			Span-80(g)							
Sample[a]	Styrene/ Divinylbenzene (mol)	AIBN (g) 0.2 wt %	1 2 wt %	2 5 wt %	3 10 wt %	4 15 wt %	5 20 wt %	6 25 wt %	Water V (mL)	Phase Ratio, vol water/ vol monomers
A	0.94/0.06	0.2111	2.111	5.278	10.556	15.834	21.112	26.39	657.9	85/15
B	0.94/0.06	0.2111	2.111	5.278	10.556	15.834	21.112	26.39	348.3	75/25
C	0.90/0.10	0.2132	2.132	5.33	10.66	15.99	21.32	26.65	664.2	85/15
D	0.90/0.10	0.2132	2.132	5.33	10.66	15.99	21.32	26.65	351.7	75/25
E	0.80/0.20	0.2184	2.184	5.46	10.92	16.38	21.84	27.30	680.2	85/15
F	0.80/0.20	0.2184	2.184	5.46	10.92	16.38	21.84	27.30	360.1	75/25
G	0.70/0.30	0.2236	2.236	5.59	11.18	16.77	22.36	27.95	696.2	85/15
H	0.70/0.30	0.2236	2.236	5.59	11.18	16.77	22.36	27.95	368.6	75/25
I	0.60/0.40	0.2288	2.288	5.72	11.44	17.16	22.88	28.60	712.1	85/15
J	0.60/0.40	0.2288	2.288	5.72	11.44	17.16	22.88	28.60	377.0	75/25

[a] Samples A1–A6 and J1–J6 have different amounts of Span-80.

[b] wt % refers to weight percent of the monomers.

temperature of the column was raised to 200°C at a rate of 20°C/min to determine the amount of diisopropyl ether. The amount of adsorbed water (μL H_2O in the support) is given by the difference between the initial amount and that remaining in the liquid phase. The aquaphilicity of the polymer support, Aq, is defined as the ratio of microliters of H_2O in the support to microliters of H_2O in the solvent.

3.2.2.4 Measurement of Specific Surface Area of Polymer

The specific surface area of the microporous polymer was determined by the Brunauer–Emmett–Teller (BET) method using a Micromeritics Accusorb 2100D physical adsorption analyzer, with N_2 as adsorbate.

3.2.2.5 Preparation of Immobilized Lipase

To investigate the adsorption of lipase on the polymer supports, a solution of 50 mg of lipase in 50 mL of 0.1 M phosphate buffer at pH 7.0 was prepared. The polymer support (500 mg), in the form of a fine powder of 0.2–0.5 mm diameter, was added to the lipase solution and stirred for 2 h at room temperature. To increase their water wettability, the hydrophobic polymer supports were prewet with 10 mL of ethanol before being dispersed into the lipase solution. After immobilization, the polymer support was separated by vacuum filtration and rinsed with 100 mL of buffer to remove the weakly adsorbed lipase. Then the immobilized lipase was dried for 30 min in a vacuum oven at room temperature and stored in a closed culture tube at 0°C. The fraction adsorbed is listed in Table 3.2.2 for the polymer supports of Table 3.2.1.

3.2.2.6 Determination of Concentration of Lipase in Buffer

The amount of unadsorbed lipase remaining in solution was determined by the Lowry photometric method.[11] The analysis was carried out using a Beckman DU-70 UV spectrophotometer at the wavelength of 750 nm.

3.2.2.7 Activity of Immobilized Lipase in Hydrolysis of Tributyrin

The activity of the immobilized lipase in the hydrolysis of tributyrin was determined as follows. Tributyrin (23.8 mmol) was dispersed as small droplets into 333.5 mL 0.1 M phosphate buffer of pH 8.0. The reaction was started by the addition of the biocatalyst containing 10.53 mg lipase and 126.47 mg polymer support F6 (Table 3.2.1) and was carried out for 24 h at 30°C with stirring (200 rpm). Periodically, 1-mL aliquots were withdrawn, mixed with an equal volume of diethyl ether, and shaken for 5 min. The concentrations of mono-, di-, and triglyceride were determined using a Hewlett Packard HP5890A gas chromatograph provided with a flame ionization detector (FID).

TABLE 3.2.2
Percentage of Lipase Activity of Various Supports

Support[a]	X%[b]	Fraction of Lipase Adsorbed (%)
Celite	16	
Cellulose	24	
Ethyl cellulose	52	
Silica gel	23	
Kieselguhr	30	
Clay	13	
Alumina	9	
CPG-100	34	
Carbon	12	
Accurel		
Nylon	18	
High-density polyethylene (HDPE)	98	
Polypropylene (PP)	88	
Celgard 2500	85	
Profax PP	14	
Microthene HDPE	12	
Poly(styrene-divinylbenzene)		
B6	84	70
D6	90	75
E3	59	54
E4	70	56
E5	86	66
E6	93	79
FI	65	56
F2	77	59
F3	85	61
F4	92	62
F5	97	70
F6	99	81
H6	94	81
J6	91	79

[a] Poly(styrene-divinylbenzene) data from this work. All others are from ref. 2.
[b] *X%* = [fatty acid produced by immobilized lipase after 24 h/fatty acid produced by free lipase after 24 h] × 100.

The organic phase was injected in a 3.2 mm × 1.8-m methyl silicone column (3% OV-101 on 100/200 chromosorb W-HP from Alltech); after 5 min at 65°C to determine the diethyl ether, the temperature was raised to 220°C at a rate of 10°C/min to determine the mono-, di-, and tributyrin. The glycerol and butyric acid were analyzed using a 3.2 mm × 1.8-m Porapak Q column at 240°C provided with an FID.

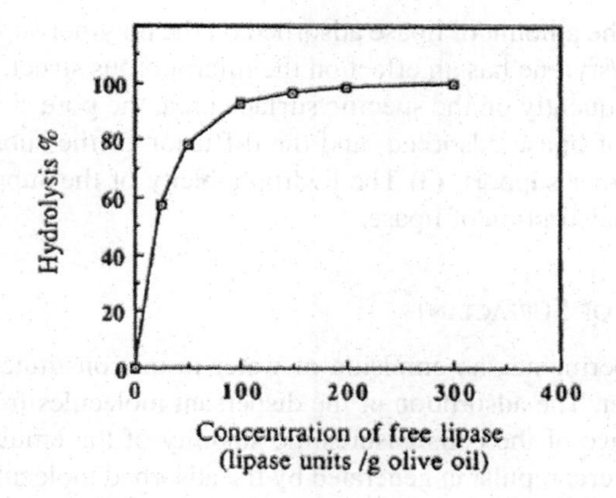

FIGURE 3.2.1 Extent of hydrolysis of olive oil as a function of free lipase concentration after 24 h at 30°C with stirring.

3.2.2.8 Activity of Lipase in Hydrolysis of Olive Oil

The activities of both the free and immobilized lipase in the hydrolysis of olive oil were determined. The hydrolysis was carried out at 30°C with stirring (200 rpm) in a 500-mL three-necked round-bottom flask containing various amounts of free or immobilized lipase and an emulsion of 50 g of olive oil and 50 g of deionized and double distilled water. The emulsion was stabilized using 1 g of the nonionic surfactant Span-80. Aliquots (1 mL) of the reaction mixture were withdrawn periodically, and in the free lipase case the enzyme was inactivated by heating at 90°C for 10 min. After centrifugation, the oil phase was dissolved in a neutralized mixture of ethanol and diethyl ether (1:1 v/v) and titrated with 0.05 M aqueous sodium hydroxide solution using phenol phthalein as indicator. The hydrolysis percent is defined as

$$\text{Hydrolysis \%} = \frac{\text{acid produced by hydrolysis}}{\text{acid produced in complete hydrolysis}} \times 100$$

Figure 3.2.1 provides the relationship between the lipase concentration and the degree of hydrolysis when the olive oil was hydrolyzed by free lipase for 24 h at 30°C. One may note that the hydrolysis is just completed in 24 h if 300 lipase units are employed per gram of olive oil. For 50 g of olive oil, 16.7 mg of lipase are needed for the hydrolysis to be completed in 24 h. The activity of the immobilized lipase will be compared to that obtained with the free lipase.

3.2.3 RESULTS AND DISCUSSION

In the preparation of the hydrophobic porous polymer supports by the concentrated emulsion polymerization, the following factors are important: (1) The surfactant (Span-80) concentration in the continuous phase, which is demonstrated later, has a marked effect on the amount of lipase adsorbed on the polymer support. (2) The ratio of divinylbenzene/styrene has an effect on the microporous structure of the polymer support and consequently on the specific surface area, the pore size and its distribution, the amount of lipase adsorbed, and the diffusion of the substrate through the microporous polymer support. (3) The hydrophobicity of the support material may greatly affect the adsorption of lipase.

3.2.3.1 Effect of Surfactant

In the present experiments, an emulsion of water in oil constitutes the precursor of the porous polymer. The adsorption of the dispersant molecules from the continuous phase on the surface of the cells ensures the stability of the emulsion. Electrostatic repulsion and/or steric repulsion generated by the adsorbed molecules are responsible for the kinetic stability of this emulsion.[12] The polymerization of the monomers of the continuous phase generates the polymer support with a microporous structure. The internal surface area of the polymer is expected to increase as the amount of dispersant becomes larger. The effect of the initial surfactant concentration in the continuous phase on the microporous structure of the polymer support can be investigated by measuring the specific surface area of the polymer support. The specific surface area of the polymer support vs. the surfactant concentration is plotted in Figure 3.2.2, which shows that the specific surface area increases with increasing surfactant concentration.

The fraction of lipase which is adsorbed from the buffer on the polymer support in 2 h is plotted in Figure 3.2.3 vs. the concentration of surfactant in the continuous phase. Our experiments indicated that adsorption saturation is achieved in about 2 h. This figure shows that the fraction of lipase adsorbed increases with increasing surfactant concentration. Comparing the two curves of Figure 3.2.3, one can see that the amount of lipase adsorbed decreases as the volume fraction of the dispersed phase increases from 0.75 to 0.85 at constant surfactant concentration. This probably occurs because the amount of surfactant available decreases as the volume fraction of the dispersed phase becomes larger. As expected, both the specific surface area

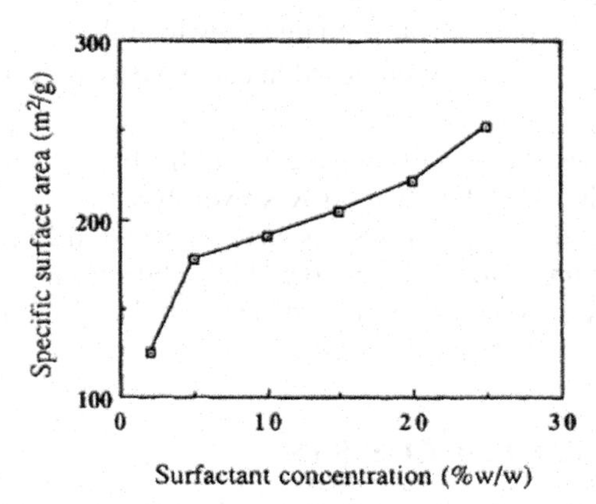

FIGURE 3.2.2 Effect of surfactant concentration (in wt % of monomers) in continuous phase on specific surface area of polymer support for F group in Table 3.2.1.

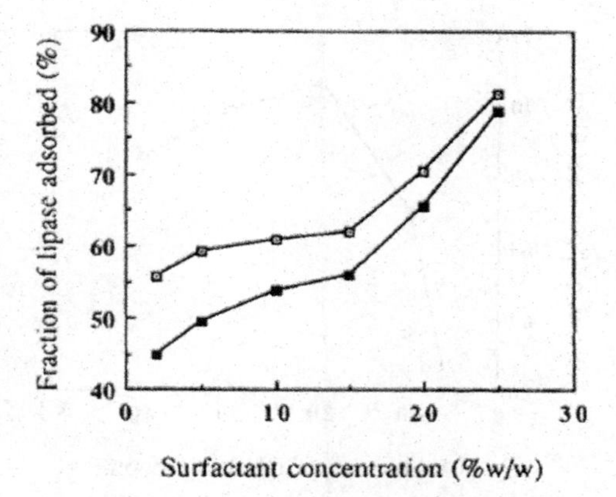

FIGURE 3.2.3 Effect of surfactant concentration (in wt % of monomers) in continuous phase on fraction of lipase adsorbed on supports F1–F6 (volume fraction of dispersed phase = 0.75) and E1–E6 (volume fraction of dispersed phase = 0.85) of Table 3.2.1. (□) F1–F6; (■) E1–E6.

and the fraction of lipase adsorbed increase as the initial surfactant concentration in the continuous phase increases.

3.2.3.2 Effect of Ratio of Divinylbenzene/Styrene

The dependence of the specific surface area of the polymer support on the mole percent of DVB in the ST-DVB mixture is presented in Figure 3.2.4, which shows that the specific surface area increases with increasing proportion of divinylbenzene. This is as expected since a greater amount of divinylbenzene generates a larger number of small pores, thus increasing the specific surface area of the polymer support. The fraction of lipase adsorbed on polymer supports prepared with different amounts of DVB in monomers is presented in Figure 3.2.5. The fraction of lipase adsorbed on the polymer support exhibits a weak maximum at 20 mol%

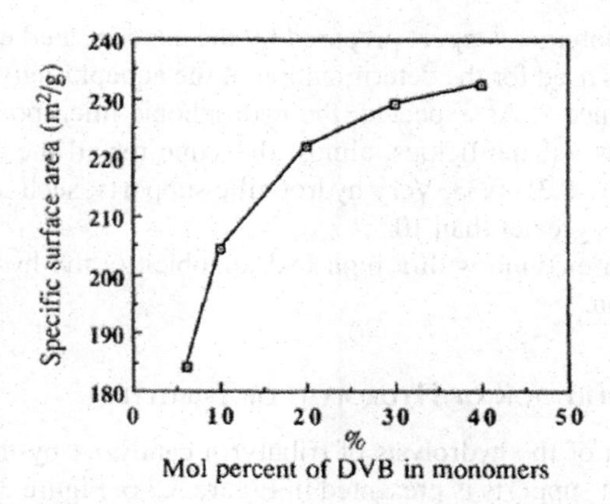

FIGURE 3.2.4 Effect of mole percent of divinylbenzene in mixture of monomers on specific surface area of polymer support (B5, D5, F5, H5, and J5 in Table 3.2.1).

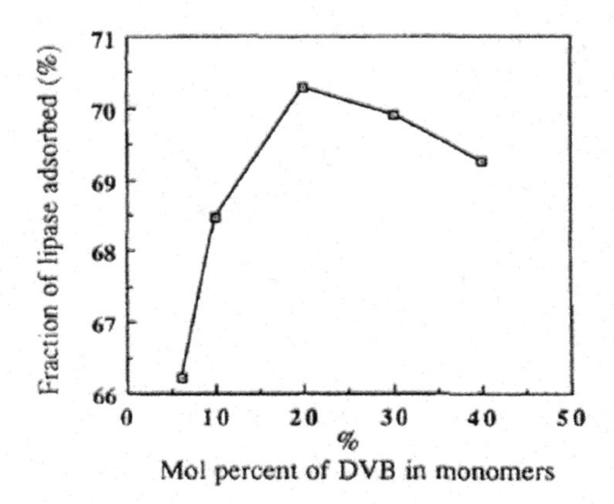

FIGURE 3.2.5 Effect of mole percent of divinylbenzene in mixture of monomers on fraction of lipase adsorbed on polymer support (B5, D5, F5, H5, and J5 in Table 3.2.1).

divinylbenzene. At low ratios, divinylbenzene is a crosslinking agent; at larger ratios, it participates also in copolymerization and the structure of the system may be somewhat different. The fact that the surface area increases with increasing amounts of divinylbenzene while the adsorption passes through a weak maximum may be due to the small pores which are accessible to N_2, which is employed in the surface area measurements, but are not accessible to lipase molecules. Thus, one can conclude that the fraction of lipase adsorbed on polymer depends not only on the initial concentration of surfactant in the continuous phase but also on the ratio of the two monomers. The crosslinking density increases the specific surface area of the polymer support and is beneficial to the adsorption of lipase for not too high contents of divinylbenzene.

3.2.3.3 Effect of Hydrophobicity of Polymer Support

The microporous polymer support prepared by the concentrated emulsion polymerization method was used for the determination of the aquaphilicity by the procedure presented in reference 9. As expected, the hydrophobic microporous polymer supports have very low aquaphilicities, almost the same for all the supports prepared in this study, namely 0.31–0.33. Very hydrophilic supports, such as Sephadex G-25, have aquaphilicities greater than 10.

Lipase being an enzyme with a high hydrophobicity,[5] the hydrophobic support favors its adsorption.

3.2.3.4 Time Evolution of Hydrolysis of Tributyrin

The time evolution of the hydrolysis of tributyrin catalyzed by the lipase immobilized on the present supports is presented in Figure 3.2.6. Figure 3.2.6a provides the concentrations of mono-, di-, and tributyrin as well as that of glycerol as a function of percentage of hydrolysis; Figure 3.2.6b shows the time dependence of the percentage

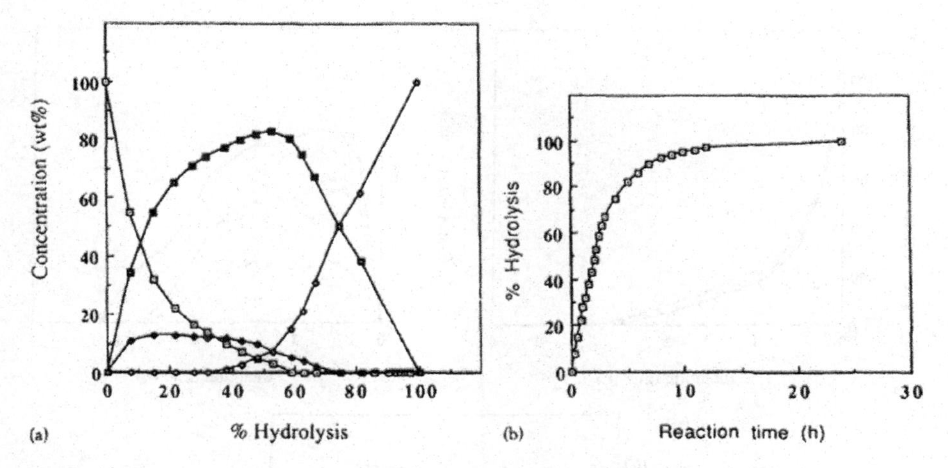

FIGURE 3.2.6 (a) Evolution of composition of reaction mixture as a function of extent of hydrolysis of tributyrin. Initial amount of tributyrin was 23.8 mmol. The biocatalyst is supported on the F6 support and contains 10.53 mg lipase and 128.47 mg polymer support. The reaction was carried out at 30°C with stirring (200 rpm). (□) Tributyrin; (◆) dibutyrin; (■) monobutyrin; (◆) glycerol, (b). (Acid produced via hydrolysis)/(acid produced via total hydrolysis) × 100 vs. the reaction time for conditions in Figure 3.2.6a.

of hydrolysis. One may note that the yield of monobutyrin has a maximum of 83% after 2 h and 15 min of reaction. The maximum occurs because the tributyrin is hydrolyzed to dibutyrin, which in turn transforms in monobutyrin, and the latter transforms in glycerol and butyric acid. Consequently, the present hydrophobic supports can be used for the preparation of the monoester.

The time evolution of mono-, di-, and tributyrin depends upon the molar ratio of the two monomers (DVB/ST) (Figure 3.2.7a–c). This is because different ratios bring about different crosslinkings between the molecular chains, and this affects the microporous network, and therefore the interactions between lipase and support, as well as the diffusion of the substrate and reaction products through the biocatalyst phase.

3.2.3.5 Time Evolution of Hydrolysis of Olive Oil

The time evolution of olive oil hydrolysis with both the free and the immobilized lipase is presented in Table 3.2.3. The evolution clearly depends on the state of the lipase (free or immobilized). The comparison between the two is made at equal amounts of free lipase. Since only a fraction of lipase is adsorbed, the amount of immobilized lipase is smaller than the amount of free lipase. The experiments have been carried out as described in the experimental section using 16.7 mg of lipase in the free lipase case and 16.7 times the fraction of adsorbed milligrams of lipase in the immobilized case. For example, for the support F6 of Table 3.2.1, for which the fraction adsorbed is 0.81 (Table 3.2.2), 180.6 mg of immobilized lipase which contains 167 mg of polymer and 13.6 mg of lipase has been employed. As expected, the initial hydrolysis rate with free lipase is higher than that with the immobilized lipase. This is due to the diffusional

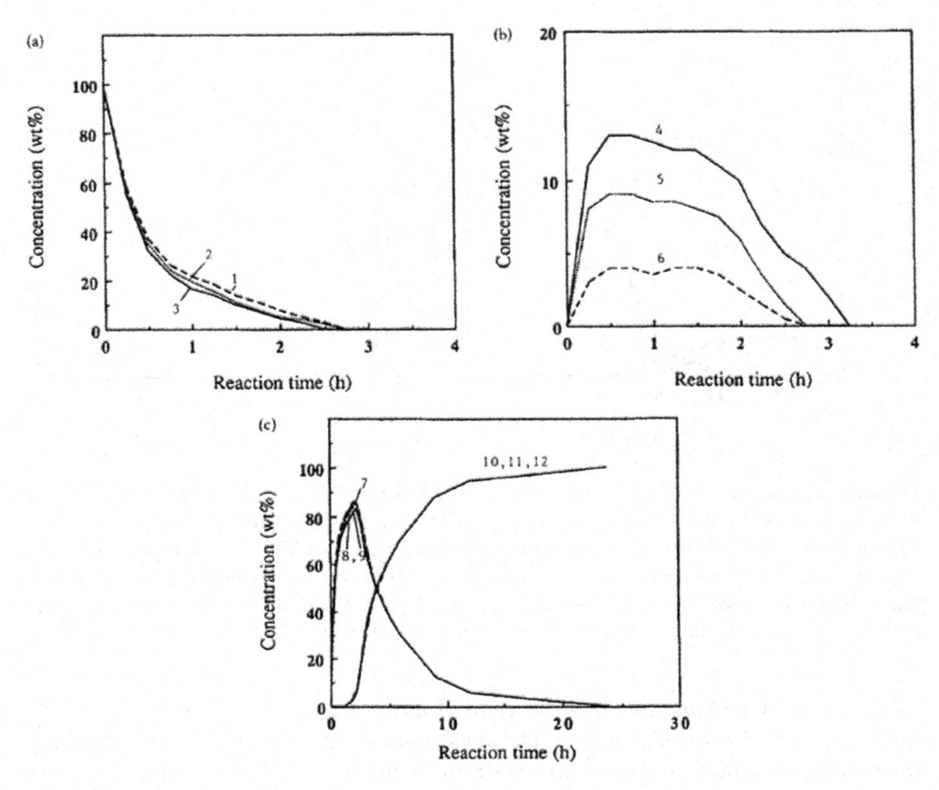

FIGURE 3.2.7 Evolution of concentrations of tributyrin, dibutyrin, and monobutyrin for various amounts of divinylbenzene in mixture of monomers for conditions in Fig. 6a. (a) 1—tributyrin, J6; 2—tributyrin, H6; 3—tributyrin, F6. (b) 4—dibutyrin, F6; 5—dibutyrin, H6; 6—dibutyrin, J6. (c) 7—monobutyrin, J6; 8—monobutyrin, H6; 9—monobutyrin, F6; 10—glycerol, F6; 11—glycerol, H6; 12—glycerol, J6.

TABLE 3.2.3
Time Evolution of Hydrolysis of Olive Oil

	Percent Hydrolysis			
		Immobilized Lipase		
Reaction Time (h)	**Free Lipase**	**F5**	**F6**	**H6**
2	60.5	19.0	20.0	19.0
4	81.5	39.0	41.0	38.0
8	91.0	74.5	75.5	72.5
12	95.0	83.5	84.5	81.5
16	97.5	87.5	89.5	86.5
20	99.0	94.0	95.0	91.0
24	100.0	97.0	99.0	94.0

resistance and to the partition effect mentioned above. It is worth noting that there is enhancement, not inactivation through immobilization of lipase molecules. Indeed, the time required for 100% hydrolysis is just 24 h for the free lipase; while for the same amount of time it is 99% for the immobilized support F6, despite the fact that the fraction of lipase adsorbed is only 0.81 (Table 3.2.2). This indicates that the rate of the enzymatic reaction per molecule is higher in the immobilized case. In fact, Table 3.2.3 shows that for F6 the higher activity of the immobilized molecules overcomes the negative effect of diffusion somewhere between 7 and 8 h. Since the particles have an average radius R of about 0.02 cm using a diffusion coefficient for the substrate of $D = 0.5 \times 10^{-7}$ cm^2/s, one can evaluate a diffusion time τ with Einstein's expression $\tau \cong R^2/D$. One obtains $\tau \cong 2$h and hence diffusion is relevant in the early stages. Similar conclusions are valid for the supports F5 and H6 as well (Table 3.2.3).

Consequently, the lipase molecule becomes more active through immobilization, perhaps because of its interactions with the head group of the surfactant molecules.

3.2.3.6 Evaluation of Activity and Stability of Immobilized Systems

To compare the supports, Table 3.2.2 lists X, the ratio of the activities in the hydrolysis of olive oil of the immobilized lipase and of the free lipase after 24 h, for various supports employed in the literature and for the supports discussed in this article. For the supports prepared in this study, the comparison is made for the conditions mentioned in the previous section. The activity after 24 h is chosen because for the experimental conditions selected the percentage of hydrolysis becomes just 100% in this time in the free lipase case.

Most of the supports decrease tremendously the lipase activity upon immobilization. The poly(styrene-divinylbenzene) supports prepared here by the concentrated emulsion polymerization method as well as Accurel powders from high-density polyethylene (HDPE) or from polypropylene (PP)[2] constitute excellent supports for lipase. Table 3.2.2 shows that X is greater than the fraction adsorbed for all the supports prepared in this article. In other words, the rate of the enzymatic reaction per lipase molecule is higher in the immobilized than in the free case.

The stability of the immobilized system is assessed by reusing 15 times the immobilized lipase in the hydrolysis of olive oil. Each enzymatic reaction was performed for 24 h, at 30°C, with stirring. Figure 3.2.8 presents the stability of the immobilized system for the poly(styrene-divinylbenzene) support F6. For comparison purposes, the stability of lipases immobilized by adsorption and covalent binding on various supports are listed in Table 3.2.4.

The hydrophobic microporous poly(styrene-divinylbenzene) support prepared by the concentrated emulsion polymerization method constitutes a most suitable adsorbent for lipase immobilization. The lipase immobilized on support F6 of Table 3.2.1 has the highest activity in the first run ($X = 99\%$) and stability ($X = 90\%$ activity after 15 reuses) (see Table 3.2.4). The high stability of the present supports may be due to the hydrophobic interactions as well as to the interactions between the enzyme and the head group of the surfactant molecule. The high activity of support F6 is due to the high partition of lipase and to the favorable interactions of the molecule of lipase with the support, which makes the immobilized enzyme molecule more active than the free one.

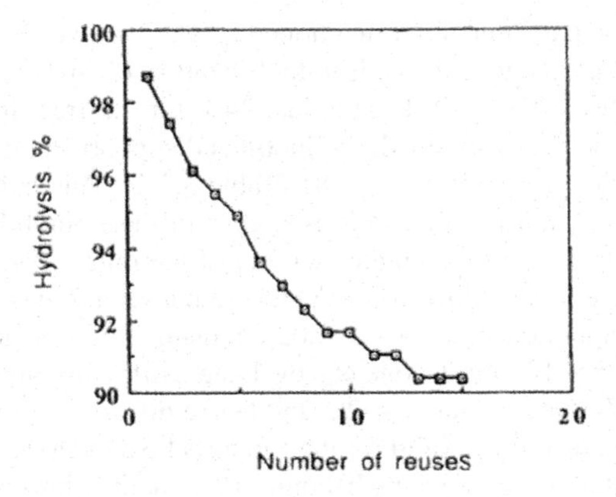

FIGURE 3.2.8 Stability of lipase immobilized on F6 (Table 3.2.1) during 15 reuses in hydrolysis of olive oil. Each enzymatic reaction was performed for 24 h at 30°C with stirring.

TABLE 3.2.4
Stability of Immobilized Lipase

	Percent Hydrolysis of Olive Oil			
Support[a]	**Batch 1**	**Batch 2**	**Batch 3**	**Batch 15**
Adsorption				
Duolite C-3	18.8	4.7	1.6	—
Duolite A-7	45.1	8.5	12.8	—
Duolite S-37	56.0	21.8	8.7	—
Duolite S-761	92.5	67.1	41.9	—
Duolite S-762	63.3	29.9	10.2	—
Celite	96.9	76.9	38.3	—
Poly(styrene-divinylbenzene)	98.7	97.4	96.2	90.4
Covalent binding				
Glutaraldehyde (alkylamine-CPG)	92.9	97.4	93.0	—
Carbodiimide (alkylamine-CPG)	96.1	96.1	93.7	—
Periodic acid (alkylamine-CPG)	33.1	33.2	20.0	—
Glutaraldehyde (alkylamine-spherosil)	59.9	71.7	70.0	—
Carbodiimide (alkylamine-spherosil) Azo	96.9	86.5	86.5	—
coupling (arylamine-spherosil)	77.6	12.9	26.6	—

[a] Poly(styrene-divinylbenzene) data from this work. All others are from ref. 5.

3.2.4 CONCLUSIONS

1. Hydrophobic, microporous poly(styrene-divinylben-zene) supports have been prepared by concentrated emulsion polymerization and used for the immobilization of lipase. The support F6 of Table 3.2.1 exhibits the highest activity (X = 99%) after 24 h of reaction and stability (X = 90% activity after 15 reuses) in the hydrolysis of olive oil. The activity and stability are among the highest obtained until now with various supports.
2. The specific surface area of the support and the fraction of lipase adsorbed on polymer support increase with increasing surfactant concentration in the continuous phase of the emulsion employed as a precursor for the polymer particles.
3. The supports prepared in this article favor the adsorption of lipase because of the hydrophobic interactions between the two, but also, perhaps, because of the interactions between lipase and the head group of the surfactant molecules. The relatively high stability of the immobilized enzymes is probably due to the strong interactions between the enzyme and the head group of the surfactant.
4. The enzymatic reaction rate per molecule of lipase in the immobilized case is higher than that of the free molecule.

REFERENCES

1. Borgstrom, B., Brockman, H. L. (eds.). 1984. *Lipases*. Elsevier, Amsterdam, the Netherlands.
2. Brady, C., Metcalfe, L., Slaboszewski, D., Frank, D. 1988. Lipase immobilized on a hydrophobic, microporous support for the hydrolysis of fats. *J. Am. Oil Chem. Soc.* 65: 917–921.
3. Hunter, J. R. 1987. *Foundation of Colloid Science*, vol. 1. Clarendon, Oxford.
4. Kennedy, J. F., Melo, E. H. M., Jumel, K. 1990. Immobilized enzymes and cells. *Chem. Eng. Prog.* 86: 81–89.
5. Kimura, Y., Tanaka, A., Sonomoto, K., Nihira, T., Fukui, S. 1983. Application of immobilized lipase to hydrolysis of triacylglyceride. *Eur. J. Appl. Microbiol. Biotechnol.* 17: 107–112.
6. Lie, E., Molin, G. 1991. Hydrolysis and esterification with immobilized lipase on hydrophobic and hydrophilic zeolites. *J. Chem. Tech. Biotechnol.* 50: 549–553.
7. Lowry, O. H., Rosebrough, N. J., Farr, A. L., Randall, R. J. 1951. Protein measurement with the Folin phenol reagent. *J. Biol. Chem.* 193: 265–275.
8. Otero, C., Pastor, E., Fernandez, V. M., Ballesteros, A. 1990. Influence of the support on the reaction course of tributyrin hydrolysis catalyzed by soluble and immobilized lipases. *Appl. Biochem. Biotechnol.* 23: 237–247.
9. Reslow, M., Adlercreutz, P., Mattiasson, B. 1988. On the importance of the support material for bioorganic synthesis. *Eur. J. Biochem.* 172: 573–578.
10. Ruckenstein, E., Kim, K. 1988. Polymerization in gel-like emulsions. *J. Appl. Poly. Sci.* 36: 907–923; Ruckenstein, E., Park, J.S. 1988. Hydrophilic-hydrophobic polymer composites. *J. Polym. Sci. Polym. Lett. (Part C)* 26: 529–536.

11. Sokolovskii, V. D., Kovalenko, G.A. 1988. Immobilization of oxydoreductases on inorganic supports based on alumina: The role of mutual correspondence of enzyme-support hydrophobic-hydrophilic characters. *Biotechnol. Bioeng.* 32: 916–919.
12. Yokozeki, K., Yamanaka, S., Takinami, K., Hirose, Y., Tanaka, A., Sonomoto, K., Fukui, S. 1982. Application of immobilized lipase to regio-specific interesterification of triglyceride in organic solvent. *Eur. J. Appl. Microbiol. Biotechnol.* 14: 1–5.

3.3 Concentrated Emulsion Polymerization Pathway to Hydrophobic and Hydrophilic Microsponge Molecular Reservoirs*

Eli Ruckenstein and Liang Hong

Department of Chemical Engineering, State University of New York at Buffalo, Buffalo, New York 14260

ABSTRACT Porous particles of micrometer size have been prepared using the concentrated emulsion polymerization method. The dispersed phase with a volume fraction greater than 0.74 contained a monomer, or a polymer, a crosslinking agent, and inert liquids as porogens. The continuous phase consisted of an aqueous solution of a surfactant when the monomer of the dispersed phase was hydrophobic, and a surfactant solution in a hydrocarbon liquid when the dispersed phase was hydrophilic. Four kinds of porous materials have been achieved: (1) hydrophobic crosslinked (with divinylbenzene (DVB) as the crosslinking agent) polystyrene particles; (2) hydrophilic crosslinked polyacrylamide (with methylenebisacrylamide (MBAA) as the crosslinking agent) particles; (3) hydrophobic porous nutshells of crosslinked polystyrene surrounding a void filled with sparse poly(vinylbenzyl chloride) (poly(VBC)); and (4) hydrophilic nutshells of crosslinked polyacrylamide encapsulating loosely entangled polyethylene oxide coils. The nutshell structure was obtained by using a two-step procedure: First, a concentrated emulsion of oil in water (o/w) with a dispersed phase consisting of monomers (VBC and DVB) dissolved in a porogen was polymerized, or a concentrated emulsion of water in oil (w/o) in which the dispersed phase contained a polymer

---

* *Chemistry Materials*, Vol. 4, 1032–1037, (1992).

polyethylene oxide dissolved in water was prepared. In the second step, styrene/DVB was introduced into the o/w emulsion and an aqueous solution of acrylamide/MBAA into the w/o emulsion. The resulted systems were allowed to polymerize, thus generating polymer shells encapsulating either sparse poly(VBC) matrixes or loosely entangled polyethylene oxide. The porous hydrophobic polymeric beads of the first kind have relatively small surface areas of 20–40 m^2/g, while those of the third kind have surface areas of about 300 m^2/g. The crosslinked particles of polyacrylamide of the second kind were prepared with water as the porogen adsorbs 68 mg of enzyme lipase per gram of polymer. The poly(VBC) encapsulated in the porous nutshell particles was functionalized with anchoring quaternary phosphonium cations. This functionalized substrate possesses a high adsorption capacity for cobalt carbonyl anions (2.78 mmol of Co/g of polymer) via ion pair formation as well as via physical adsorption.

3.3.1 INTRODUCTION

Porous polymer particles are used in many applications, such as chromatography, ion exchange, reactive polymer matrixes, and more recently in controlled release of drugs.[1] The synthesis of these macroreticular particles is, in general, based on the suspension polymerization of a dispersed phase that consists of a monomer, a crosslinking agent, an initiator, and a suitable inert solvent that functions as porogen.[2] The size of the particles depends on the suspending medium as well as on the nature and amount of dispersant it contains. The porosity of the particles is controlled by varying the volume fraction of the porogen and by adjusting the concentration of the crosslinking agent.[3]

In the present paper a novel method for the preparation of porous polymer particles which is based on the polymerization of concentrated emulsions[4] is proposed. The concentrated emulsion is obtained by dispersing with stirring a hydrophobic (or hydrophilic) liquid into a surfactant-containing hydrophilic (or hydrophobic) liquid. The volume fraction of the dispersed phase is larger than 0.74, which represents the volume fraction of the most compact arrangement of spheres of equal radii and can be as large as 0.99. The concentrated emulsions have the appearance and behavior of gels and, for this reason, can be easily handled.[5] The concentrated emulsion is stabilized by the adsorption of the surfactant on the surface of the dispersed droplets. In contrast to suspension polymerization, the concentrated emulsion method employs a small volume fraction of continuous phase, and this offers some advantages, such as easier surface modifications[6] and, as demonstrated below, the generation of nutshell particles.

Four kinds of polymeric beads have been prepared: (1) porous highly crosslinked hydrophobic polystyrene particles; (2) lightly crosslinked hydrophilic polyacrylamide particles; (3) porous nutshells of highly crosslinked polystyrene encapsulating sparse poly(VBC) matrixes; and (4) hydrophilic particles with lightly crosslinked polyacrylamide nutshells encapsulating loosely entangled polyethylene oxide molecules. All four kinds of particles can constitute reservoirs for chemically active species and medicines. As examples, such particles are used for trapping cobalt carbonyl complexes and immobilizing the enzyme lipase.

3.3.2 EXPERIMENTAL SECTION

3.3.2.1 Reagents Used

Styrene (Aldrich, 99%), divinylbenzene (Aldrich, tech., 55%), acrylic acid (Aldrich, 99%), and vinylbenzyl chloride (Kodak, *p* and *m* isomeric mixture) were distilled under vacuum before use. Acrylamide (Aldrich, 99%), tributylphosphine (Aldrich, 92%), cobalt carbonyl (Fluka, purum 90%–95%), polyethylene oxide (Aldrich, 99%, $MW = 600{,}000$), sodium dodecyl sulfate (SDS, Aldrich, 99%), sorbitan monooleate (Span-80, Fluka), *N,N'*-methylenebisacrylamide (MBAA, Aldrich, 99%), Lipase (Sigma, EC3.1.1.3., Type VII, 900 units/mg solid), and solvents were used without further purification. The aqueous solution of sodium acrylate (1.81 M, pH = 7.20) was prepared by neutralizing acrylic acid with a solution of sodium hydroxide.

3.3.2.2 Instruments Employed

Energy-dispersive spectroscopy surface analysis (EDS) was performed on PGT/IMIX field emission electron microscopy equipment (15 keV). Porous polymer particles were investigated with scanning electron microscopy (SEM, Hitachi S-800). The specific surface area and the pore size distribution were measured with a physical adsorption analyzer (Accusorb 2100E, Micromeritics). The infrared absorption spectrum of the encapsulated cobalt carbonyl anions was obtained with a Mattson Alpha Centauri FT-IR instrument. The elemental analyses for Cl, P, and Co were carried out by Quantitative Technologies, Inc. (Bound Brook, NJ).

3.3.2.3 Measurement of the Specific Surface Area and Pore Size Distribution

The polymer samples (the weights used have been in the range 0.02–0.2 g) were first degassed at room temperature overnight under 1–5 µmHg and then at 75°C–80°C for 2 h under the same pressure. To estimate the error of the instrument, a silica catalyst support (Aldrich, $S_w = 600\ m^2/g$) was employed to measure the surface area and a value of 430 m^2/g was obtained. As for the estimation of the pore size distribution, our instrument has allowed us to determine only pore sizes smaller than 50 A. However, larger pores of submicron sizes could be measured from the SEM pictures.

3.3.2.4 Preparation of Porous Particles by Concentrated Emulsion Polymerization Method

An aqueous solution of 0.5 g of SDS in 4 mL of water was prepared in a 150-mL round-bottom flask. Then a mixture containing 1.5 mL of DVB, 2 mL of styrene, 12.5 mL of toluene (or decane), and 25 mg of azobisisobutyronitrile (AIBN) as initiator was introduced dropwise, under stirring with a Teflon blade (about 600 rmp), over about 20 min, at room temperature. The obtained concentrated emulsions were seeded in glass tubes and polymerized at 40°C for 3 days. The porogen and unreacted monomer were removed from the porous polymer particles by extraction with methanol in a Soxhlet funnel for 24 h. Finally, the particles were vacuum dried.

3.3.2.5 Preparation of Hydrophobic Particles with a Porous Nutshell Structure by a Two-Step Concentrated Emulsion Polymerization Method

A solution containing 0.5 g of SDS in 4 mL of water was introduced into a 150-mL round-bottom flask. Then a solution of VBC (1 g, 6.65 mmol) in 13 mL of toluene-decane mixture (volume ratio = 1/1) containing 26 mg of AIBN was added dropwise, under stirring with a Teflon blade (about 600 rpm), over about 20 min, at room temperature. The resultant gel-like concentrated emulsion (with a volume fraction of the dispersed organic phase of 0.78) was polymerized at 40°C for 16 h. After polymerization, the concentrated emulsion was cooled to room temperature, and DVB (3.5 mL) was added slowly under vigorous stirring (about 700 rpm). The emulsions loaded with DVB were packed tightly by shaking, sealed in glass tubes, and polymerized for 48 h at 40°C–45°C. The aggregates of fine particles of polymer thus obtained were dispersed by stirring in methanol. After filtration, the polymer powders were purified in a Soxhlet funnel with methanol for 24 h, followed by vacuum drying. The polymer particles thus obtained have a porous nutshell encapsulating a sparse matrix of poly(VBC) (see Results and Discussion).

3.3.2.6 Conversion of the Encapsulated Poly(VBC) into Poly[(vinylbenzyl) tributylphosphonium chloride] (poly(VBPC)), via Quarternization

Tributylphosphine (2.0 g, 10 mmol) was introduced into a suspension of hydrophobic particles with porous nutshell structure (2.5 g, containing 3.7 mmol of chlorine) in 20 mL of DMF. The mixture was stirred under a N_2 atmosphere, at 60°C–65°C for 72 h. After cooling, the mixture containing the quaternized polymer powder was introduced into a 60-mL mixture of ethyl ether and petroleum ether (V/V = 1/4). The filtered polymer powder was introduced into 20 mL of THF, precipitated with 60–80 mL of petroleum ether (this purification step was repeated three times), and finally dried under vacuum at 50°C.

3.3.2.7 Entrapping $Co(CO)_4^-$ Anions and $NaCo(CO)_4$ Complexes

Dark red crystals of $Co_2(CO)_8$ (0.588 g, 1.62 mmol) were added to a suspension of polymer beads which encapsulate poly(VBPC) in a mixture of an aqueous solution of NaOH (0.5 M, 50 mL) and toluene (30 mL). This slurry was stirred at room temperature for 1.5 h and filtered in a Buchner funnel. The $Co(CO)_4^-$-loaded powder was then purified in a Soxhlet with CH_2Cl_2 for 24 h and finally dried under vacuum.

3.3.2.8 Preparation of Hydrophilic Particles [Polyacrylamide and Poly(sodium acrylate)] with Water as Porogen

To a solution of Span-80 (1.0 g) in 4 mL of *m*-xylene held in a 100-mL round-bottom flask, 20 mL of an aqueous solution of acrylamide (4 g, 56 mmol), MBAA (0.5 g, 3.5 mmol), and initiator $K_2S_2O_8$ (10 mg) were added dropwise, under stirring with a Teflon blade (about 600 rpm), over about 200 min, at room temperature. The resultant gel-like concentrated emulsion of water in oil (w/o, with a volume fraction of dispersed aqueous phase of 0.83) was sealed in glass tubes and polymerized at 40°C for 3 days. To precipitate the polymer beads, the polymerized emulsion containing polyacrylamide was introduced into THF. The surfactant and the nonpolymerized monomer were removed from the polymer by extraction with THF in a Soxhlet for 4 h. The particles were finally vacuum dried. For the preparation of poly(sodium acrylate), a gel-like concentrated emulsion of w/o was prepared using 20 mL of an aqueous solution of sodium acrylate (0.17 g/mL, pH = 8.5) containing 10 mg of $K_2S_2O_8$; the other conditions were the same as for polyacrylamide.

3.3.2.9 Preparation of Hydrophilic Particles with Polyacrylamide Nutshell and Encapsulated Polyethylene Oxide

To a solution of Span-80 (1.0 g) in 4 mL of decane held in a 100-mL round-bottom flask, a solution of polyethylene oxide (0.2 g, MW = 600 000) and $K_2S_2O_8$ (8 mg) in 15 mL of water was added dropwise, under stirring (about 600 rpm), over about 20 min, at room temperature. To the gel-like w/o concentrated emulsion thus obtained, an aqueous solution of acrylamide (2 g, 28 mmol) and MBAA (0.125 g, 1.75 mmol) in 5 mL of water was added slowly under vigorous stirring (about 700 rpm). The obtained emulsion was poured and sealed into glass tubes and subjected to polymerization at 40°C–45°C for 48 h. The aggregates of fine particles were precipitated by stirring the polymerized emulsion in THF. Further, the polymer was purified with THF in a Soxhlet for 8 h and vacuum dried.

3.3.2.10 Lipase Immobilized to the Hydrophilic Substrate

Lipase (25 mg) was dissolved in 50 mL of a phosphate buffer solution (0.89 M, KH_2PO_4/Na_2HPO_4, pH = 7.0) in a 150-mL flask. Polyacrylamide particles (200 mg) thus obtained were suspended into the solution by stirring at room temperature for 2 h. The system was subsequently centrifuged to precipitate the enzyme-containing substrate. The liquid was decanted and used for the determination of the enzyme which remained in the solution, on the basis of the visible absorption spectrum at 740 nm.[7]

3.3.3 RESULTS AND DISCUSSION

3.3.3.1 Preparation of Porous Particles by the Concentrated Emulsion Polymerization Method

Two kinds of crosslinked polystyrene porous particles (with 21 mol % DVB) have been prepared by the concentrated emulsion polymerization method, using either toluene or decane as the porogen. Because toluene is a good solvent for polystyrene, while decane is a "nonsolvent," the pore morphologies obtained were different (Figure 3.3.1). The particles based on toluene (with a volume fraction of dispersed phase of 78%) have very small pores which cannot be detected in the SEM picture (Figure 3.3.1a). The pore size distribution between 20 and 50 Å determined with the adsorption analyzer is given in Figure 3.3.2. The results almost coincide with those of a previous study[8] in which porous polystyrene beads were prepared by suspension

FIGURE 3.3.1 Scanning electron micrograph of crosslinked polystyrene particles: (a) particles prepared by using toluene as porogen (the length of the scale is 3.7 μm); (b) particles prepared by using decane as porogen (the length of the scale is 4.0 μm).

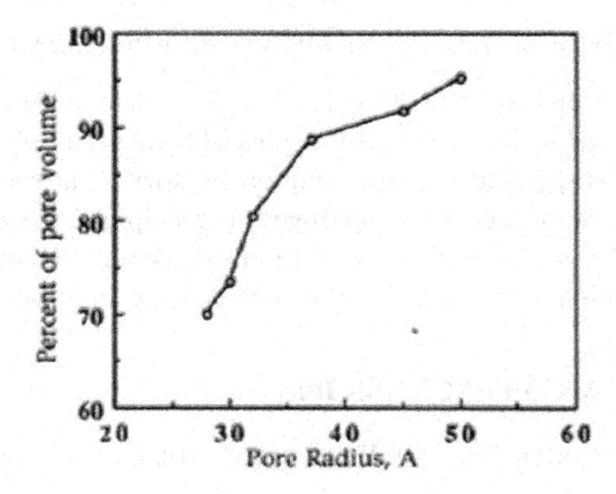

FIGURE 3.3.2 Pore sizes (below 50 Å) distribution of crosslinked polystyrene particles prepared by using toluene as porogen.

polymerization. In contrast, the porous particles based on decane have pore sizes as large as 0.2–0.3 μm, which can be detected in the SEM pictures (Figure 3.3.1b), and also larger surface areas (47 m^2/g) than those based on toluene (25 m^2/g). The main difference between the concentrated emulsion polymerization and the suspension polymerization consists in the much smaller volume fraction of continuous phase in the former than in the latter. The gel-like emulsion that constitutes the precursor in the former case contains polyhedral cells separated by thin films of continuous phase.

The polymerization of the cells does not appreciably change their size, which is in the range 0.1–10 μm. In suspension polymerization the sizes are larger, in the range 3–300 μm.[1] The fact that the volume fraction of the continuous phase is very small in the concentrated emulsion method is particularly advantageous in the preparation of porous nutshell particles discussed in the next section.

3.3.3.2 Porous Nutshell Particles and Their Use as Hosts for Cobalt Carbonyl Anions

A two-step concentrated emulsion polymerization procedure is suggested to prepare porous nutshell particles. The detailed compositions are listed in Table 3.3.1. In the first step, a gel-like concentrated emulsion of a mixture of solvent/nonsolvent for the polymer (as porogen), containing small amounts of VBC and DVB and an initiator AIBN, dispersed in water was prepared and polymerized at 40°C for 16 h. In the second step, a mixture of styrene and DVB was introduced into the polymerized gel-like emulsion, and the resulting system was polymerized for 3 days. During the first polymerization, the crosslinked poly(VBC) matrix is swollen by the solvent (toluene), and a nucleus is generated surrounded by a liquid layer containing the nonsolvent (decane) as the major ingredient. The mixture of styrene-DVB, introduced in the second step, mixes with decane but cannot penetrate into the already polymerized nucleus and then polymerizes to generate a porous nutshell of crosslinked polystyrene surrounding a "void" filled with a sparse poly(VBC) matrix (this mechanism is depicted in Figure 3.3.3). Since the volume fraction of the continuous phase is small, it is easier for the mixture of styrene-DVB to surround the prepolymerized particles than when the volume fraction is large.

The SEM picture (Figure 3.3.4) shows that the nutshell resembles a shell of marblelike pieces, with submicron-size pores among them. The pores are of submicrometer size,

TABLE 3.3.1
Preparation of Porous Nutshell Particles by a Two-Step Concentrated Emulsion Procedure

Entry	VBC-DVB (mol % of DVB)	Porogen (12.5 mL)	ST-DVB (mol % of DVB)	Elem Anal, (mmol/g of polymer)		Specific Surface Area (m^2/g)[b]
				Cl	P	
1	1.0–0.11 g (6.3%)	decane/ toluene ($V/V = 1$)	1.8–1.4 g (21%)	1.61	0.93	225
2	as above	as above	3.2 g[a] (55%)	1.49	0.83	332
3	as above	toluene alone	as above	1.54		70
4	as above	decane alone	as above	1.42		100
5	1.0–0.60 g (28%)	decane/ toluene ($V/V = 1$)	as above	1.58		284

[a] Only DVB was introduced, which contains 45 mol% of 3- or 4-ethyl vinylbenzene.
[b] The surface area of porous particles was measured before quaternization.

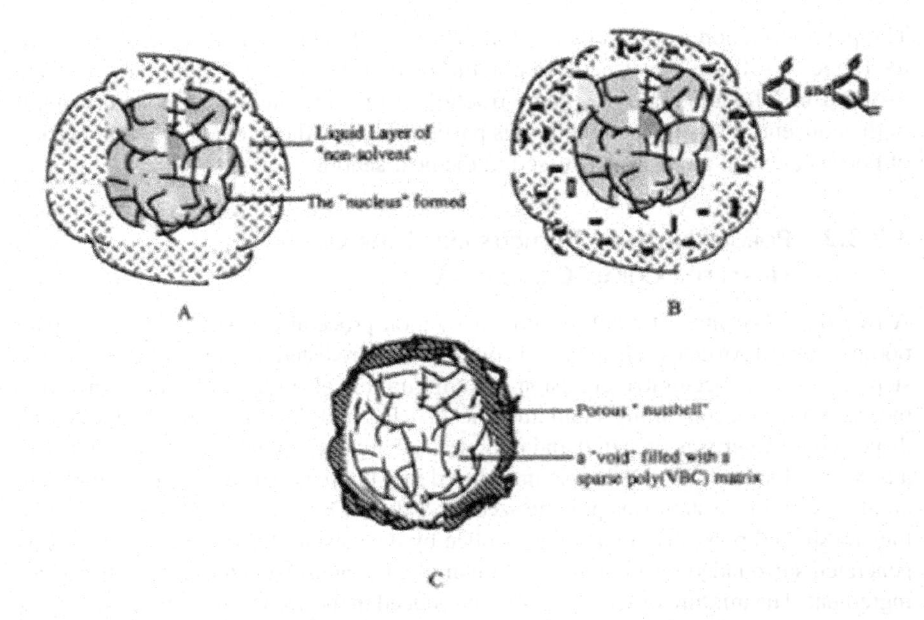

FIGURE 3.3.3 Schematic illustration of the process for generating porous nutshell particles: A, cell containing a nucleus; B, polymerization of styrene/DVB in the nonsolvent layer; C, formation of the porous nutshell particle.

FIGURE 3.3.4 Scanning electron micrographs of porous nutshell particles containing poly(VBC): (a) material from 1 of Table 3.3.1; (b) material from 2 of Table 3.3.1. The lower part of both micrographs represent a 5-fold magnification of the rectangular area marked in the upper part.

can be estimated directly from the pictures, and are a result of the nonsolvent effect of decane for polystyrene. The nature of the porogen[3] and the extent of crosslinking of poly(VBC) are expected to affect the structure of the nutshell particles. When a single porogen is used (3 and 4 of Table 3.3.1), the porous particles have relatively low specific surface areas. When a solvent/nonsolvent porogen mixture is employed, the surface area is much larger. Their effect on the surface area of the particles is presented in Table 3.3.1. In addition, if the extent of crosslinking of the encapsulated poly(VBC) increases (5 of Table 3.3.1), the density of the polymer particle increases, and the surface area and pore volume diminish. The comparison of N_2 adsorption volumes by the porous particles (2 and 5 of Table 3.3.1; Figure 3.3.5) is in agreement with this conclusion.

The EDS experiments indicate a low surface concentration of phosphorus in the quaternized samples (Figure 3.3.6 and Table 3.3.1). Since the elemental analysis indicates a high content of phosphorus in all the samples, one can infer that the quaternary onium cations are hidden inside the polystyrene "cages."

There are several potential applications of these porous materials. They include the controlled release of drugs and the entrapment of highly active and toxic catalysts or reagents. Here we provide as an example the adsorption of cobalt carbonyl anions. This was accomplished by impregnating the dry quaternized substrate (containing poly(VBPC) inside the particle; entry 2 in Table 3.3.1) with an aqueous solution of tetracarbonylcobalt anions prepared by treating a $Co_2(CO)_8$ solution in benzene (5.4×10^{-2} M) with an aqueous solution of NaOH.[9] After purification by extraction with CH_2Cl_2 for 24 h, the elemental analysis indicated a content of Co of 2.78 mmol/g of polymer, which exceeds the content of encapsulated phosphonium cations (0.83 mmol/g of polymer). This implies that a large fraction of $NaCo(CO)_4$ complexes is physically adsorbed on the substrate. The FT-IR spectrum displays the characteristic absorption frequencies of the $Co(CO)_4^-$--Bzl P^+Bu_3 and $NaCo(CO)_4$ at 1746, 1736, 555, and 524 cm^{-1} (Figure 3.3.7). The deviations of the stretching frequencies of the carbonyl from the reported data,[10] 1888 and 555 cm^{-1} ($NaCo(CO)_4$ in DMF), may be attributed to the formation of ion pairs and to the different medium.

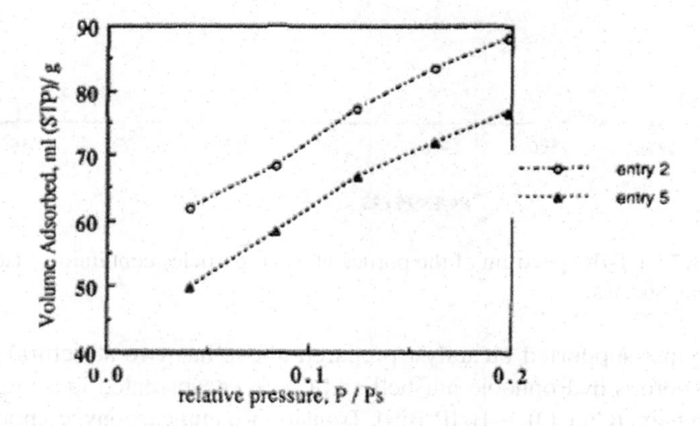

FIGURE 3.3.5 N_2 adsorption isotherm of porous nutshell particles at 77 K (2 and 5 of Table 3.3.1).

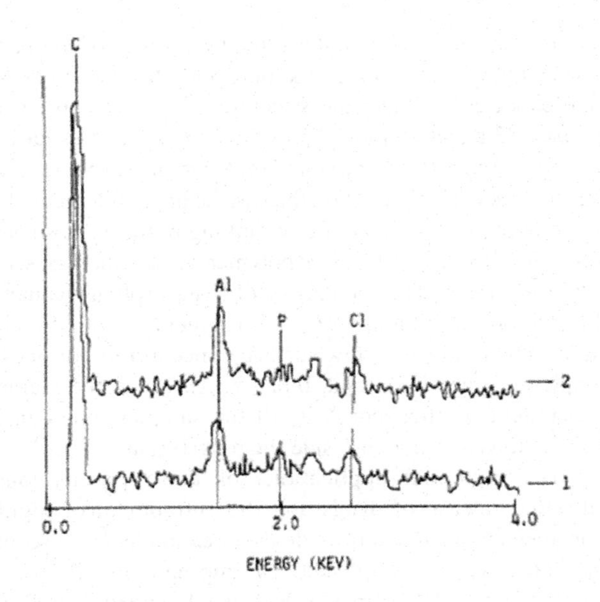

FIGURE 3.3.6 Energy-dispersive spectroscopy of porous nutshell particles containing pendant phosphonium groups (1 and 2 of Table 3.3.1).

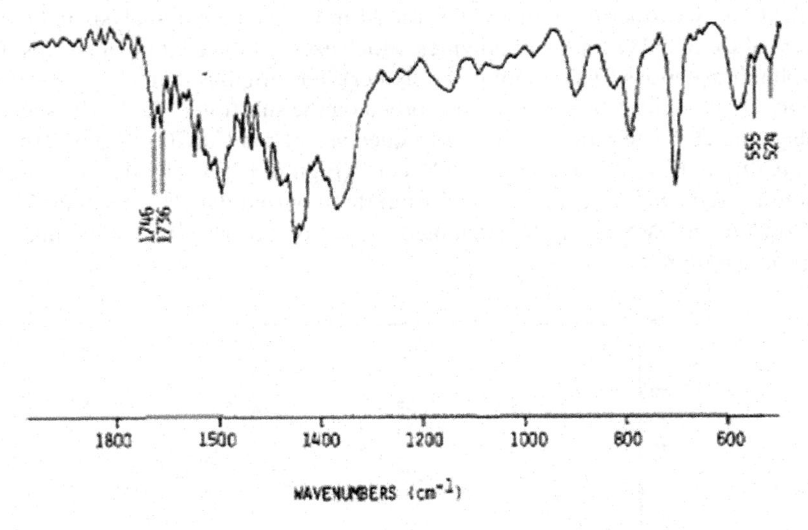

FIGURE 3.3.7 FT-IR spectrum of the porous nutshell particles containing adsorbed tetracarbonylcobalt species.

The polymer-supported bicatalyst prepared above has two structural characteristics: (1) a porous hydrophobic nutshell and (2) an encapsulated loose texture with pendant ion pairs ($Co(CO)_4^-$--$BzlP^+Bu_3$). Transition-metal carbonyl compounds constitute the most useful active species in catalyzing versatile organic reactions, especially under phase-transfer catalysis conditions.[11]

3.3.3.3 Porous Hydrophilic Particles

The preparation of porous hydrophilic particles by suspension polymerization has been reported in patent form.[12] In the present paper a concentrated emulsion of water (containing a hydrophilic monomer and a crosslinking agent) in oil with Span-80 as emulsifier was employed to prepare hydrophilic particles. While three hydrophilic monomers have been tried (Table 3.3.2), only acrylamide and sodium acrylate formed stable concentrated emulsions.

The SEM picture (Figure 3.3.8a) shows that the hydrophilic porous spherical particles of polyacrylamide are of micron size and have a smooth surface. This morphology may be a result of a low extent of crosslinking, due to the small proportion of crosslinking agent and of the good solvating ability of water for polyacrylamide. The hydrophilic particles can be used as biocompatible substrates for the solid-phase synthesis of peptides[13] as well as for immobilization of enzymes. We observed, for instance, that these hydrophilic particles can adsorb 68 mg of lipase/g of polymer from an aqueous solution containing 0.5 mg of lipase/mL.

If the preparation procedure for hydrophilic particles is modified by first preparing a concentrated emulsion of w/o, in which the dispersed medium is an aqueous solution of polyethylene oxide, followed by the addition of acrylamide and MBAA and polymerization, a polyacrylamide shell that encapsulates the polyethylene oxide can be generated (Figure 3.3.8b). This kind of material could be used in the controlled release of drugs, since if a biodegradable polymer (which is often water soluble) and drug molecules are encapsulated into the polyacylamide shell particles, the releasing process can be controlled through the degrading action of a specific enzyme present in the body.

TABLE 3.3.2
Preparation of Hydrophilic Particles by Polymerizing Concentrated Emulsions of Water in Oil

Continuous Phase	Dispersed Phase[a] (20 mL of aqueous solution)	Stability of the Gel-Like Emulsion at 40°C
decane	AA (4 g) + MBAA (0.25 g)	stable
m-xylene	as above	stable
decane	HEA (5 g) + MBAA (0.50 g)	cannot be formed at room temp
m-xylene	HEA (4 g) + MBAA (0.25 g) + PEO (0.5 g)	separates into two phase after about 1 h
decane	SA (3.4 g) + MBAA (0.50 g)	about 30 vol% separates after 12 h
m-xylene	as above	about 10 vol% separates after polymerization
decane	AA (2 g) + MBAA (0.125 g)/ PEO (0.2 g)	stable

[a] AA, acrylamide; HEA, 2-hydroiyethyl acrylate; SA, sodium acrylate; PEO, polyethylene oxide), *MW* = 400; MBAA, *N,N'*-methylenebisacrylamide.

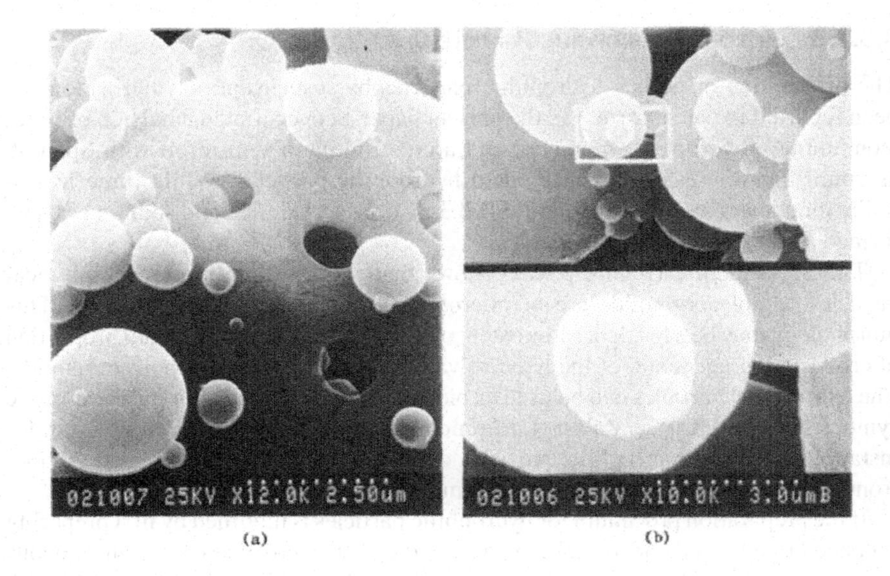

FIGURE 3.3.8 Scanning electron micrograph of polyacrylamide particles: (a) lightly crosslinked polyacrylamide particles; (b) particles with lightly crosslinked polyacrylamide nutshell encapsulating polyethylene oxide.

ACKNOWLEDGMENT

This work was supported by the National Science Foundation.

REFERENCES

1. Eury, R.; Patel, R.; Longe, K.; Cheng, T.; Nacht, S. *Chemtech* **1992**, *22*, 42.
2. Sherrington, D. C. In *Macromolecular Syntheses,* Pearce, E. M., Ed.; Wiley: New York, 1982; Vol. 8, p 30.
3. Howard, G. J.; Midgley, C. A. J. *Appl. Polym. Sci.* **1981**, *26*, 3845.
4. Ruckenstein, E.; Kim, K.-J. *J. Appl. Polym. Sci.* **1988**, *36*, 907.
5. Ruckenstein, E.; Park, J.-S. *J. Polym. Set., Part C: Polym. Lett.* **1988**, *26*, 529.
6. Ruckenstein, E.; Hong, L. *Chem. Mater.* **1992**, *4*, 122.
7. Clark, J. M. *Experimental Biochemistry,* W. H. Freeman: San Francisco, CA, 1964.
8. Guyot, A,; Bartholin, M. *Prog. Polym. Sci.* **1982**, *8*, 277.
9. Alper, H.; Aroumainian, H.; Petrignani, J.-F.; Manul, J. *J. Chem. Soc., Chem. Commun.* **1985**, 340.
10. Edgell, W. F.; Lyford, IV, J. *J. Chem. Phys.* **1970**, *52*, 4329.
11. (a) Alper, H. *Aldrichimica* **1991**, *24*, 3. (b) Brunet, J.-J. *Chem. Rev.* **1990**, *90*, 1041.
12. Mauz, O.; Sauber, K.; Noetzel, S. U.S. Patent, 1985, 4,542,069.
13. Bayer, E. *Angew. Chem., int. Ed. Engl.* **1991**, *30*, 113.

3.4 Polymer-Supported Quaternary Onium Salts Catalysts Prepared via Concentrated Emulsion Polymerization*

Liang Hong and Eli Ruckenstein

Department of Chemical Engineering, State University of New York at Buffalo, Buffalo, NY 14260, USA

ABSTRACT A concentrated emulsion has a very large volume fraction of dispersed phase (0.74–0.95 in this case) and the appearance of a gel. Three procedures based on concentrated emulsion polymerization are suggested for the preparation of polymer-supported quaternary onium salts. (1) Concentrated emulsions of vinyl benzyl chloride (VBC) in water are subjected to polymerization. The polymer resins thus obtained are composed of particles in the micrometre range. A large fraction of the pendant benzyl-chloride groups present in the poly(VBC) particles are converted to onium chloride by a quaternization reaction. (2) A small amount of VBC is added to a partially polymerized concentrated emulsion of styrene (containing a crosslinking agent) in water under vigorous stirring. The system is subsequently subjected to complete polymerization. The obtained polystyrene-poly(VBC) is found to consist of particles having a non-uniform poly(VBC) shell that covers a crosslinked polystyrene core. This polymer is then subjected to a quaternization reaction in order to generate a polymer substrate with bound quaternary onium chloride. (3) A concentrated emulsion of styrene in an aqueous solution of a quaternary onium chloride monomer is prepared. The onium chloride adsorbed on the surface of the dispersed phase polymerizes simultaneously with styrene when the concentrated emulsion is subjected to polymerization. The polymer-supported onium salts thus

* *POLYMER*, 1992, Volume 33, 1968–1975, (1992).

prepared were used as phase transfer catalysts in the alkylation of isopropylidene malonates. The catalysts containing larger pores had a higher activity than those with a more compact structure. The catalyst prepared by the second approach had a greater catalytic activity than that prepared by the first approach. The catalyst prepared by the third procedure had a low activity because of the small amount of supported onium salt it contains.

3.4.1 INTRODUCTION

Insoluble polymer-supported phase transfer catalysts have been widely used in organic synthesis[1] because they can avoid the elimination of the catalyst with the product. However, when conventional methods are employed to prepare polymer-supported catalysts, the slow diffusion of reagents into and out of the polymer matrix often limits their utility. In order to avoid this difficulty, colloidal polymer supports[2,3] were prepared by conventional emulsion polymerization because of their high surface area per unit volume. In the present paper, polymer-supported phase transfer catalysts are prepared using the concentrated emulsion polymerization method.[4] In these emulsions, the volume fraction of the dispersed monomer (the internal phase ratio) exceeds the critical value of 0.74 (which represents the volume fraction of the most compact arrangement of spheres of equal radius) and can become as large as 0.99. When its volume fraction is sufficiently small, the continuous phase is present in the form of a network of thin films separating polyhedral droplets of the dispersed phase. In the present paper, the volume fraction of the dispersed phase is in the range of 0.74–0.95. The concentrated emulsions have the appearance of a gel and, as demonstrated in previous papers of this laboratory,[5,6] can form at room temperature and remain stable at the polymerization temperature only if one of the phases is sufficiently hydrophobic and the other is sufficiently hydrophilic. The adsorption of an emulsifier, which is dissolved in the continuous phase, on the surface of the droplets ensures the stability of this gel-like emulsion.

In the present investigation, hydrophobic monomers are employed as the dispersed phase and water as the continuous phase. A suitable initiator is included in the dispersed phase and the continuous phase contains an appropriate emulsifier. The concentrated emulsion is prepared at room temperature, and its polymerization occurs by heating at 45°C. A porous polymer composed of a three-dimensional array of small particles with diameters in the micrometre range is thus obtained. The size of the polymer particles can be altered by changing the volume fraction of the dispersed phase and the nature of the dispersant. The main advantage of the concentrated emulsion method is that it allows the preparation of agglomerates of micrometre size particles of polymer-supported catalysts, which, in contrast to the individual particles produced by the conventional emulsion polymerization, can be more easily handled and recovered.

In this paper three procedures have been employed to prepare polymer-supported quaternary onium salts. (1) A concentrated emulsion of vinylbenzyl chloride (VBC) (containing a crosslinking agent) in water was prepared at room temperature. This emulsion was subjected to polymerization by heating at 45°C. The porous polymer thus obtained is brittle and can be easily ground to a fine powder. This powder was quatemized with a tertiary phosphine or amine in ethanol at 70°C–80°C. (2) Because of diflusional limitations, only the region near the surface of the particles is actively involved in the chemical reaction. For this reason the above technique was modified as follows: a concentrated emulsion of styrene (containing a crosslinking agent) in water was prepared at room temperature and was subjected to mild polymerization at 45°C. Then, a small amount of VBC was added under vigorous stirring and the system was subjected to complete polymerization. This was followed by a quaternization reaction by treating the polymer powder with a tertiary phosphine or amine in ethanol at 70°C–80°C. (3) A quaternary phosphonium or ammonium salt monomer was dissolved in the continuous aqueous phase, from which a part was adsorbed because of its hydrophobic moiety upon the surface of the droplets of styrene (containing a crosslinked agent) of a concentrated emulsion. These adsorbed molecules polymerize with styrene when the concentrated emulsion is subjected through heating to polymerization.

The polymer-supported quaternary onium salts thus prepared were employed as phase transfer catalysts in an alkylation reaction. The emphasis in this paper is on the relations among the preparation method, the morphology of the prepared polymers and the phase transfer catalytic activities.

3.4.2 EXPERIMENTAL

3.4.2.1 REAGENTS

Styrene (Aldrich), divinylbenzene (DVB) (containing 45% 3- and 4-ethylvinylbenzene) (Aldrich) and VBC (Kodak) were used after distillation under reduced pressure. The dispersant, sodium dodecylsulphate (SDS) (Aldrich), was used as received. The initiator, azobis(isobutyronitrile) (AIBN) (Alfa), was recrystallized from methanol. Tributylphosphine (TBP) (99%, Aldrich), triethylphosphine (TEP) (99%, Aldrich), tributylamine (TBA) (99%, Aldrich), triethylamine (TEA) (99%, Aldrich), trimethyl-amine (TMA) (99%, Aldrich), ethyl bromoacetate (98%, Aldrich), 2,2-dimethyl-l,3-dioxane-4,6-dione (98%, Aldrich), potassium chromate (98%, ACS grade, Aldrich), silver nitrate (>99%, ACS grade, Aldrich) and various solvents were used without further purification.

3.4.2.2 SYNTHESIS OF THE QUATERNARY PHOSPHONIUM AND AMMONIUM SALTS

Tributylvinylbenzyl phosphonium chloride (VBPC) was synthesized using the method reported by Nishikubo et al.[7] and tributylvinylbenzyl ammonium chloride (VBAC) was synthesized as suggested by Brandstrom and Ldamm.[8] Triethylvinylbenzyl phosphonium chloride (VEPC) was prepared from a mixture of VBC (6.7 g, 44 mmol) and TEP (5 g, 42 mmol) dissolved in 10 mL absolute ethyl alcohol. After the solution was heated at 40°C under a nitrogen atmosphere for 6 days, the ethyl alcohol was evaporated using an aspirator. The white solid that remained in the flask was recrystallized twice by dissolving it in chloroform and precipitating subsequently with ethyl ether. A final drying in vacuum yielded 5.5 g of pure product. Triethylvinylbenzyl ammonium chloride (VEAC) was synthesized using the method indicated by Fieser.[9] Trimethylvinylbenzyl ammonium chloride (VMAC) was prepared as follows: TMA (4.0 g, 68.3 mmol) was introduced at –5°C into 10 mL acetonitrile containing 11.04 g (70 mmol) of VBC. The white solid that precipitated after 5 min was recrystallized by dissolving in acetone, and precipitating with ethyl ether. A final drying in vacuum yielded 9.5 g (66 mol%) of VMAC.

3.4.2.3 Preparation of Poly(VBC)

The poly(VBC) crosslinked with 1.3 mol% DVB was prepared by the polymerization of a concentrated emulsion of VBC (containing DVB) in water. Various volume fractions of dispersed phase were employed. A typical procedure for the preparation of the concentrated emulsion is as follows. A small amount of aqueous solution of SDS was placed in a single-necked flask (100 mL capacity) equipped with a mechanical stirrer. A mixture of VBC and DVB containing the initiator AIBN was added dropwise with a syringe, under stirring, at room temperature. The obtained gel-like emulsion was packed in a tube, and subjected to polymerization in a water bath at 45°C for 48 h.

3.4.2.4 Preparation of Poly(VBC)-Supported Onium Chloride

A typical procedure for the preparation of poly(VBC)-supported onium chloride is as follows. A powder of poly(VBC) polymer (3 g, 0.02 mol of chlorine) and 5.56 g (0.027 mol) of TBA were introduced into 30 mL ethyl alcohol and subjected to heating at 80°C for 48 h under stirring. After reaction, 20 mL acetone was poured into the flask and the obtained system was stirred and subsequently filtered. The quaternized polymer powder was then introduced in acetone and subjected to heating under reflux for 0.5 h in order to eliminate the unreacted reactants. Finally, the solid was dried under vacuum at 50°C.

3.4.2.5 Preparation of the Polystyrene Core-Poly(VBC) Shell-Bound Benzyltributyl Phosphonium or Ammonium Chloride

A mixture of styrene (12.49 g, 120 mmol) and DVB (2.73 g, 11.6 mmol) containing AIBN (0.030 g) was added dropwise at room temperature to an aqueous solution of SDS (0.34 g of SDS dissolved in 2 mL water) under stirring. The flask containing the gel-like emulsion was immersed in a water bath (40°C–45°C) for 6 h for mild polymerization. VBC (1.85 g, 12.1 mmol) was subsequently added, under vigorous mechanical stirring (about 700 rev min^{-1}), to this partially polymerized concentrated emulsion. Vigorous stirring was necessary because of the relatively high viscosity of the partially polymerized concentrated emulsion. The polymerization was then continued for another 40 h at 40°C–50°C. The polymer thus obtained was washed twice with ethyl alcohol and subsequently heated in methanol under reflux for 2 h to remove the unpolymerized VBC. After drying in vacuum at 50°C overnight, a polymer containing polystyrene particles (crosslinked with 9.7 mol% DVB) covered with a non-uniform shell of poly(VBC) was obtained. The material thus obtained was ground to a fine powder.

Further, 2.48 g (1.88 mmol chlorine) of the polymer powder was introduced together with 2.85 g (14 mmol) of TBP or 2.40 g (14 mmol) of TBA in 20 mL of ethyl alcohol and allowed to react at 80°C for 24 h under magnetic stirring. The polymer powder containing pendant benzyltributyl phosphonium chloride or benzyltributyl ammonium chloride was filtered, washed twice (first in refluxing methanol and subsequently in refluxing acetone) and finally dried under vacuum at 50°C.

3.4.2.6 Preparation of Polystyrene Porous Media with Surface-Bound Quaternary Onium Salts

A concentrated emulsion was prepared at ambient temperature by dropwise addition under stirring of a mixture of styrene and DVB to an aqueous solution containing SDS and a quaternary onium monomer such as VBPC. The obtained gel-like emulsion was packed into a tube by gentle tapping and subjected to polymerization by heating the tube in a water bath at 40°C for 72 h. After polymerization, the polymer was ground and the powder was introduced several times (for a total of 2–3 h) in warm water, under stirring, was filtered, and finally was dried in vacuum.

3.4.2.7 Determination of the Quaternary Ammonium Chloride Content in the Poly(VBC)-Supported PTC Catalysts

The Mohr method[10] was employed to determine quantitatively the chloride anion, which is the counterion of the onium cation. To a flask (100 mL) containing 10 mL of aqueous solution of K_2CrO_4 (about 0.4 M), 0.1 g poly(VBC)-supported onium salt powder was added. The system was subjected to ultrasonic cleaning for 15 min to ensure ion exchange between Cl^- and CrO_4^{2-} ions. Then, the system containing the fully swollen polymer was titrated with a 0.020 N solution of silver nitrate.

3.4.2.8 Electronic Spectroscopy for Chemical Analysis (ESCA)

Polystyrene powder (50 mg) containing bound onium salts was dissolved in 2–3 mL chloroform. The sticky solution was poured into an aluminium pan (2 cm in diameter), the solvent (chloroform) was evaporated and the obtained disc was employed for the ESCA.

3.4.2.9 Monoalkylation Reaction in the Triphase Reaction System

The alkylation reaction of propylidene malonate (the Meldrum acid[11]) was carried out under triphase (polymer-supported phase transfer catalyst/inorganic base/reagent and solvent) conditions. In a typical run, propylidene malonate (0.77 g, 5 mmol), a powder of potassium carbonate (0.69 g, 5 mmol) and a powder of poly(VBC)-supported benzyltributyl ammonium chloride catalyst (0.42 g, 1.0 mmol) were introduced into a flask (25 mL). Chloroform (10 mL) followed, under magnetic stirring, by a solution of ethyl bromoacetate (1.18 g, 7.0 mmol) in 4 mL chloroform were added to the flask. The heterogeneous system was heated at 60°C–65°C under stirring for 4 h. After reaction, the catalyst and the inorganic salt were filtered and the filtrate was subjected to evaporation under a water aspiratory pump. Acetonitrile (5 mL) was added to the residual liquid and the crude product, which precipitated when 40 mL of distilled water was poured into the acetonitrile solution, was purified by recrystallization from benzene with petroleum ether. The isolated yield of the adduct was 0.64 g (molar yield 55.6%). ^{1}H NMR ($CDCl_3$) 1.30 (t, 3H), 1.91 (s, 3H), 1.89 (s, 3H), 2.40 (d, 2H), 4.01 (t, 1H), 4.20 (q, 2H).

3.4.2.10 Instruments

Infrared spectra were obtained with a Mattson Alpha Centauri FT-IR instrument, and the ^{1}H NMR analyses were performed with a GEM-300 spectrometer. The morphologies of the polymer samples were investigated with a scanning electron microscope (Amray 100A). The elemental analysis of Cl, N and P was carried out by Quantitative Technologies, Inc. (Bound Brook, NJ). Angular dependent ESCA (Physical Electronics PHI 5100) and energy dispersive spectroscopy (EDS) (PGT/IMIX) were employed to identify the surface-bound phosphonium salt and chloromethyl groups.

3.4.3 RESULTS AND DISCUSSION

3.4.3.1 Morphology and Phase Transfer Catalytic Reactivity of Poly(VBC)-Supported Quaternary Benzyltributyl Ammonium Chloride

Because VBC has a reactive chloromethyl group, it has frequently been used as a precursor for functionalization.[1,2] In the present concentrated emulsion method various volume fractions of VBC were employed to prepare poly(VBC) resins (Table 3.4.1). The SEM micrographs of the quaternized resins show that their morphology changes from macropores (samples 1a and 1b, Figure 3.4.1a and b) to a more compact structure (sample 1d, Figure 3.4.1d) with increasing volume fraction of VBC. The morphology of sample 1c is intermediary (Figure 3.4.1c).

A comparison between Figures 3.4.1a and 3.4.1a′ indicates that the quaternization reaction (Scheme 3.4.1) does not change the morphology.

In Scheme 3.4.1 y and z represent the content of VBC and onium salt units, respectively, in mol%: for Z = N, y:z is about 33:65 (samples 1a, 1b, 1c and 1d); for Z = P, y:z is 41; 57 (sample 1e).

The content of pendant benzyltributyl ammonium chloride in this poly(VBC)-supported catalyst was determined both by the titration of Cl^- and by the elemental

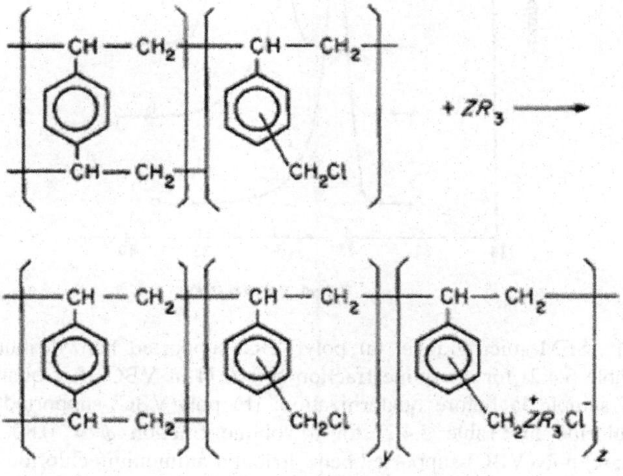

$$Z = N, P;\ R = n\text{-}C_4H_9$$

SCHEME 3.4.1

TABLE 3.4.1
Compositions of Various Concentrated Emulsions of VBC in Water Used for the Preparation of the Substrates of the Catalysts 1a, 1b, 1c and 1d

	1a	1b	1c	1d
Dispersed phase				
$\left[\text{VBC+DVB}\left(13.2\times10^{-3}\ \frac{\text{mol DVB}}{\text{mol VBC}}\right)\right]$ (mL)	17.0	16.0	15.0	18.0
AIBN g/mL (VBC + DVB)		2.6×10^{-3}		
Continuous phase				
Water (mL)	6	4	2	1
SDS g/mL (VBC + DVB)		0.02		
Volume fraction of VBC, ϕ	0.74	0.80	0.88	0.95

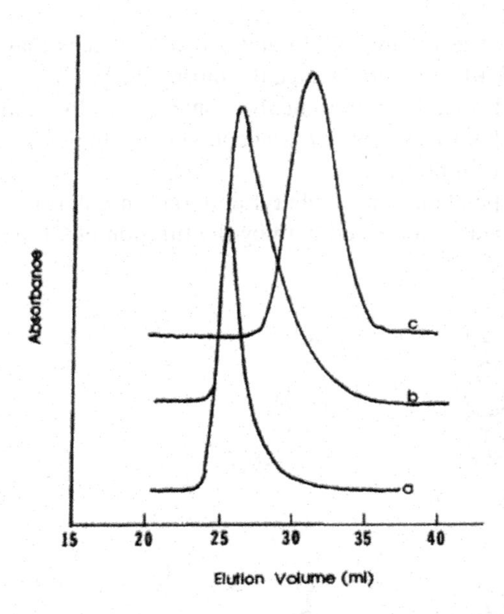

FIGURE 3.4.1 SEM micrographs: (a) poly(VBC)-supported benzyltributyl ammonium chloride la (Table 3.4.2) for a volume fraction ϕ = 0.74 of VBC after quaternization; (a′) poly(VBC) of sample la before quaternization; (b) poly(VBC)-supported benzyltributyl ammonium chloride lb (Table 3.4.2) for a volume fraction ϕ = 0.80 of VBC after quaternization; (c) poly(VBC)-supported benzyltributyl ammonium chloride lc (Table 3.4.2) for a volume fraction ϕ = 0.88 of VBC after quaternization; (d) poly(VBC)-supported benzyltributyl ammonium chloride ld (Table 3.4.2) for a volume fraction ϕ = 0.95 of VBC after quaternization.

analysis of N and P as described in the Experimental section; the results of these independent analyses are listed in Table 3.4.2. Since the yield of quaternization is in the range of 65%–68%, the bound-benzyltributyl ammonium chloride is present in a large portion of the poly(VBC) particles. The FT-IR. spectrum of sample lc shows the characteristic absorption bands of bound benzyltributyl ammonium chloride (1636 cm^{-1}, 1379 cm^{-1} and 1099 cm^{-1}) and of bound benzyl chloride (707 cm^{-1}) (Figure 3.4.2).

The relation between the phase transfer catalytic activity and the morphology of the poly(VBC)-supported quaternary benzyltributyl ammonium chloride was investigated for the alkylation reaction of isopropylidene malonate *1* (Scheme 3.4.2).

The yields of monoalkyl-isopropylidene malonate *3* are used to compare the catalytic activities of the poly(VBC)-supported quaternary benzyltributyl ammonium chloride catalysts with their unsupported counterpart VBAC (Table 3.4.3). Table 3.4.3 demonstrates that the catalysts 1a and 1b provide higher yields than 1d. It is interesting to note that the catalysts 1a and 1b have numerous macropores (Figure 3.4.1a and b), while catalyst 1d has a more compact morphology (Figure 3.4.1d). It is clear that

TABLE 3.4.2
Titration for Cl and Elemental Analysis for N and P in the Poly(VBC)-Supported Quaternary Onium Salts and the Diameters of Their Particles

	1a	1b	1c	1d	1e[a]
Volume fraction, ϕ	0.74	0.80	0.88	0.95	0.88
CP$^-$ (mmol/g polymer)	2.41	2.37	2.43	2.43	2.10
N (mmol/g polymer)			2.44		
P (mmol/g polymer)					2.13
Range of diameters in Figure 3.4.1 (μm)					
Before quaternization	2.5–25				
After quaternization	12–29	3.4–17	0.97–1.1		

[a] 1e has the same substrate as 1c but has a phosphonium salt instead of an ammonium salt

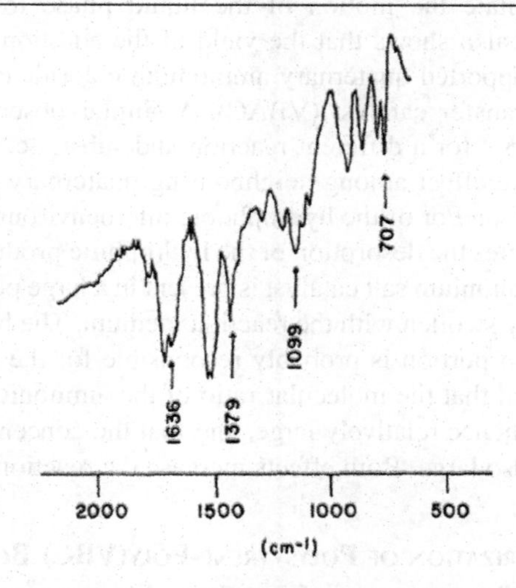

FIGURE 3.4.2 FT-IR spectrum of poly(VBC)-supported ammonium salt of sample 1c (Table 3.4.2).

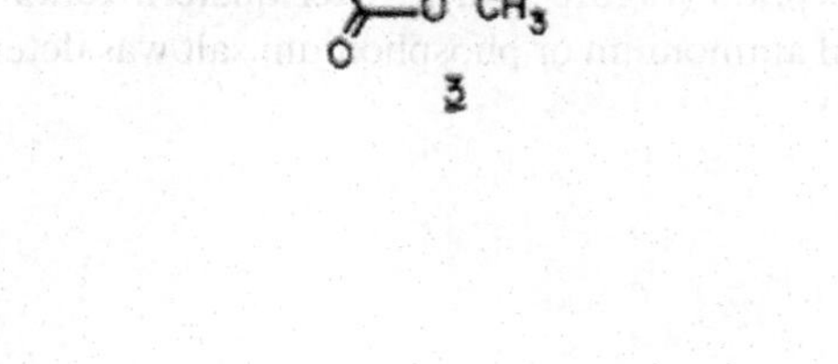

SCHEME 3.4.2

TABLE 3.4.3
Comparison of the Yields of Monosubstituted Isopropylidene Malonate 3 (Scheme 3.4.2) in Phase Transfer Catalyst Reaction with Various Supported and Unsupported Ammonium Salts[a]

Catalyst	Amount (g)	Ammonium Salt (mmol)	Time (h)	Yield (mol%)
1a	0.50	1.21	4	64.3
1b	0.50	1.20	4	63.5
1c	0.50	1.22	4	61
1d	0.51	1.24	4	55
VBAC	0.41	1.22	4	54

[a] Ammonium salt (mmol): malonate *1* (mmol) = 1.20–1.24:5.0

the macropores facilitate the motion of the liquid phase toward the individual particles. Table 3.4.3 also shows that the yield of the addition product *3* is higher for the poly(VBC)-supported quaternary ammonium chloride catalysts than for the unsupported phase transfer catalyst (VBAC). A similar observation was recently made by other authors[13] for a different reaction and substrate. This can be a result of either a cooperative effect among neighbouring quaternary ammonium sites on the polymer substrate, and/or of the hydrophobic microenvironment on the polymer support, which facilitates the desorption of the hydrophilic product. It is important to note again that the ammonium salt catalyst is present in a large portion of the particle, which is also probably swollen with the reaction medium. The high concentration of catalyst in the swollen portion is probably responsible for the cooperativity effect. It should also be noted that the molecular ratio of the ammonium salt to reactant is 1.22:5 (Table 3.4.3), hence relatively large, and that the concentration of reagent in the swollen layer is also large. Both effects increase the reaction rate.

3.4.3.2 Characterization of Polystyrene-Poly(VBC) Bound Phase Transfer Catalysts with Core-Shell Morphology

Polymer substrates with a core-shell morphology have been prepared by adding, under vigorous stirring, a precursor monomer (VBC in this case) to mildly polymerized concentrated emulsions of styrene in water (Table 3.4.4) and subjecting the systems to complete polymerization. The SEM micrograph (Figure 3.4.3) shows that in such cases a non-uniform shell of poly(VBC) coats a core of polystyrene. The EDS analysis (Figure 3.4.4), indicates the presence of Cl in the superficial layer of the particles. The FT-IR spectrum of the polystyrene-poly(VBC) presents a characteristic absorption band of the C-Cl group at 710 cm^{-1} over the background of the polystyrene absorption (Figure 3.4.5). After quaternization (Scheme 3.4.5), the content of substituted ammonium or phosphonium salt was determined by elemental analysis (Table 3.4.4).

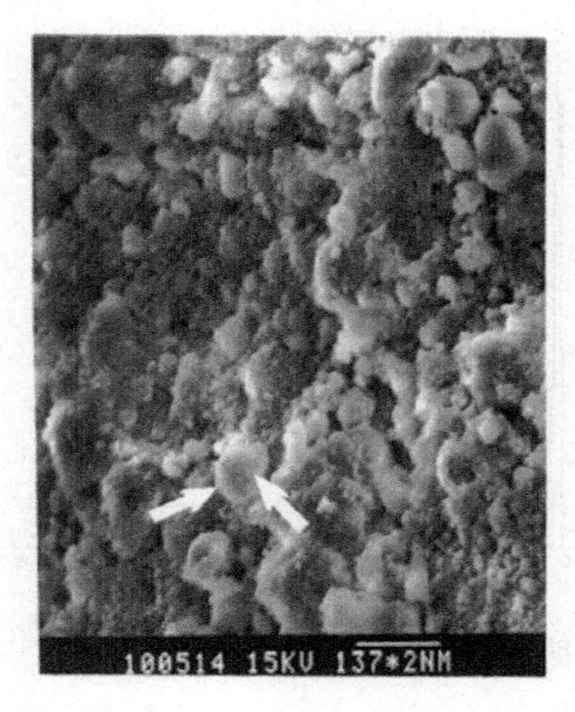

FIGURE 3.4.3 SEM micrograph of polystyrene (core)-poly(VBC) (shell) substrate. The arrows indicate the core and shell. The composition of the substrate is given in Table 3.4.4.

TABLE 3.4.4
The Compositions of the Concentrated Emulsions Used to Prepare the Core-Shell Polystyrene-Poly(VBC) Substrates and the Elemental Analysis of the Substrates and of Their Quaternization Products

Feed Composition			**Elemental Analysis (mmol/g polymer)**		
Dispersed phase			Cl	N	P
Styrene	12.49 g	Polystyrene-poly(VBC) substrate	0.75		
DVB	2.73 g				
VBC	1.85 g				
AIBN	0.030 g	Polystyrene-poly(VBC)-supported ammonium salt (catalyst 4a)		0.47	
Continuous phase					
SDS	0.34 g				
Water	2 mL	Polystyrene-poly(VBC)-supported phosphonium salt (catalyst 4b)			0.27

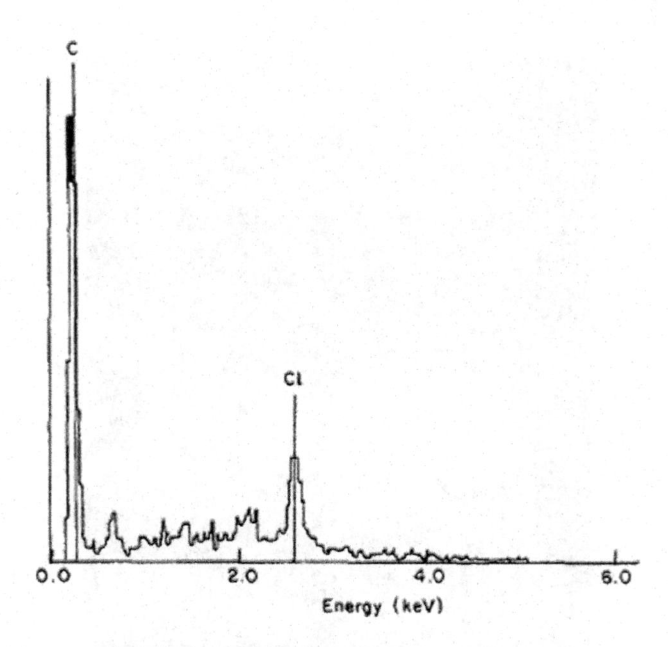

FIGURE 3.4.4 EDS spectrum of the substrate from Figure 3.4.3.

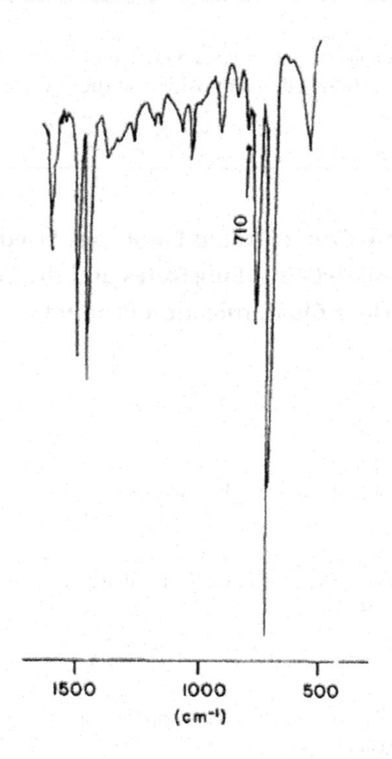

FIGURE 3.4.5 FT-IR spectrum of the polystyrene (core)-poly(VBC) (shell) substrate.

In Scheme 3.4.3, *y* and *z* represent the content of VBC and onium salt units, respectively, in mol%: for Z = N, *y*:*z* = 31:69; for Z = P, *y*:*z* = 62:38.

The polystyrene-poly(VBC)-supported quaternary benzyltributyl ammonium chloride 4a or benzyltributyl phosphonium chloride 4b catalysts listed in Table 3.4.4 were employed as phase transfer catalysts in the alkylation reaction of isopropylidene malonate *1* (Scheme 3.4.2). Two other catalysts, poly(VBC)-supported quaternary benzyltributyl phosphonium chloride le (Table 3.4.2) and the unsupported catalyst VBPC were also used for comparison (Table 3.4.5). The latter table shows that although the molar amounts of phosphonium chloride contained by catalysts 4b, le and VBPC are the same, the yields of the product (5-substituted isopropylidene malonate *3*) are in the sequence VBPC > 4b > le. Because VBPC is a low molecular weight catalyst which can dissolve in the reaction solvent, each of the molecules of VBPC can act as a reactive site; if the concentration of catalyst on the substrate is low and, consequently, the cooperativity effect is absent, the system which uses unsupported VBPC as catalyst should have the highest yield. This appears to be the case for the experiments presented in this section. One may note that the supported phosphonium chloride catalyst 4b provides a yield which is near that of VBPC and much higher than that of supported ammonium catalyst 4a. The low yield of the supported phosphonium chloride catalyst le (which has the same morphology as catalyst 1c of Figure 3.4.1c) is probably due to the deep penetration of the phosphonium catalyst in the polymer. Therefore a part of the catalyst is no longer available for reaction. It is important to note that in the present section, because of the smaller amount of catalyst, the molar ratio between the catalyst and reagent is 1:23, whereas in the previous section it was much larger, namely 1.22:5. Since there the amount of catalyst on the support was high, the cooperativity effect resulted in a higher activity of the supported catalyst than of the unsupported catalyst. The transition to cooperativity is now under investigation in our laboratory.

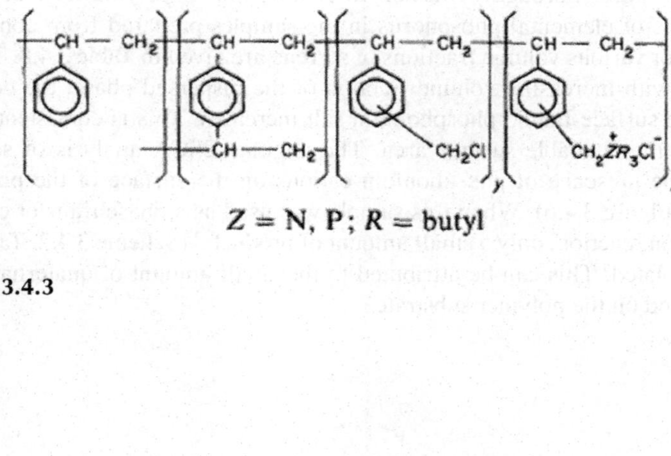

$Z = N, P; R = \text{butyl}$

SCHEME 3.4.3

TABLE 3.4.5
Comparison of the Yields of Monosubstituted Isopropylidene Malonate 3 (Scheme 3.4.2) in Phase Transfer Catalyst Reactions with Various Supported and Unsupported Onium Salts[a]

Catalyst	Amount Polymer Substrate (g)	Quaternary Onium Salt (mmol)	Time (h)	Yield (mol%)
4a (Table 3.4.4)	0.65	0.306	4	35
4b (Table 3.4.4)	0.80	0.219	4	52.8
le (Table 3.4.1)	0.1	0.213	4	17.4
VBPC	0.078	0.219	4	59.2
7c (Table 3.4.7)	1.0	5.5×10^{-2}	4	Very small

[a] Phosphonium salt (mmol): malonate *1* (mmol) = 0.219:5.0.

3.4.3.3 Polystyrene-Supported Onium Salts Prepared Via Their Adsorption from the Continuous Phase on the Surface of Styrene Droplets of a Concentrated Emulsion

In the third approach used for the preparation of phase transfer catalysts, a concentrated emulsion of styrene in water was prepared after the quaternary onium salt (Scheme 3.4.4) was dissolved in water. The onium salt is expected to be adsorbed upon the surface of styrene droplets with their vinylbenzyl group oriented towards the inside of the droplets. It is therefore possible that they will polymerize simultaneously with styrene. It is important to note that experiment has indicated that concentrated emulsions do not form for concentrations of quaternary onium salt greater than some critical values (Table 3.4.6). The instability of the concentrated emulsions containing quaternary onium salts is probably a result of the neutralization or screening of the net charge of the surface of the styrene droplets, by the formation of ion-pairs between the anionic surfactant and cationic quaternary onium ion. VBPC was used as a typical quaternary onium salt in experiments (Table 3.4.7). After polymerization, the samples were thoroughly washed to remove the unpolymerized onium salt. The amounts of elemental phosphorus in the samples prepared from concentrated emulsions for various volume fractions of styrene are given in Table 3.4.8. This table shows that with increasing volume fraction of the dispersed phase, the conversion of VBPC to surface-bound phosphonium salt increases. This is consistent with the increase in the available surface area. The angular ESCA analysis of sample 7b identified the presence of phosphonium cations on the surface of the polystyrene substrates (Figure 3.4.6). When this sample was used as a phase transfer catalyst in the alkylation reaction, only a small amount of product *3* (Scheme 3.4.2, Table 3.4.5) could be isolated. This can be attributed to the small amount of quaternary onium salt supported on the polymer substrate.

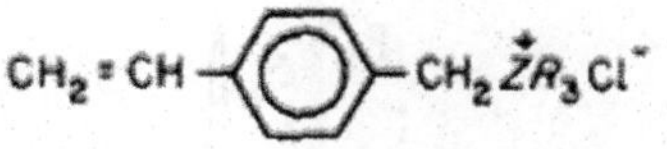

$$CH_2{=}CH{-}C_6H_4{-}CH_2\overset{+}{Z}R_3\,Cl^-$$

$Z = N$, R = butyl, ethyl, methyl
$Z = P$, R = butyl, ethyl

SCHEME 3.4.4

TABLE 3.4.6
Critical Concentrations at Room Temperature of Quaternary Onium Salts Above Which No Concentrated Emulsions Form[a]

	VBPC	VBAC	VEPC	VEAC	VMAC
Concentration (M)	0.23–0.25	0.18–0.20	0.15–0.17	0.20–0.23	0.15–0.17

[a] Styrene (15 mL) was dispersed in the aqueous phase (0.30 g SDS + onium salt + 2 mL H_2O)

TABLE 3.4.7
Concentrations in the Concentrated Emulsions of Styrene in Water Containing the Quaternary Onium Salt VBPC

	7a	7b	7c	7d
Dispersed phase				
Styrene (g)	13.8	15.3	5.2	7.3
DVB (g)	0.28	0	0.105	0.148
AIBN	2.2×10^{-4} g/g monomer for all systems			
Continuous phase				
VBPC (g)	0.142	0.113	0.142	0.142
Surfactant (SDS) (g)	0.3 g for all systems			
Water	2 mL for all systems			

TABLE 3.4.8
Elemental Analysis of the Surface-Supported Phosphonium Salts

Catalyst	VBPC ((mmol × 10^2)/g styrene)	Volume Fraction, ϕ	P((mmol × 10^2)/g polymer)
7c	7.70	0.74	2.87
7d	5.48	0.80	2.75
7a	2.90	0.88	2.20

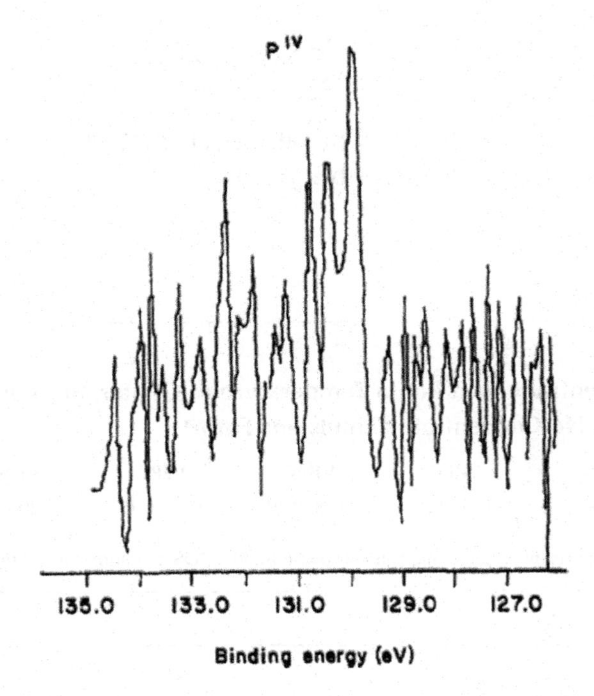

FIGURE 3.4.6 ESCA spectrum of sample 7b (Table 3.4.7). Atomic concentrations (%) of P^{IV} = 2.18. Binding energy of P^{IV} = 130.9 eV.

3.4.4 CONCLUSION

Polymer substrates of poly(VBC) and polystyrene-poly(VBC) have been prepared by the concentrated emulsion polymerization method. The active quaternary phosphonium and ammonium chloride groups were subsequently introduced via quaternization reactions. Two kinds of phase transfer catalysts have been thus obtained: (1) poly(VBC)-supported onium salt, in which thick layers of bound onium salts were generated, and (2) polystyrene-poly(VBC)-bound onium salt, in which the substrate has a core-shell morphology with poly(VBC) covering a core of polystyrene. These two kinds of polymer-supported onium salts have been employed as phase transfer catalysts in the alkylation of isopropylidene malonate. Surface-bound quaternary onium salts were also prepared via the adsorption of the onium salt from the continuous phase on the surface of the dispersed phase of the concentrated emulsion. The adsorbed onium salts can be polymerized simultaneously with the styrene present in the droplets. The amount of supported onium salt thus obtained has been small. The latter approach needs further investigation.

REFERENCES

1. Svec, F. *Pure Appl. Chem.* 1988, **60,** 377.
2. Ford, W. T., Chandran, R. and Turk, H. *Pure Appl. Chem.* 1988, **60,** 396.
3. Guyot, A. *React. Polym.* 1989, **10,** 113.

4. Ruckenstein, E. and Kim, K. J. *J. Appl. Polym. Sci.* 1988, **36,** 907.
5. Chen, H. H. and Ruckenstein, E. *J. Colloid Interface Sci.* 1990, **136,** 473.
6. Ruckenstein, E. and Park, J. S. *Polymer* 1992, **33,** 405.
7. Nishikubo, T., Uchida, J., Matsui, K. and Iizawa, T. *Macromolecules* 1988, **21,** 1583.
8. Brandstrom, A. and Ldamm, B. *Acta Chem. Scand.* 1974, **B28**, 590.
9. Fieser, M. and Fieser, L. F. (Eds) *Reagents for Organic Synthesis*, Vol. 7, John Wiley & Sons, New York, 1979, p. 18.
10. *Kirk-Othmer Encyclopedia of Chemical Technology*, 2nd Edn, Vol. 5, John Wiley & Sons, New York, 1964, p. 5.
11. McNab, H. *Chem. Soc. Rev.* 1978, **7,** 345.
12. Alami, S. W., LeMaguer, D. and Caze, C. *React. Polym.* 1987, **6,** 213.
13. Kondo, S., Yasui, H. and Tsuda, K. *Makromol. Chem.* 1989, **190,** 2079.

3.5 Preparation of Latex Carriers for Controlled Release by Concentrated Emulsion Polymerization*

Kyu-Jun Kim and Eli Ruckenstein

Department of Chemical Engineering, State University of New York at Buffalo, Buffalo, New York 14260

ABSTRACT Spherical latexes of submicron diameter containing a model hydrophobic herbicide, namely, 2-(2,4-dichlorophenoxy)-propionic acid, were prepared by the concentrated emulsion polymerization method. A mixture of styrene, herbicide, and initiator was dispersed into a small amount of water (6% by volume based on the entire emulsion) containing surfactant, at room temperature. The heating at 40°C of the foam-like gel thus obtained led to the formation of latexes containing the herbicide. It was found that the model herbicide dispersed in the polymer latex was released to water over a period of several weeks and that the amount released was strongly dependent upon temperature and latex concentration in water.

3.5.1 INTRODUCTION

The controlled release technology has recently received widespread attention because of its high efficiency in the use of active chemicals, such as drugs, pesticides, and fertilizers, without the need for multiple applications.[1–3] One of the methods to achieve controlled release is the physical incorporation of the bioactive agent into an inert matrix by encapsulation, blending, or embedding. The molecular diffusion of the active agent in the matrix is generally controlling the release rate. Although natural or synthetic polymers have been widely utilized as vehicles for bioactive agents,[1–3] no investigation has yet been reported in which polymer latexes

* *Journal of Applied Polymer Science*, Vol. 38, 441–452, (1989).

of submicron size were employed as inert media. The colloidal size of such a vehicle and the possible stabilization of the colloidal dispersions in aqueous environments could allow the polymer latexes to become valuable candidates for various controlled release applications where hydrophobic active agents are released into hydrophilic environments. However, when the solubility of the active agent in water is not small, neither the conventional emulsion polymerization (in which polymerization takes place in micelles) nor the miniemulsion polymerization (in which polymerization occurs in the monomer droplets)[4] are appropriate procedures for the preparation of polymer latexes containing active ingredients. In both systems, the ratio of the volumes of water to monomer is large and, therefore, a significant amount of the active compound escapes from the monomer phase to the water phase. In the conventional emulsion polymerization, there is an additional, more important, negative factor since polymerization occurs in the micelles and hence both the monomer and the active agent have to migrate from the monomer droplets to the micelles. In contrast, in concentrated emulsion polymerization,[5] a very small amount of water, which can be smaller than 1% (based on the entire emulsion), is required to prepare a foam-like emulsion (gel) and the polymerization occurs in the monomer droplets. The stability of this gel is ensured by the adsorption of a surfactant, which is dissolved initially in the continuous phase, on the interface between the network of thin liquid films of the continuous phase and the polyhedral drops of the dispersed phase. An efficient loading of the active agent in the polymer latexes thus becomes possible, since the amount of active agent dissolved in the water phase is very small.

In this paper, we report the preparation of polymer latexes containing a model herbicide, namely, 2-(2,4-dichlorophenoxy)-propionic acid, (2,4-DP), by the concentrated emulsion polymerization method, as well as the results obtained concerning the release behavior in water of these latexes. The solubility of this compound in water is very low (350 ppm at 20°C)[6]. In this case, the miniemulsion method may also be appropriate for preparing latexes containing 2,4-DP. It is found that polymerization proceeds without a chain transfer process between the growing polymer radical and model herbicide, leading to the formation of latexes of 0.32 μm average diameter. The herbicide is released from the latexes over a period of several weeks with an amount released that strongly depends upon temperature and amount of water.

3.5.2 EXPERIMENTAL

Styrene (Aldrich) was distilled and azobisisobutyronitrile (AIBN, Alfa) was recrystallized from methanol. Sodium dodecylsulfate (Aldrich) and 2-(2,4-dichlorophenoxy)-propionic acid (2,4-DP, Sigma) were used as received. Water was deionized and distilled.

A small amount of water (2.5 mL) containing sodium dodecylsulfate (0.4 g) was placed in a single-neck flask of 100 mL capacity equipped with a mechanical stirrer. Styrene (36 g) in which 2,4-DP (2.67 g) and AIBN (0.27 g) were dissolved by heating at 40°C was added dropwise to the stirred aqueous solution through an addition funnel, at room temperature. A gel-like emulsion is thus generated, where the dispersed monomer phase is composed of cells separated among them by thin films of water.

The concentrated emulsions thus prepared were transferred to centrifuge tubes (15 mL capacity), which were then sealed with rubber septa. A mild centrifugation was employed to achieve better packing of the polyhedral drops. Polymerization was carried out at 40°C in the presence of air. It is difficult to disperse the latexes in water when polymerization approaches completion because they agglomerate. Therefore, additional water (typically twice as much as the initial weight of the emulsion) was added into the tubes when the conversion reached about 50% (after about 22 h). During polymerization, a small fraction of the cells of the gel collapsed and formed a bulk monomer phase in which bulk polymerization occurred. The undispersed polymer was removed by filtration through a coarse filtering funnel. The filtered aqueous solution was further polymerized. After complete conversion, the concentration of the latexes in water was measured by weighing the polymer after filtering through a fine membrane (0.1 μm porosity).

The molecular weight of the polymer and its distribution were determined by employing a GPC composed of 500, 10^4, 10^5, and 10^6 Å pore size columns (Waters, Ultrastyragel) in series. Methylene chloride was used as the mobile phase and the flow rate was 1 mL/min. The GPC was calibrated using monodisperse polystyrene standards (Polymer Laboratories, U.K.) with molecular weight of 10,000–12,000,000. For the GPC measurements, the aqueous latex solution was poured into methanol and the precipitated polymer was collected, dried, and dissolved in chloroform. The latter solution was injected into the mobile phase.

The polymer latexes were examined with a JEOL 100U transmission electron microscope. Copper grids coated with carbon films were used as substrates and dilute latex solutions (approximately 1×10^{-6} g/mL) were deposited on the substrates.

For the release investigations, the latexes dispersed in water were kept in a constant temperature bath and aliquots of the sample were taken after various lengths of time. The release experiments were conducted by measuring the concentration of 2,4-DP in water by HPLC. Polymer latexes were separated from water by filtering through a nitrocellulose membrane of 0.1 μm porosity (Microfiltration Systems), and the filtered aqueous solutions were injected into the HPLC. μ-Bondapak C_{18} and Shodex RSpak columns were connected in series and CH_3CN/water (50/50 vol%) was used as the mobile phase. The wavelength of the detector was 278 nm and the retention volume for 2,4-DP was 3.2 mL at a flow rate of 0.8 mL/min.

3.5.3 RESULTS AND DISCUSSION

3.5.3.1 Preparation of Polystyrene Latexes Containing 2,4-DP

The polymerization of concentrated emulsions whose dispersed phase consisted of styrene and 2,4-DP was first investigated in order to study the effect of the active agent.

Figure 3.5.1 presents GPC curves for the polystyrene prepared by the concentrated emulsion method in the presence (curve b) and absence (curve a) of 2,4-DP. It also includes the GPC curve for the polystyrene prepared by the bulk polymerization

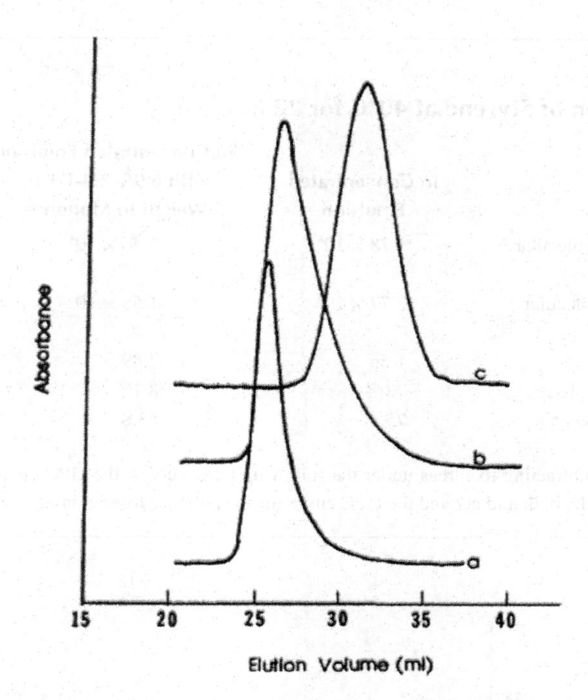

FIGURE 3.5.1 GPC curves of polystyrenes produced at 40°C after 22 h of reaction: (a) prepared by concentrated emulsion polymerization; (b) prepared by concentrated emulsion polymerization in the presence of 6.9% 2,4-DP by weight to monomer; (c) prepared by bulk polymerization.

method (curve c). The curves show that, with the addition of 2,4-DP to the dispersed phase, the molecular weight distribution broadens and the contribution of the molecular weights in the range between those produced in concentrated emulsions without 2,4-DP and those produced in bulk becomes significant. The data regarding polymerization in various conditions are summarized in Table 3.5.1.

As shown in Table 3.5.1, the percent conversion in concentrated emulsion decreases in the presence of 2,4-DP. The fractions of the cells of the foam that collapse during polymerization with the formation of a bulk monomer phase are comparable and relatively small in both cases. This indicates that the stability of the gel-like emulsion is not dramatically deteriorated by the presence of 2,4-DP.

In order to understand the results of Table 3.5.1, bulk polymerization was conducted in the presence of various amounts of 2,4-DP. Figure 3.5.2 is a plot of polymer conversion against the weight percent of 2,4-DP in the monomer phase and shows that the percent conversion is practically unaffected by the presence of 2,4-DP. The effect of 2,4-DP on the molecular weight of the resulting polymer was also investigated, and the corresponding GPC curves are presented in Figure 3.5.3.

TABLE 3.5.1
Polymerization of Styrene at 40°C for 22 h

	In Concentrated Emulsion	In Concentrated Emulsion with 6.9% 2,4-DP by Weight to Monomer	In Bulk
Number average molecular weight	1.78×10^6	5.46×10^5	1.27×10^5
Weight average molecular weight	2.77×10^6	1.58×10^6	2.27×10^5
Polydispersity	1.56	2.89	1.79
Formation of bulk phase[a]	7.0%	8.1%	–
Polymer conversion (%)	75.4	53.8	15.2

[a] Calculated by subtracting the areas under the following GPC curves: the GPC curve for the polymer formed both in the bulk and gel and the GPC curve for the polymer formed in gel.

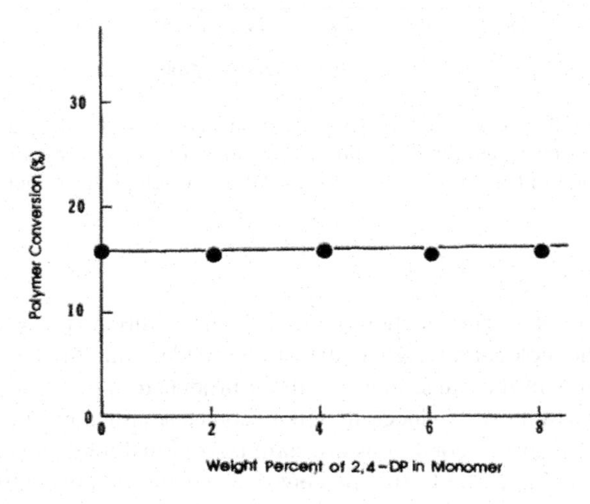

FIGURE 3.5.2 A plot of conversion in bulk polymerization of styrene as a function of weight percent of 2,4-DP in the monomer (40°C, 22 h).

Despite the addition of a substantial amount of 2,4-DP, no appreciable difference between the two curves could be detected. Therefore, it is fair to conclude that no noticeable chain transfer occurs with the 2,4-DP during polymerization, since such a process would lower the molecular weight by the premature kinetic chain-breaking of the growing polymer radical. This is an important observation, since the chain-transfer process could result in the loss of active agent by its incorporation into the polymer.

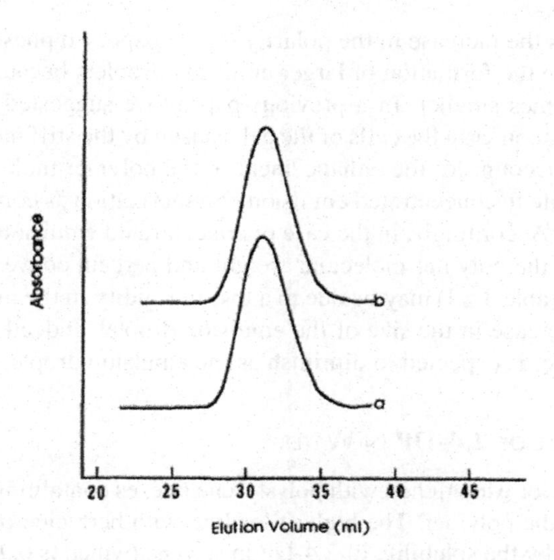

FIGURE 3.5.3 GPC curves of the polystyrenes prepared by bulk polymerization (40°C, 22 h): (a) without 2,4-DP; (b) with 8% 2,4-DP by weight to monomer.

Figure 3.5.4 presents transmission electron micrographs of the polymer latexes produced in the absence (a) and presence (b) of an herbicide in the dispersed phase. It is noticeable that the latexes containing 2,4-DP have a diameter of about 0.32 μm, which is significantly larger than those of the pure polystyrene latexes of 0.23 μm

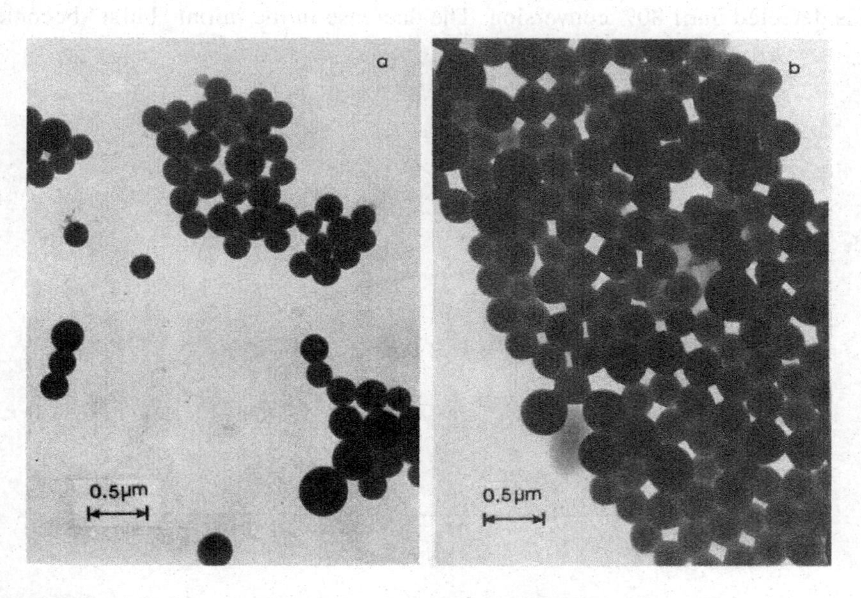

FIGURE 3.5.4 Transmission electron micrographs of polymer latexes: (a) prepared without 2,4-DP; (b) prepared with 6.9% 2,4-DP by weight to monomer.

diameter. Perhaps the increase in the polarity of the dispersed phase in the presence of 2,4-DP leads to the formation of larger emulsion droplets because the interfacial free energy becomes smaller. In a previous paper,[5] we suggested that the smaller mobility of the monomer in the cells of the gel, caused by the stiff nature of the emulsifier layer, may account for the enhancement in the polymer molecular weight and polymerization rate in concentrated emulsion polymerization as compared to polymerization in bulk. Accordingly, in the case of concentrated emulsion polymerization, the depression in the polymer molecular weight and percent conversion in the presence of 2,4-DP (Table 3.5.1) may be due to a loss in rigidity in the monomer medium caused by the increase in the size of the emulsion droplet. Indeed, the influence of the emulsifier layer is expected to diminish as the emulsion droplet becomes larger.

3.5.3.2 Release of 2,4-DP in Water

The release behavior was studied with polystyrene latexes containing 6.9% 2,4-DP by weight based on the polymer. The highest loading with herbicide of the polystyrene latexes is limited by the solubility of 2,4-DP in styrene (which is 6.9% by weight).

When diluted with a sufficient amount of water (7.8×10^{-5} g/mL latex concentration) after 22 h of polymerization (the conversion is about 50%), the polymer latexes released instantaneously half of the 2,4-DP embedded. It is likely that the looseness of the matrix due to unpolymerized monomer is responsible for this sudden, large initial release of active agent. It is, therefore, useful to examine the effect of the extent of polymerization on the initial release. Figure 3.5.5 plots the percent of 2,4-DP released immediately (about 5 min) on dilution against polymer conversion. No significant decrease in the initial release was detected until 80% conversion. The decrease in the initial "burst" becomes

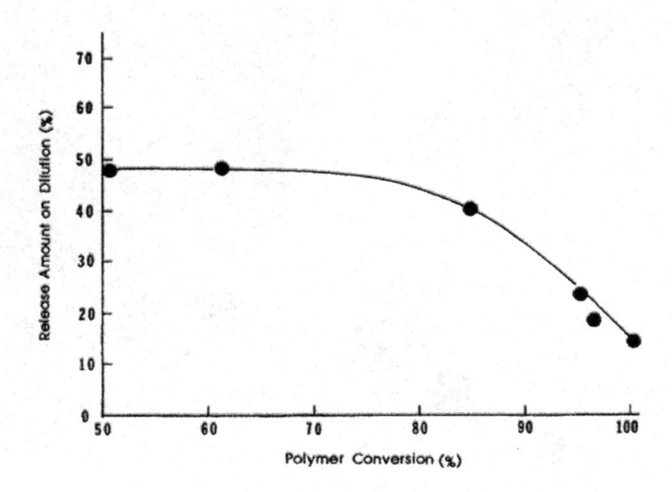

FIGURE 3.5.5 A plot of the percent of 2,4-DP released on dilution as a function of conversion (latex concentration after dilution; 7.8×10^{-5} g/mL).

notable only when the conversion exceeds 85%. Accordingly, polymerization should be carried out close to completion in order to minimize the loss of active agent on dilution. However, it should be noted that, in spite of high conversion, the latexes still released initially about 15% 2,4-DP when diluted with water (7.8×10^{-5} g/mL latex concentration). In order to clarify this observation, the initial instantaneous release when the latexes are diluted with various amounts of water was measured, and the results are plotted in Figure 3.5.6. The initial release increases with the amount of water until the latex concentration becomes 1.0×10^{-3} g/mL and remains constant for smaller values. This indicates that the dissolution of 2,4-DP located on the latex surface and/or in the vicinity of the latex surface is responsible for most of the initial release in Figure 3.5.5. The initial release would be independent of the amount of water used for dilution if it resulted solely from the escape of 2,4-DP from the monomer phase during the preparation of the concentrated emulsion.

Figure 3.5.7 shows the cumulative release of 2,4-DP after 3 days in a wide range of latex concentrations. As expected, the release is significantly inhibited at high latex concentrations because the amount of water available per particle is small. This is quite beneficial for practical applications since the premature release can be prevented by keeping the latex concentration high. While the decrease of the cumulative release with increasing particle concentration at sufficiently large values of the latter is understandable, since the amount of water becomes smaller, the occurrence of the weak maximum is less understandable. If real, it probably has a kinetic origin, the rate of dissolution being increased by the stirring produced by the Brownian motion of the particles. This may be important when the number of particles is sufficiently large but not too large to interfere with one another.In Figure 3.5.8 the cumulative release of 2,4-DP is plotted against time at various temperatures. It shows that

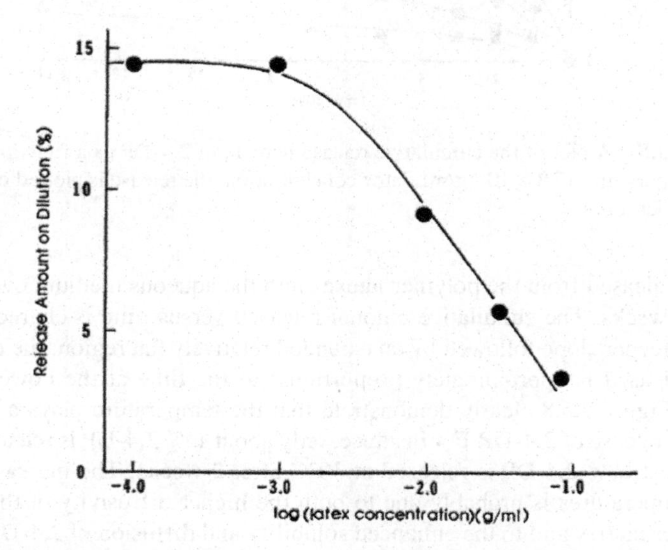

FIGURE 3.5.6 Amount of 2,4-DP released on dilution with various amounts of water (the values in the abscissa signifies the latex concentrations after dilution).

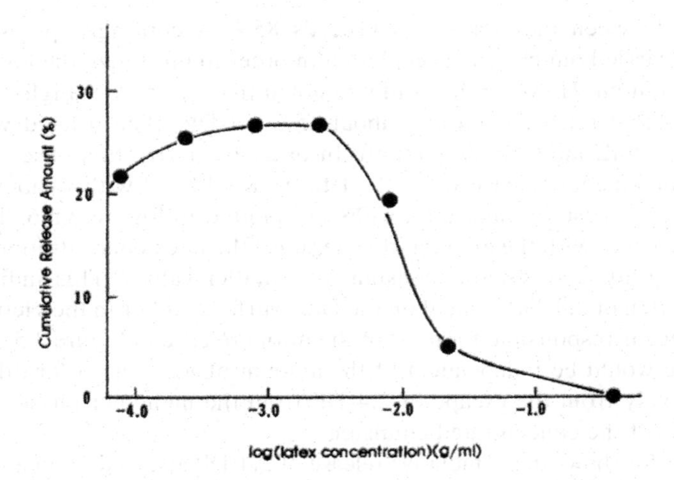

FIGURE 3.5.7 A plot of the cumulative release percent of 2,4-DP as a function of latex concentration (22°C, 3 days; the release is plotted by excluding the initial release).

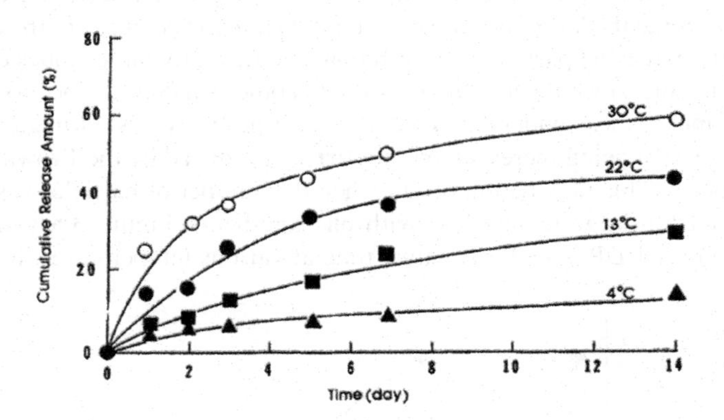

FIGURE 3.5.8 A plot of the cumulative release percent of 2,4-DP as a function of time at various temperatures (7.8×10^{-5} g/mL latex concentration; the release is plotted by excluding the initial release).

2,4-DP is released from the polymer latexes into the aqueous medium over a period of several weeks. The cumulative amount released versus time is characterized by an initial steeper slope followed by an extended relatively flat region; the cumulative amount released is approximately proportional to the time at the power 1/2. The results of Figure 3.5.8 clearly demonstrate that the temperature plays a significant role on the release of 2,4-DP. For instance, only about 14% 2,4-DP is released at 4°C, while almost 60% 2,4-DP is released at 30°C after 2 weeks. The increased release at high temperatures is probably due to both the higher diffusivity of the solute in the polymer matrix and to the enhanced solubility and diffusion of 2,4-DP in water.

The initial release rate against temperature can be described by an Arrhenius plot with an activation energy of 62.7 KJ/mol (Figure 3.5.9).

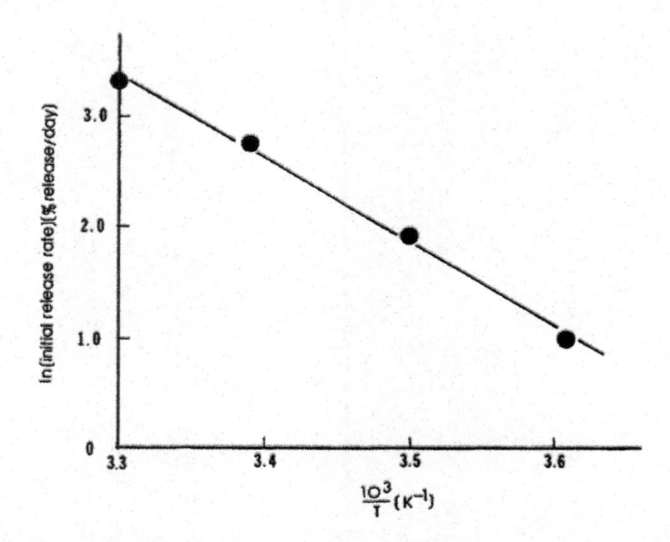

FIGURE 3.5.9 Arrhenius plot for the initial release rate against temperature (the points have been calculated on the basis of Figure 3.5.8).

3.5.4 CONCLUSION

The concentrated emulsion polymerization method can be used for the preparation of polymer latexes of submicron size in which active compounds are dispersed. Complete conversion is recommended in order to minimize the initial sudden release of active agent on dilution with water. The strong dependence of the release on temperature and amount of water suggests that the polymer latexes could be employed as valuable carriers in temperature and moisture sensitive release systems.

Dr. I. Sushumna has carried out the electron microscopy work.

REFERENCES

1. *Controlled Release Technologies: Methods, Theory, and Applications,* A. F. Kydonieus, Ed., CRC Press, Boca Raton, FL, 1980.
2. *Controlled Release Systems,* S. K. Chandrasekaran, Ed., AIChE Symposium Series 206, AIChE, New York, 1981.
3. *Controlled Release Technology,* K. G. Das, Ed., Wiley-Interscience, New York, 1983.
4. Y. T. Choi, M. S. El-Aassar, E. D. Sudel, and J. W. Vanderhoff, *J. Polym. Sci.,* 23, 2973 (1985).
5. E. Ruckenstein and K. J. Kim, *J. Appl. Polym. Sci.,* 36, 907 (1988).
6. *Encyclopedia of Chemical Technology,* Wiley-Interscience, New York, 1980, Vol. 12, p. 312.

4 Plastics Toughening and Compatibilization

CONTENTS

Plastics represent the most important application of synthetic polymer materials, because most (~80% by mass) of the polymers produced worldwide are used as plastics. Among commercial plastics, some of them, such as polystyrene, are relatively brittle. It is important to modify these plastics through toughening technologies, in order to optimize their mechanical properties and expand their applicability. The toughening of plastics can be accomplished by either copolymerization of different monomers or melt blending of different polymers.

Copolymerization can be carried out in various media, such as bulk, solution, suspension, and emulsion. However, when crosslinking is involved in the copolymerization process, emulsion becomes a highly valid option of medium, because crosslinking can be limited within the resulting latex particles with well-maintained processability. In particular, concentrated emulsion polymerization has emerged as a very attractive approach for the preparation of crosslinked latex particles when macromonomers are involved. A remarkable advantage of concentrated emulsion polymerization is that the site of polymerization is in the droplets. It is different with conventional emulsion polymerization in which the site of polymerization is the micelles, and it is concerned with whether the macromonomers can diffuse readily into the micelles with small monomers. Another advantage of concentrated emulsion polymerization is the synthetic efficiency regarding the volume of reaction system. This advantage becomes critical when it is compared with miniemulsion polymerization which typically also has polymerization sites within the droplets but allows only a small volume fraction of dispersed phase.

In general, concentrated emulsion polymerization is a powerful technique to prepare toughened polymeric blends/composites. There are two appropriate pathways leading to toughened materials. First, they can be obtained via polymerization in a single concentrated oil-in-water emulsion with the oil droplets containing monomers and elastomer modifiers. Second, they can be synthesized by mixing two partially polymerized concentrated emulsions each containing different components, followed by continuing the polymerization to completion.

Concentrated emulsion possesses kinetic stability rather than thermodynamic stability. Its long-term kinetic stability can largely be ascribed to its high viscosity derived from the high concentration of the dispersed phase. Sometimes, its stability is further strengthened by partial polymerization of the droplets in the dispersed phase. For this reason, droplets from different concentrated emulsions can be mixed to form polymer blends, thus eliminating the complexity of melt blending. Moreover, some reactions, such as compatibilization reactions, can be performed on the surface of these droplets to boost the stability of the immiscible blends through covalent stabilization.

Sections 4.1 and 4.2 illustrate macromonomer-containing copolymerizations in single concentrated emulsions. Different products can be obtained by using different experimental settings. The product reported in Section 4.1 is a semi-interpenetrating polymer network (SIPN), while the product described in Section 4.2 is an AB binary network.

Section 4.3 presents the general strategy of the droplets-mixing approach of concentrated emulsion polymerization. Two concentrated emulsions containing different monomers are prepared, and each of them is subjected to partial polymerization.

Subsequently, the two systems are mechanically mixed, and the mixture is subjected to complete polymerization.

Section 4.4 continues focusing on the droplets-mixing approach, including introducing a vinyl-terminated macromonomer into one of the two concentrated emulsions. Some AB binary networks from the macromonomers are formed, which lead to the self-compatibilization of the resulting blends. More powerful compatibilization approaches are presented in Sections 4.5 and 4.6.

Section 4.5 illustrates an interesting case of mixing concentrated emulsions that are based on epoxy resins and various monomers, followed by polymerization and heterogeneous crosslinking among the droplets. The heterogeneous crosslinking proceeds via the reactions between epoxy/anhydride and epoxy/amine. The single T_g values of the systems indicate that the components of the composites are satisfactorily compatibilized.

Section 4.6 describes an alternative but much simpler system. Functional groups are introduced to different droplets, and the compatibilization is achieved directly through surface reactions among the droplets.

It should be highlighted that one more advantage of the concentrated emulsion is its high rate of polymerization. Since initiator is present in every droplet, the polymerization in the individual droplets can be initiated almost simultaneously. Because diffusion of free radicals across droplets for termination is restricted, concentrated emulsion polymerization may involve slower termination rates as compared with the conventional free radical polymerization process. Sections 4.7 and 4.8 describe explorative studies of such high-rate polymerization systems. Specifically, Section 4.7 focuses on an adiabatic process and Section 4.8 addresses the effect of external heating for high-rate concentration polymerization.

4.1 Semi-Interpenetrating Polymer Network Latexes via Concentrated Emulsion Polymerization*

Eli Ruckenstein and Hangquan Li

Department of Chemical Engineering, State University of New York at Buffalo, Buffalo, New York 14260

ABSTRACT Latexes of semi-interpenetrating polymer networks (SIPN) of polyurethane (PU) and poly(methyl methacrylate) (PMMA) were prepared via the concentrated emulsion polymerization. In this procedure, a partially cross-linked PU was first prepared in a low polarity solvent from the appropriate precursors. Subsequently, MMA and an initiator were introduced into the solution, and the solution was used as the dispersed phase of a concentrated emulsion in water. Finally, SIPN latexes were obtained via the polymerization of the concentrated emulsion. For comparison purposes, SIPN materials have been also prepared via bulk polymerization. The studies with differential scanning calorimetry and transmission electronic microscopy showed that partial interpenetration was achieved in the SIPN latexes. The tensile behavior and particle morphology of the SIPN materials were investigated by changing the proportion of PU, the molar ratio of NCO/OH, the theoretical cross-link density, and the concentration of the initiator. The SIPN latexes prepared possess a high toughness.

4.1.1 INTRODUCTION

Interpenetrating polymer networks (IPNs) constitute a special type of polymer blends that combine two networks. If only one of the two components is in the form of a network while the other one is in the form of a linear polymer,

* *Journal of Applied Polymer Science*, Vol. 55, 961–970, (1995).

the composite is called a semi-IPN or SIPN. Sperling[1,2] and Frisch[3] and their coworkers demonstrated that in many cases the interpenetration of the two phases yields materials with new or enhanced properties. A number of methods have been developed to produce IPNs, such as latex blending,[4,5] sequential polymerization,[6,7] and simultaneous polymerization.[8,9] A large number of useful IPN materials have been produced by coupling various polymers. Among the coupled polymers, IPNs or SIPNs based on polyurethane (PU) and poly(methyl methacrylate) (PMMA) have received a lot of attention because they can find applications as structural,[10] energy damping,[11,12] and biomaterials.[13] Allen[14–18] and Meyer[10,11,19–21] and their coworkers have thoroughly investigated the PU/PMMA IPNs and shown that they exhibit improved impact and shear resistance. However, the PU/PMMA IPN materials were exclusively prepared by the bulk polymerization method, which produces bulk materials difficult to process. It is clear that if IPN materials could be prepared in the form of latexes, which are more easily processable,[4,22] their usefulness will be increased. However, the conventional pathway to IPN latexes, namely the seed latex method, is not suitable for the PU/PMMA system because the isocyanate, which is a necessary precursor in the formation of PU, reacts rapidly with water. Hence water cannot be used as a reaction medium. Furthermore, the polymerizations of PU and PMMA occur via different mechanisms, namely by step addition and free radical polymerization, respectively, and consequently the seed method cannot be employed.

In this study, a recently developed procedure, based on concentrated emulsions,[23,24] was employed to prepare PU/PMMA SIPN latexes. A concentrated emulsion differs from the conventional emulsion in that the volume fraction of the dispersed phase is larger than 0.74, which represents the most compact arrangement of spheres of equal size, and can be as large as 0.99. At high volume fractions, the dispersed phase is in the form of polyhedral droplets separated by a network of thin films of the continuous phase. PU/PMMA SIPN materials were also prepared via simultaneous bulk polymerization for comparative purposes.

In the proposed procedure a partially cross-linked PU polymer was first prepared in a solvent (chloroform). Then, the MMA monomer and an initiator were introduced and the system thus obtained was used as the dispersed phase of a concentrated emulsion in water. SIPN latexes were prepared via the polymerization of the concentrated emulsion. It is important in the first step, when the PU is prepared, that the cross-linking of the PU is moderate. In this manner only, the viscosity of the system also remains moderate, and the dispersion in water, to generate a concentrated emulsion, is possible.

4.1.2 EXPERIMENTAL

4.1.2.1 Materials

The chemicals employed are listed in Table 4.1.1. MMA was distilled before use. Azobisisobutyronitrile (AIBN) was recrystallized from methanol. The other compounds were used as received. Water was deionized and distilled.

TABLE 4.1.1
Chemicals Used

Chemicals	Description	Code
Diols	Poly(propylene glycol) diol MW	
	725	PPG725 II
	1000	PPG1000 II
	2000	PPG2000 II
	Polybutadiene diol MW = 4900	
Triols	Poly(propylene glycol) triol MW	
	725	PPG725 III
	1500	PPG1500 III
	3000	PPG3000 III
Isocyanate	Toluene-2,4-diisocyanate, 80% remainder 2,6 isomer	TDI
Catalyst	Dibutyltin dilaurate, 95%	DBTL
Monomer	Methyl methacrylate, 99%	MMA
Initiator	Azobisisobutyronitrile	AIBN
Surfactant	Dodecyl sulfate sodium salt, 70%	SDS
Solvents	Chloroform, 99%; Toluene, 99%	

Purchased from Aldrich.

4.1.2.2 Preparation Procedure

4.1.2.2.1 Concentrated Emulsion Polymerization

The preparation was carried out in two steps. The PU precursors [poly(propylene glycol) (PPG) diol and triol, and [toluene-2,4-diisocyanate (TDI)] were mixed with a solvent (usually chloroform) in a test tube at room temperature. In order to increase the hydrophobicity of the system and provide contrast in the transmission electron microscopy (TEM) study, 2 g polybutadiene diol/100 g of the PPGs was incorporated. After the catalyst dibutyltin dilaurate (DBTL) (1 g/100 g of PU precursors) was added to the system, the tube was introduced into a water bath at 60°C to allow the PU formation, and kept for a time shorter by half an hour than the time needed for gelation to occur. This ensures that enough cross-linking has taken place, without being too advanced for the viscosity of the system to become exceedingly high. After the PU solution was cooled to room temperature, the MMA monomer as well as the initiator (AIBN) were introduced in various amounts, and the solution was kept at room temperature for more than 12 h. While its viscosity greatly increased, it remained in a range suitable to be dispersed as the dispersed phase of a concentrated emulsion. The concentrated emulsion was prepared as follows. An aqueous solution of sodium dodecyl sulfate (SDS) (10 wt%) was placed in a single neck 100 mL flask provided with a mechanical stirrer. The above solution was added dropwise with a syringe into the flask with vigorous stirring in the latter. The volume fraction of the SDS aqueous solution was 0.15 of the whole system. The whole addition process lasted about 15 min, and took place at room temperature. The gel-like concentrated emulsion thus formed was additionally stirred under a flow of nitrogen for 15 min and finally transferred to a tube of 30 mL capacity. The tube was sealed with a rubber septum and shaken under nitrogen flow for 10 min. The tube was finally introduced into a water bath at 60°C to carry out the polymerization of MMA, which lasted 48 h. The product thus obtained was then washed with methyl alcohol in an extractor for 24 h and dried in a vacuum oven for another 24 h.

4.1.2.2.2 Simultaneous Bulk Polymerization

The reactants (PU precursors, MMA, and initiator, free of DBTL and solvent) were mixed in a tube, which was introduced into a Cole Parmer ultrasonic mixer, at 60°C, for 3 h. After 3 h the viscosity of the system became so high that further mixing was no longer possible. Consequently, the tube was transferred into a water bath of 60°C where it was kept for 48 h.

4.1.2.3 Thermal Transition Temperatures

The thermal transitions of the samples were measured by differential scanning calorimetry (DSC) with a Perkin-Elmer DSC instrument. Each sample was heated from –70°C to 210°C, with a heating rate of 10°C/min.

4.1.2.4 Tensile Testing

Powders of the products of the concentrated emulsion polymerization, or bulk material for the products of bulk polymerization, were hot pressed in a Laboratory Press (Fred S. Carver, Inc.) at 180°C for 3–5 min, and then cooled to room temperature. The sheets thus obtained were cut to the size required by the ASTM D.638-58T. The tensile testing was conducted at room temperature, with an Instron Universal Testing Instrument (model 1000). The elongation speed of the instrument was 20 mm/min.

4.1.2.5 TEM

The phase morphology was examined by TEM (Philips T-400). The samples were first thermo-pressed in the same way as for tensile testing, then sectioned with an ultramicrotome, and stained with OsO_4.

4.1.2.6 Scanning Electron Microscopy (SEM)

The particle morphology was examined by SEM (Hitachi S-800).

4.1.3 RESULTS AND DISCUSSION

4.1.3.1 Gelation Time of PU

As already noted, the polymerization of PU must be halted below gelation, otherwise the system acquires such a high viscosity that it cannot be dispersed to generate a concentrated emulsion. The effects of the concentrations of the PU precursors as well as the temperature on the gelation time is summarized in Table 4.1.2. Gelation occurred when the solution lost its flowability. As expected, the time of gelation increases with an increasing amount of solvent. The gelation time should be long enough to allow for the preparation of the concentrated emulsion. On the basis of the gelation experiments we selected a volume of the solvent that was twice as large as the volume of the precursors, and 60°C as the reaction temperature. Under these conditions the system exhibits a suitable viscosity, which is not too high to impede the generation of a concentrated emulsion.

TABLE 4.1.2
Gelation Time (min) for Various PU Polymerization Systems

Vol. Ratio of Solvent/PU Precursors	Chloroform (R = 0.9)		Chloroform (R = 1.0)		Toluene (R = 1.0)	
	Room Temp.	60°C	Room Temp.	60°C	Room Temp.	60°C
1.0	630	58	550	55	540	54
1.5	1050	97	960	94	910	90
2.0	2580	270	2340	252	2280	245
3.0	NG	1500	NG	1350	NG	1300

PU precursors: PPG725II/PPG3000III weight ratio = 6/4. R = molar ratio of isocyanate/hydroxyl groups. NG, no gelation.

4.1.3.2 DSC Measurements

Instead of the sharp transitions of the individual polymer species, a diffuse transition region was observed on each of the DSC diagrams. This is typical for the interpenetration or semi-interpenetration structures. Because of the interpenetration, the segments of different polymers are entangled and move together. Consequently, a transition range between the glass-transition temperatures (T_gs) of the two polymers will be generated. Indeed, the T_g of PPG[10] is −60°C and that of PMMA[25] is 115°C and the transition ranges, listed in Table 4.1.3, are located between 50°C and 115°C. Although these transitions cannot be ascribed to a particular phase structure, the temperature shift indicates some interpenetration. The effect of some of the parameters should be noted. The PU/PMMA weight ratio has a major effect on the transition temperature. The starting temperature of the transition for samples with a PU/PMMA weight ratio of 1/3 is about 25°C higher than that with a ratio of 1/2. Because the PMMA segments are much stiffer than those of PPG, the greater the number of PMMA segments included, the higher the transition-starting temperature. The NCO/OH molar ratio has a minor effect. The samples with NCO/OH = 1.0 have a somewhat lower transition-starting temperature than those with NCO/OH = 1.1. From the investigation of the mechanical properties, we concluded that the samples with a ratio NCO/OH = 1.0 provide the highest toughness, hence probably the best interpenetration. Consequently, a lower starting transition temperature is associated with a better interpenetration. The presence of a transition

TABLE 4.1.3
Thermal Transition for PU/PMMA SIPNs

PU/PMMA Wt Ratio	NCO/OH Molar Ratio	PPG Composition 725II/3000III Wt Ratio	M_c	Transition Range (°C)
1/3	1.0	6/4	5000	75–110
1/3	1.0	4/6	3333	70–105
1/3	1.1	6/4	5000	80–110
1/3	1.1	4/6	3333	85–115
1/2	1.0	6/4	5000	50–75
1/2	1.0	4/6	3333	50–90
1/2	1.1	6/4	5000	55–90
1/2	1.1	4/6	3333	55–95

range was also reported by Klempner and Frisch[26] who attributed it also to the physical interpenetration of the two polymer chains. The theoretical cross-linking density, expressed in terms of the average molecular weight, M_c, between two successive cross-linked points, seems not to play a major role regarding the transition temperature. The data in the lower part of Table 4.1.3 show a somewhat broader transition range for a lower M_c.

4.1.3.3 Phase Morphology

The TEM micrographs of the SIPN latexes are presented in Figure 4.1.1, where the black regions represent the polybutadiene which was included in PU, because in the present systems only the polybutadiene can be stained by OsO_4. Considering that the white regions are largely occupied by PMMA, the size of the PMMA domains for a sample with NCO/OH = 1.0 (Figure 4.1.1a) is 50–100 nm, but is somewhat larger for NCO/OH = 0.8 (Figure 4.1.1b). As previously reported,[27] in such a range of domain sizes, a pronounced toughening is expected to occur.

4.1.3.4 Stress-Strain Curves

The stress-strain behavior was examined for PU/PMMA SIPN with various PU/PMMA weight ratios (Figure 4.1.2), and also for SIPNs with a constant PU/PMMA weight ratio, but various molar ratios (R) of isocyanate to hydroxyl groups (NCO/OH) (Figure 4.1.3). The curves of Figure 4.1.2 show that with increasing elastomer (PU) content, SIPN changes from a rather brittle material to a ductile one. The samples with a PU/PMMA weight ratio between 1/4 and 1/2 behave like leather. Whatever the PU content, all the specimens break without necking. Very light stress whitening was observed that suggests that the physical entanglement of the two phases prevents the

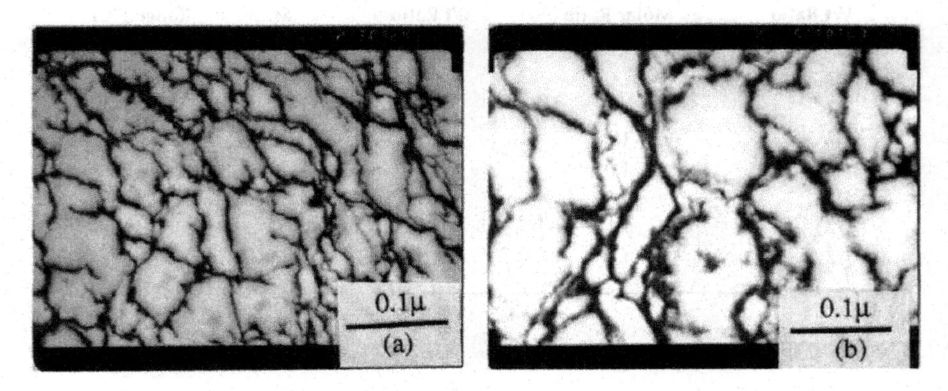

FIGURE 4.1.1 Phase morphology of PU/PMMA SIPN latexes, NCO/OH molar ratio: (a) 1.0 and (b) 0.8. PU composition: PPG725II/PPG3000III weight ratio = 6/4, PU/PMMA weight ratio = 1/2, [AIBN] = 1 g/100 g MMA.

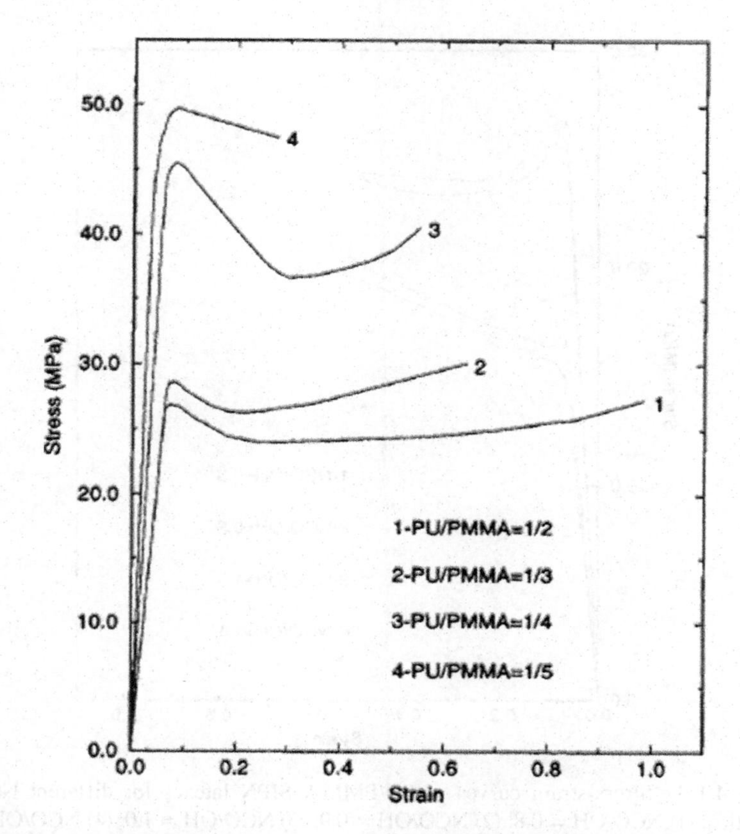

FIGURE 4.1.2 Stress-strain curves of PU/PMMA SIPN latexes for different PU/PMMA weight ratios: (1) PU/PMMA = 1/2; (2) PU/PMMA = 1/3; (3) PU/PMMA = 1/4; (4) PU/PMMA = 1/5. PU composition: PPG725II/PPG3000III weight ratio = 6/4, NCO/OH molar ratio = 1.0, [AIBN] = 1 g/100 g MMA.

interfacial decohesion and that the interpenetration is satisfactory. Figure 4.1.3 shows that for values of *R* between 0.9 and 1.1, the tensile curves exhibit a yield point. For lower values of *R*, the yield point disappears.

4.1.3.5 Effect of *R* of Isocyanate to Hydroxyl Group

Table 4.1.4 shows that the optimum values of *R* for concentrated emulsion and bulk polymerizations are different. For the bulk polymerization, the highest toughness (measured as the area under the stress-strain curve) and the highest elongation are for $R = 0.9$; for the concentrated emulsion polymerization they are for $R = 1.0$. For bulk polymerization, the increase of *R* above 0.9 does not change the tensile strength, but the elongation at break point and the toughness decrease sharply. For the samples prepared via the concentrated emulsion method, the change in *R* moderately changes the tensile strength and somewhat more strongly the elongation and toughness.

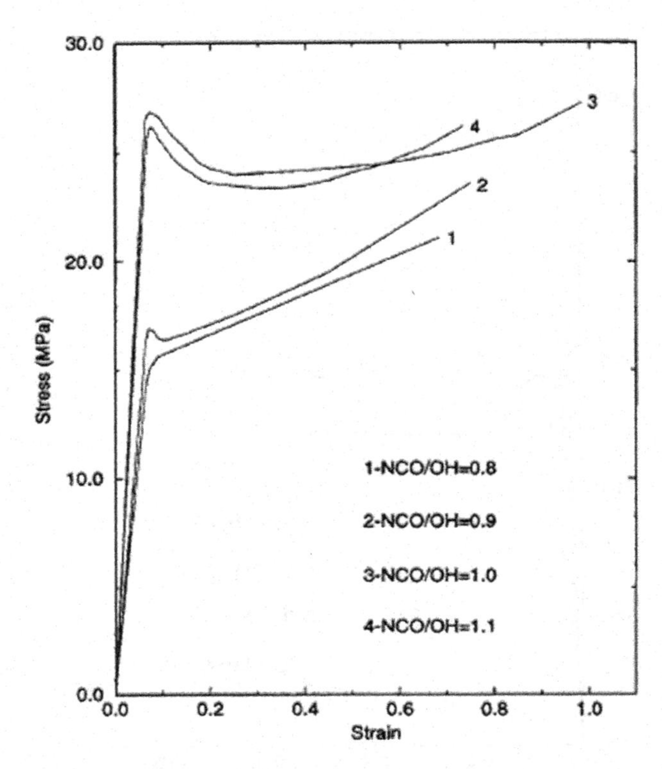

FIGURE 4.1.3 Stress-strain curves of PU/PMMA SIPN latexes for different NCO/OH molar ratios: (1) NCO/OH = 0.8; (2) NCO/OH = 0.9; (3) NCO/OH = 1.0; (4) NCO/OH = 1.1. PU composition: PPG725II/PPG3000III weight ratio = 6/4, PU/PMMA weight ratio = 1/2, [AIBN] = 1 g/100 g MMA.

Consequently, the effect of R in the concentrated emulsion method is less strong than in bulk polymerization. Because part of the isocyanate groups are consumed in side reactions with water, the optimum value of R is a little higher for the concentrated emulsion method.

It was reported[11,14] that, for bulk polymerization, a ratio NCO/OH of 1.1–1.2 constitutes the optimum value, while the value determined by us was $R = 0.9$. Some differences in the PU precursors are probably responsible for the different results.

4.1.3.6 Effect of Concentration of Initiator (AIBN)

The variation of tensile strength, elongation at break, and toughness for a set of SIPN prepared with various amounts of AIBN is presented in Table 4.1.5. The tensile strength remained substantially constant, over the range investigated, for both kinds

TABLE 4.1.4
Tensile Properties of PU/PMMA SIPN for Various NCO/OH Molar Ratios

NCO/OH	Concentrated Emulsion Polymerization						Simultaneous Bulk Polymerization					
	Tensile Strength (MPa)		Elongation (%)		Toughness (MJ/m³)		Tensile Strength (MPa)		Elongation (%)		Toughness (MJ/m³)	
Molar Ratio	Ave. Value	SE	Ave. Value	SE	Ave. Value	SE	Ave. Value	SE	Ave. Value	SE	Ave. Value	SE
0.8	23.1	0.8	68.3	5.3	13.6	1.4	18.6	1.2	120.3	5.0	18.8	1.7
0.9	23.6	1.2	75.0	7.1	15.2	1.4	20.7	0.9	125.7	6.5	21.9	2.0
1.0	27.3	2.4	98.4	7.7	20.8	2.1	20.4	0.6	75.2	1.7	16.2	0.7
1.1	26.2	1.9	73.4	5.5	16.9	1.1	20.6	0.8	51.2	5.8	10.4	0.9

PU composition: PPG725II/PPG3000III wt ratio = 6/4, PU/PMMA wt ratio = 1/2, [AIBN] = 1 g/100 g MMA. SE, standard error.

TABLE 4.1.5
Tensile Properties of PU/PMMA SIPN for Various AIBN Concentrations

[AIBN] (g/100 g MMA)	Concentrated Emulsion Polymerization ($R = 1.0$)						Simultaneous Bulk Polymerization ($R = 0.9$)					
	Tensile Strength (MPa)		Elongation (%)		Toughness (MJ/m^3)		Tensile Strength (MPa)		Elongation (%)		Toughness (MJ/m^3)	
	Ave. Value	SE	Ave. Value	SE	Ave. Value	SE	Ave. Value	SE	Ave. Value	SE	Ave. Value	SE
0.6	32.7	3.1	68.3	5.0	17.9	2.4	26.6	2.5	71.8	3.7	16.0	3.0
0.8	34.6	2.9	72.5	6.6	20.8	1.8	26.9	3.4	98.3	4.8	21.8	4.8
1.0	33.7	1.7	74.2	4.9	20.7	2.7	22.7	1.6	110.8	6.1	22.2	2.6
1.2	35.0	3.3	73.8	4.4	20.7	3.1	23.1	0.9	88.0	7.5	19.3	1.9

PU composition: PPG1000II/PPG3000III wt ratio = 2/1, PU/PMMA wt ratio = 1/3. SE, standard error.

of polymerization, the elongation at break of the materials based on the concentrated emulsion method remaining also constant but passing through a maximum for those based on bulk polymerization.

A full interpretation of the above results is difficult, because, even for the simpler linear blends, the contribution of each phase to the mechanical properties is not easy to evaluate. Theoretically, higher content of AIBN may result in higher polymerization rates and lower molecular weights. However, the polymerization of MMA is complicated by the autoacceleration (gel) effect[28] in which the viscosity of the medium plays a role. This effect is additionally enhanced in concentrated emulsions. The maximum that appears in the elongation of the samples prepared via bulk polymerization is probably due to the fact that the proper concentration of AIBN provides the proper initial polymerization rate of MMA which matches the rate of PU formation, thus leading to a system with smaller PMMA domains more uniformly distributed. Because in a simultaneous bulk polymerization the formation of polymers is accompanied by phase separation,[1] only when the rate of MMA polymerization and PU formation are comparable can one avoid the formation of PMMA domains that are too large.

4.1.3.7 Effect of Theoretical Cross-Link Density (M_c)

The effect of the variation of the M_c of the PU on the tensile strength, elongation, and toughness is presented in Table 4.1.6. Both the elongation and the toughness are strongly dependent on M_c, but the tensile strength is almost independent. When the PU network is highly cross-linked (low M_c), the SIPNs have a low elongation and low toughness, and the materials are rather brittle. For lightly cross-linked networks (high M_c), both the elongation and the toughness are again low. However, for moderate cross-linking, there are maxima in elongation and toughness. The samples prepared by different procedures have somewhat different optimum values for M_c. The samples based on simultaneous bulk polymerization exhibit maxima near $M_c = 5{,}000$, whereas those based on concentrated emulsion polymerization have maxima between 4,000 and 5,000. One may note that for $M_c > 5{,}000$, both the elongation and the toughness show a sharp fall in both cases. This is probably due to the increase in the density of hydrogen bonds, which make the PU chains stiffer.

4.1.3.8 Effect of PU/PMMA Ratio

The effect of the PU/PMMA ratio upon the tensile strength, elongation at break, and toughness of the SIPNs is presented in Table 4.1.7. The tensile strength exhibits a sharp decrease from the value 70 MPa[27] of the unmodified PMMA. The elongation at break increases from its value for PMMA (1%–2%) rapidly. Comparing these mechanical properties of the samples prepared via the concentrated emulsion

TABLE 4.1.6
Tensile Properties of PU/PMMA SIPN for Various Values of M_c

PU Composition Wt Ratio of PPG725II/PPG3000III	M_c	Concentrated Emulsion Polymerization (R = 1.0)						Simultaneous Bulk Polymerization (R = 0.9)					
		Tensile Strength (MPa)		Elongation (%)		Toughness (MJ/m³)		Tensile Strength (MPa)		Elongation (%)		Toughness (MJ/m³)	
		Ave. Value	SE	Ave. Value	SE	Ave. Value	SE	Ave. Value	SE	Ave. Value	SE	Ave. Value	SE
3/7	2850	24.7	2.1	35.6	2.7	7.5	1.3	20.5	1.8	43.5	3.2	8.7	1.8
4/6	3333	25.2	1.2	50.7	4.4	10.6	1.9	20.9	1.4	58.1	5.4	11.4	2.5
5/5	4000	24.0	1.5	97.2	7.0	19.7	2.6	17.8	0.9	85.7	6.9	14.3	3.1
6/4	5000	23.6	1.2	98.4	7.7	20.8	2.1	20.7	1.0	125.7	6.5	21.9	2.0
7/3	6666	24.5	1.2	30.5	2.4	6.0	1.1	21.6	1.2	31.7	1.7	6.2	1.3

PU/PMMA wt ratio = 1/2, [AIBN] = 1 g/100 g MMA. SE, standard error.

TABLE 4.1.7
Tensile Properties of PU/PMMA SIPN for Various PU/PMMA WT Ratios

	Concentrated Emulsion Polymerization ($R = 1.0$)						Simultaneous Bulk Polymerization ($R = 0.9$)					
	Tensile Strength (MPa)		Elongation (%)		Toughness (MJ/m^3)		Tensile Strength (MPa)		Elongation (%)		Toughness (MJ/m^3)	
PU/PMMA Wt Ratio	Ave. Value	SE	Ave. Value	SE	Ave. Value	SE	Ave. Value	SE	Ave. Value	SE	Ave. Value	SE
1/2	23.6	1.2	98.4	7.7	20.8	2.1	20.7	0.9	125.7	6.5	21.9	2.0
1/3	30.1	1.4	64.2	4.0	16.4	1.7	25.5	2.3	76.4	3.8	17.8	2.3
1/4	40.5	0.9	52.0	4.2	16.8	2.1	33.0	2.2	63.8	3.9	16.8	2.5
1/5	48.8	1.9	27.8	2.6	10.8	1.5	42.2	1.5	32.1	2.6	11.6	1.6

PU composition: PPG725II/PPG3000III wt ratio = 6/4, [AIBN] = 1 g/100 g MMA. SE, standard error.

polymerization with those prepared via bulk polymerization, one can note that the former has a higher tensile strength, but a lower elongation at break, with the toughness being roughly the same for both. The somewhat different behaviors in mechanical properties can be explained as follows. Because of the presence of water in the concentrated emulsion, a fraction of NCO groups will react with water, and some OH groups at the free ends of the PU molecules will remain unreacted, thus generating defects in the PU network. Because of these defects, additional physical entanglements between the PMMA molecules occur and, as a result, the elongation becomes somewhat lower and the tensile strength larger. In addition, in the concentrated emulsion polymerization, because of the gel effect, the average molecular weight of PMMA is higher than in the bulk polymerization, resulting in a higher physical entanglement. Consequently the tensile strength is somewhat higher and the elongation somewhat smaller. The differences in the tensile properties in the polymers obtained by the two procedures are however small because: the polymerizations of the precursors that lead to PU and PMMA are independent processes; and the polymerization of the concentrated emulsion occurs in solution, the PU formed is swollen in the solvent, and MMA penetrates inside and polymerizes there. Thus an SIPN is formed in a manner similar to that generated in simultaneous bulk polymerization.

4.1.3.9 Morphology of Particles

The micromorphologies of the samples are presented in the SEMs of Figure 4.1.4. All samples appear to be composed of microparticles bound together. In a concentrated emulsion, the volume fraction of the dispersed phase, ϕ, is large (in the present case $\phi = 0.85$). The cells of the dispersed phase are therefore tightly compacted together, and the intercell binding and reaction are inevitable. Perhaps the OH end groups of PU in adjacent cells are bridged by the diisocyanate molecules. Several chemical parameters have an influence on the morphology. The effect of different solvents can be seen from Figure 4.1.4a–c, which represent samples prepared with chloroform, toluene, and a 50/50 v/v mixture of chloroform and toluene, respectively. One can see that the particles in (a) are finer than in (b). Chloroform is a good solvent for both PU and PMMA, toluene is a good solvent for PU and MMA monomer, but a less good solvent for PMMA. For this reason, in toluene more PU segments tend to bind together than in chloroform, resulting in a coarser morphology. The sample based on the 50/50 v/v mixture of chloroform and toluene has an intermediary morphology between (a) and (b). A comparison of (a) and (d) shows the effect of the amount of solvent used. In a more diluted system, the size of the particles is larger than in a less diluted one.

4.1.4 CONCLUSION

SIPN latexes of PU/PMMA were prepared via the concentrated emulsion polymerization method. DSC and TEM investigations revealed that semi-interpenetration was achieved in the obtained material. As the relative proportion of PU/PMMA increases, the elongation at the break point and the toughness increase, but the tensile strength falls. The molar ratio of isocyanate to hydroxyl groups (R) affects the elongation

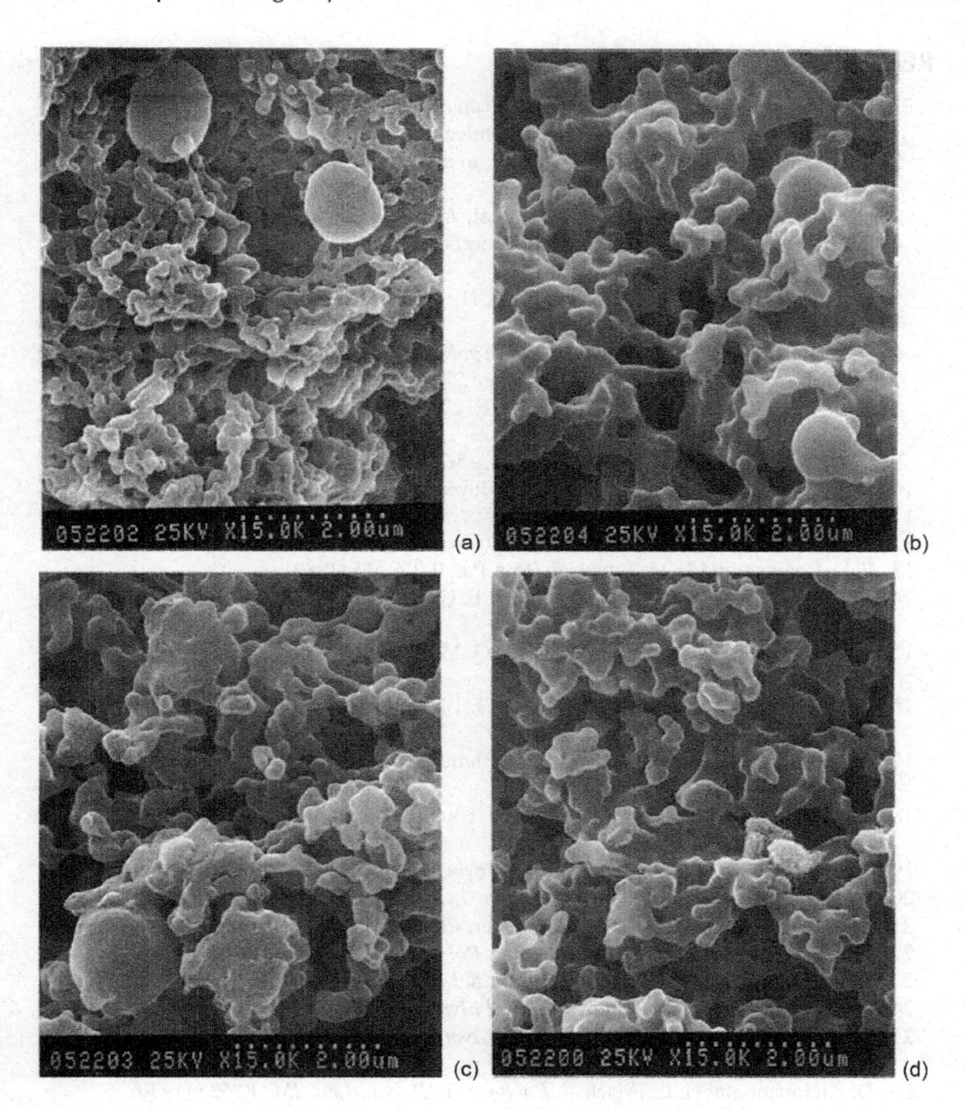

FIGURE 4.1.4 Particles morphologies of SIPN latexes prepared with various solvents: (a) volume ratio of chloroform to PU precursors = 2; (b) volume ratio of toluene to PU precursors = 2; (c) volume ratio of 50/50 v/v chloroform/toluene mixture to PU precursors = 2; (d) volume ratio of chloroform to PU precursors = 3. PU composition: PPG725II/PPG3000III weight ratio = 6/4, PU/PMMA weight ratio = 1/2, [AIBN] = 1 g/100 g MMA, NCO/OH molar ratio = 1.0.

at the break point and the toughness of the SIPN. There is an optimum value of R, 1.0 for concentrated emulsion polymerization and 0.9 for simultaneous bulk polymerization, for which the elongation at the break point and the toughness are maxima. There is also an optimum theoretical crosslink density (M_c) for optimum toughness.

REFERENCES

1. L. H. Sperling, in *Interpenetrating Polymer Networks*; D. Klempner, L. H. Sperling, and L. A. Utracki., Eds.; American Chemical Society, Washington DC, 1994; p. 1.
2. T. Hur, J. A. Manson, R. W. Hertzberg, and L. H. Sperling, *J. Appl. Polym. Sci.*, 39, 1933 (1990).
3. K. C. Frisch, D. Klempner, and S. Migdal, *J. Polym. Sci. Chem.*, 12, 885 (1974).
4. R. H. Sheu, M. S. El–Aasser, J. W. Vanderhoff, *J. Polym. Sci., Polym. Chem. Ed.*, 28, 629 (1990).
5. D. J. Hourston, R. Satgurunathan, and H. C. Varma, *J. Appl. Polym. Sci.*, 34, 901 (1987).
6. D. J. Hourston and J. A. McCluskey, *J. Appl. Polym. Sci.*, 25, 959 (1980).
7. B. Mcgarey and R. W. Richards, *Polymer*, 27, 1315 (1986).
8. K. C. Frisch, D. Klempner, H. X. Xiao, E. Cassidy, and H. L. Frisch, *Polym. Eng. Sci.*, 25, 12 (1985).
9. S. Singh, H. L. Frisch, and H. Ghiradella, *Macromolecules*, 23, 375 (1990).
10. I. Hermant, M. Damyanidu, and G. C. Meyer, *Polymer*, 24, 1419 (1983).
11. A. Morin, H. Djomo, and G. C. Meyer, *Polym. Eng. Sci.*, 23, 394 (1983).
12. P. Tan, and H. Xie, *Hecheng Xiangjiao Gongye*, 1, 180 (1984).
13. F. D. Roemer, and L. H. Tateosian, Eur. Pat. 0,014,515 (1984).
14. G. Allen, M. J. Bowden, D. J. Blundell, F. G. Hutchnson, G. M. Jeffs, and J. Vyvoda, *Polymer*, 14, 597 (1973).
15. G. Allen, M. J. Bowden, D. J. Blundell, G. M. Jeffs, J. Vyvoda, and T. White, *Polymer*, 14, 604 (1973).
16. G. Allen, M. J. Bowden, G. Lewis, D. J. Blundell, and G. M. Jeffs, *Polymer*, 15, 13 (1974).
17. G. Allen, M. J. Bowden, G. Lewis, D. J. Blundell, G. M. Jeffs, and J. J. Vyvoda, *Polymer*, 15, 19 (1974).
18. G. Allen, M. J. Bowden, G. Lewis, D. J. Blundell, G. M. Jeffs, and W. E. A. J. Davis, *Polymer*, 15, 28 (1974).
19. H. Djomo, J. M. Widmaier, and G. C. Meyer, *Polymer*, 24, 1415 (1983).
20. I. Hermant and G. C. Meyer, *Eur. Polym. J.*, 20, 85 (1984).
21. D. Jehl, J. M. Widmaier, and G. C. Meyer, *Eur. Polym. J.*, 19, 597 (1983).
22. D. J. Hourston, and J. Romaine, *J. Appl. Polym. Sci.*, 39, 1587 (1990).
23. E. Ruckenstein and J. S. Park, *J. Polym. Sci., Chem. Lett. Ed.*, 26, 529 (1988).
24. E. Ruckenstein and K. J. Kim, *J. Appl. Polym. Sci.*, 36, 907 (1988).
25. J. Brandrup and E. H. Immergut, Eds. *Polymer Handbook*, 3rd ed., John Wiley & Sons, New York, 1989.
26. D. Klempner and H. L. Frisch, *J. Polym. Sci., Polym. Lett. Ed.*, 8, 525 (1970).
27. J. A. Brydson, *Plastics Materials*, 3rd ed., Newnes-Butterworths, London, UK, 1979.
28. A. M. North, and G. A. Reed, *Trans. Faraday Soc.*, 57, 859 (1961).

4.2 AB Crosslinked Polymer Latexes via Concentrated Emulsion Polymerization*

Hangquan Li and Eli Ruckenstein

Department of Chemical Engineering, State University of New York at Buffalo, Amherst, NY 14260, USA

ABSTRACT A series of polycaprolactone (PCL)/poly(methyl methacrylate) (PMMA) AB crosslinked polymers (ABCP) in the form of latexes were prepared via the concentrated emulsion polymerization method. PCL diols were first reacted with acryloyl chloride in toluene to form a solution of vinyl-terminated PCL. Dissolving the methyl methacrylate monomer and a suitable initiator in this solution, the solution was employed to prepare a concentrated emulsion in water. After completing the polymerization of the concentrated emulsion, latexes of ABCP were obtained, which, depending on their composition, can be either elastomers or tough plastic materials. The toughness of PMMA is greatly improved in both cases. The effects of the PCL chain-length, composition and self-crosslinking of PCL on the mechanical properties were investigated.

4.2.1 INTRODUCTION

In order to improve the properties of polymers, two or more different polymer chains are often combined. There are several ways to combine two kinds of polymer molecules: bonding the end of polymer B to the backbone of polymer A results in a graft copolymer; different chains bonded end to end generate block copolymers. If polymer A is crosslinked by polymer B and not by itself, an AB crosslinked polymer (ABCP)[1–4] is obtained. In both the ABCP and in the interpenetrating polymer network (IPN), two kinds of polymer chains are combined in

* *Polymer*, Vol. 36, No. 11, pp. 2281–2287.

a network structure. The difference between the two is that in an ABCP the two kinds of chains form one single network and in an IPN they form two separate networks. Like IPN, ABCP has received attention in the past 20 years.[4] However, because of crosslinking, the processability of the material prepared by bulk polymerization is poor and consequently its commercial importance has been limited. If ABCP could be prepared as latexes, it could become employed in melt-processing and find wider applications.

In the present work, a novel system, polycaprolactone (PCL)/poly(methyl methacrylate) (PMMA), was prepared in the form of latexes via a recently developed method, the concentrated emulsion polymerization method.[5,6] A concentrated emulsion differs from the conventional one in that the volume fraction ϕ of the dispersed phase is greater than 0.74 (which represents the volume fraction of the most compact arrangements of spheres of the same size) and may be as high as 0.99. The continuous phase has a low volume fraction and for large values of ϕ is in the form of a network of thin liquid films that separate polyhedral drops. Because of its structure, the concentrated emulsion polymerization method possesses a number of advantages:

1. High yield per unit volume: since it is 'concentrated', the volume fraction is higher than 0.74; in contrast, for a conventional emulsion, it is lower than 0.57 (ref. 7).
2. Easy to handle: since the volume fraction of the dispersed phase is very high, the cells from which the emulsion is formed are tightly compacted; this gives the emulsion the appearance and behaviour of a paste.
3. The system is more stable and more uniform than a conventional emulsion.

In previous publications,[8,9] rubber toughened polystyrene and polyvinylidene composites were successfully prepared via the concentrated emulsion polymerization method. For the preparation of ABCP latexes, the same method seems to be the proper one to follow.

The modification of PMMA, which is brittle, was tried for years. The attempts were mainly focused on using polyurethanes based on polyglycol ethers as modifiers.[10–13] However, because of their hydrophilicity, those ethers cannot be employed in the presence of water (as the continuous medium) to prepare concentrated emulsions. In the present work, PCL, which has proven to be a good modifier of poly(vinyl chloride),[14,15] was employed, and a series of ABCP latexes from vinyl-terminated PCL and methyl methacrylate (MMA) monomer were prepared. Since the crosslinking is limited to each latex, the material possesses flowability and can therefore be melt-processed. The relationships between their properties and various structural parameters, e.g., crosslinking density, the length of the soft segments (PCL) and the PCL/PMMA weight ratio, were investigated.

4.2.2 EXPERIMENTAL

4.2.2.1 Materials

The chemicals employed are listed in Table 4.2.1. MMA was distilled before use. Azobisisobutyronitrile (AIBN) was recrystallized from methanol. The other compounds were used as received. Water was deionized and distilled.

4.2.2.2 Preparation Procedure

Preparation of vinyl-terminated PCL. A solution of PCL diol (MW = 2000, 1250 or 530) in toluene (20 g per 100 mL), was placed in a single-necked flask equipped with a magnetic stirring bar. Triethylamine (TEA) (1 mol for each mole of OH groups of the dissolved PCL diol) was added to the solution. The flask was sealed with a rubber septum and the solution subjected to magnetic stirring. Acryloyl chloride (AC) (1 mol for each mole of OH groups) was introduced dropwise with a syringe through the rubber septum, at room temperature. The OH group reacts with AC to form a vinyl group:

$$ROH + Cl{-}CO{-}CH = CH_2 \rightarrow RO{-}CO{-}CH{=}CH_2 + HCl$$

TABLE 4.2.1
Chemicals Used (Purchased from Aldrich)

Chemical	Description	Code
Diols	Poly(caprolactone) diol MW = 530	PCL530
	1250	PCL1250
	2000	PCL2000
Isocyanate	Tolylene 2,4-diisocyanate, 80% remainder 2,6 isomers	TDI
	Trimethylolpropane, carbamate with tolylene diisocvanate, 70 wt%	TMDI
Triethylamine	99%	TEA
Acryloyl chloride	96%	AC
Propionyl chloride	97%	—
Monomer	Methyl methacrylate, 99%	MMA
Initiator	Azobisisobutyronitrile	AIBN
Surfactant	Sodium dodecyl sulfate salt, 70%	SDS
Solvents	Toluene, 99%	

The HCl molecules produced in the above reaction react with TEA:

$$HCl + N(Et)_3 \rightarrow N(Et)_3HCl$$

and the formed quaternary ammonium salt remains suspended in the system as a white powder. Removing the white powder by filtering, a toluene solution of vinyl-terminated PCL (VTPCL) was obtained.

VTPCLx (with x = 530, 1250, 2000) was prepared by this procedure, x denoting the average molecular weight of the PCL diol employed.

If the molar amounts of TEA and AC used are only half the molar amount of OH groups, semi-vinyl-terminated PCL (SVTPCL) can be obtained. In this case, AC must be diluted to 0.1 g ml^{-1} toluene before being introduced, and the addition of the AC solution must be as slow as one drop per 2 s in order to increase the uniformity of distribution of the remaining OH groups.

VTPCL3250 and 4000. Introducing the proper amount of tolylene 2,4-diisocyanate (TDI) into a solution of SVTPCL2000 in toluene, pairs of SVTPCL2000 become connected by individual TDI molecules and VTPCL4000 is obtained. Of course, 4000 represents the average molecular weight of VTPCL. VTPCL 3250 was obtained in the same way, starting from a toluene solution of SVTPCL2000 and 1250.

VTPCL6000 and 5250. These compounds were prepared in two steps. In the first step, a solution of SVTPCL2000 in toluene was allowed to react with TDI (NCO/OH = 2/1), for 4 h. Because of the excess of TDI, it is likely that the OH groups of SVTPCL will react with different TDI molecules. In the second step, PCL diol 2000, in a molar amount half that of SVTPCL2000, was introduced. Each molecule of PCL diol 2000 can react with two molecules of isocyanate-terminated SVTPCL2000 to generate VTPCL6000. The reaction lasted 24 h. When PCL diol 1250 replaced PCL diol 2000, VTPCL5250 was obtained.

It should be emphasized that the two steps are absolutely necessary, since if TDI and PCL diol had been introduced at the same time, polyurethane of ultra-high MW would have been formed and the solution would have had too high a viscosity to be usable in preparing concentrated emulsions. More importantly, the expected molecular weight could not have been obtained.

Uni-vinyl-terminated PCL2000. Reacting VTPCL2000 with propionyl chloride, the OH groups were replaced by ethyl groups and the uni-vinyl-terminated PCL2000 (UVTPCL2000) was obtained.

Concentrated emulsion polymerization. In the modified PCL polymer toluene solutions (0.2 g/mL toluene), MMA monomers in various proportions and a suitable initiator (AIBN, 0.008 g/g MMA) were introduced. The solution thus prepared was used as the dispersed phase of a concentrated emulsion. An aqueous solution of sodium dodecyl sulfate (SDS) (10 wt%) was first placed in a single-necked

100 mL flask provided with a mechanical stirrer. Then, the solution containing the modified PCL and MMA was added dropwise with vigorous stirring into the flask with a syringe, until the volume fraction of the SDS aqueous solution became 0.2. The whole addition process lasted about 15 min and took place at room temperature. The gel-like concentrated emulsion thus formed was additionally stirred for 15 min under a flow of nitrogen, and finally transferred to a tube of 30 mL capacity. The tube was sealed with a rubber septum and introduced into a water bath at 60°C to carry out the copolymerization of MMA and modified PCL, which lasted 48 h. The product thus obtained was washed with methyl alcohol in an extractor for 24 h and dried in a vacuum oven for another 24 h.

Preparation of semi-simultaneous interpenetrating network (semi-IPN). The semi-IPN was prepared by simultaneous bulk polymerization. The reactants (PCL diol 2000, trimethylolpropane with TDI (TMDI), MMA and initiator, free of solvent) were mixed in a tube which was introduced into a Cole Parmer ultrasonic mixer at 60°C for 3 h. After 3 h the viscosity of the system became so high that further mixing was no longer possible. Consequently, the tube was transferred into a water bath of 60°C, where it was kept for 48 h.

4.2.2.3 Tensile Testing

The powders of the products of the concentrated emulsion polymerization, or the bulk material for the products of bulk polymerization, were hot-pressed in a Laboratory Press (Fred S. Carver Inc.) at 180°C for 3–5 min, and then cooled to room temperature. The sheets thus obtained were cut to the size required by ASTM D.638–58T. The tensile testing was conducted at room temperature with an Instron Universal Testing Instrument (model 1000). The elongation speed of the instrument was 20 mm min^{-1}.

4.2.2.4 Molecular Weight (MW) and Molecular Weight Distribution (MWD)

These properties were determined by gel permeation chromatography (GPC.; Waters 150°C). The determination was carried out at 30°C with tetrahydrofuran as the mobile phase.

4.2.2.5 Thermal Transition Temperatures

The thermal transitions of the samples were measured by differential scanning calorimetry (DSC) with a Perkin-Elmer DSC instrument. Each sample was heated from –70 to 210°C, with a heating rate of 10°C min^{-1}.

4.2.2.6 Scanning Electron Microscopy

The particle morphology was examined by scanning electron microscopy (SEM; Hitachi S-800).

4.2.3 RESULTS AND DISCUSSION

4.2.3.1 The MW and MWD of the VTPCL

The MW and MWD of VTPCL are listed in Table 4.2.2. Since the range of MWs of the VTPCLs is low (500–6000) and their Mark-Houwink constants are not available, the polystyrene standards could not be used for calibration. However, the PCL diols provided by the supplier were prepared via anionic ring-opening polymerization and have, for this reason, a narrow MWD (M_w/M_n = 1.4–1.6); consequently they have been used as standards. One can see from Table 4.2.2 that the number average MW values of the prepared VTPCLs do not deviate appreciably from the designed values, being only somewhat smaller. VTPCL530 constitutes an exception, because the MW (530) is smaller than the lower limit of GPC sensitivity (about 800). One may also note the wider distributions of VTPCL3250 and VTPCL5250. Since they were prepared by connecting species of different MWs, broader distributions are expected.

TABLE 4.2.2
MW and MED of VTPCLx

		Determined MW		
x	Designed MW	Number Average MW	Weight Average MW	M_w/M_n
530	580	610	1160	1.9
1250	1300	1220	2070	1.7
2000	2050	2000	3200	1.6
3250	3470	2900	7250	2.5
4000	4220	4090	6140	1.5
5250	5470	4850	14060	2.9
6000	6390	5970	11940	2.0

4.2.3.2 DSC Measurements

The PCL homopolymer has a crystalline structure, and using DSC we determined its melting point to be near 40°C. The PMMA homopolymer is amorphous, with a glass transition temperature (T_g) of about 115°C (ref. 16). However, the DSC diagrams of the ABCP samples exhibited neither of these thermal transitions; instead they showed a diffuse thermal transition region between about 50°C and 75°C. In the PCL/PMMA ABCP, different segments are bonded to each other. One may consider that the chains in the network constitute a kind of copolymer with very long sequences. Consequently, the PCL segments can no longer crystallize and the T_g of PMMA is lowered by its combination with the PCL segments. The different long sequences move together and as a result a common thermal transition region is generated. The presence of a single thermal transition also constitutes the proof that an AB network was indeed formed. The absence of a T_g near 115°C indicates that very little (if any) PMMA homopolymer was formed.

4.2.3.3 Stress-Strain Curves

Two sets of stress-strain curves are presented in Figures 4.2.1 and 4.2.2. The samples of Figure 4.2.1 have the same PCL/PMMA weight ratio but different MWs of PCL, and those of Figure 4.2.2 were prepared with the same PCL (PCL1250) but with various PCL/PMMA weight ratios. In Figure 4.2.1 one can see that the curve for PCL1250 has a pronounced yield point and that for PCL2000 has a smeared yield point. The curves for the PCLs with higher MWs have no yield point. The yield point is characteristic of a plastic material with moderate physical crosslinking. In the present systems, PCL has only non-polar, soft, relatively short segments and the physical crosslinking can be generated only by the polar segments of PMMA. By introducing the PCL segments among the PMMA chains, the inter-PMMA chain distance is increased and the density of physical crosslinks is reduced. One can conclude that, for a weight ratio PCL/PMMA = 4/6, if the MW of PCL is higher than about 2000, the physical crosslinking entirely disappears and the material behaves like a rubber. From Figure 4.2.2 one can see that the amount of PCL introduced affects the physical crosslinking. When the PCL/PMMA weight ratio is 5/5, too many PCL molecules keep the PMMA chains separated from each other, and the stress-strain curve exhibits only a smeared yield point. When the PCL/PMMA weight ratio is 2/8 or lower, there is strong physical crosslinking and the samples break before the yield point is reached. For the intermediary ratios of 4/6 or 3/7, clear yield points can be observed.

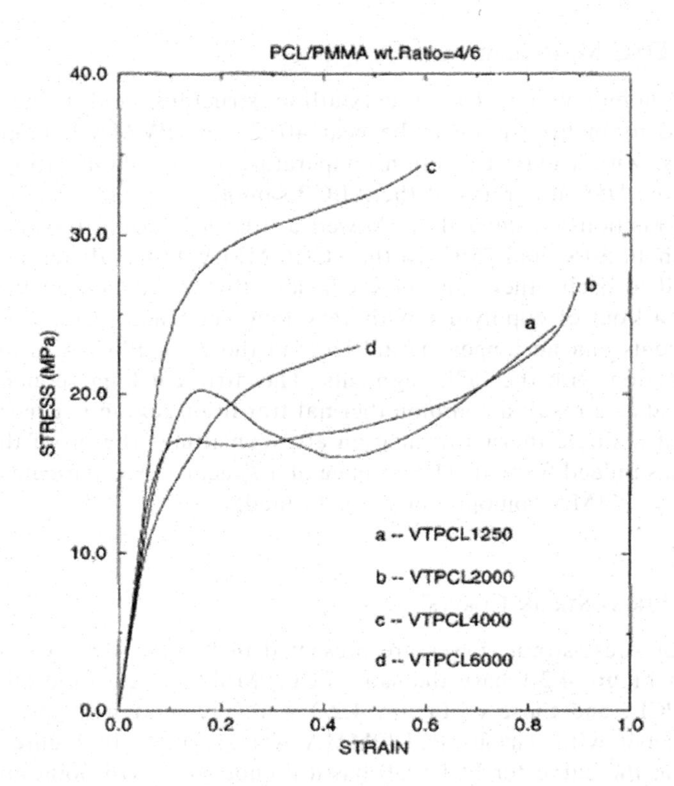

FIGURE 4.2.1 Stress-strain curves of ABCP (PCL/PMMA weight ratio = 4/6): (a) VTPCL 1250; (b) VTPCL2000; (c) VTPCL4000; (d) VTPCL6000.

From these stress-strain curves one can conclude that, depending on composition, the PCL/PMMA ABCPs can be either tough plastic materials or relatively soft materials.

4.2.3.4 Effect of PCL Free Ends

By free ends we understand the ends of PCL chains which are not bonded to the PMMA chains. The samples listed in Table 4.2.3 were prepared via the copolymerization of MMA with various proportions of bi-vinyl-terminated PCL (VTPCL) and uni-vinyl-terminated PCL (UVPCL). Sample no. 1 constitutes a full ABCP with no free ends, and sample no. 5 has each of the PCL chains with one end free. (It should be noted that the full ABCP is only an ideal limit that involves a conversion of 100%). Table 4.2.3 shows that the presence of free ends plays a negative role regarding, in particular, the elongation at break. Sample no. 2 has the optimum tensile strength. Sample no. 5, the graft copolymer, has the lowest tensile strength and is the most brittle. From the effect of the free ends one can conclude that without chemical crosslinking the PCL and PMMA grossly phase separate and the mechanical properties of the composite are poor.

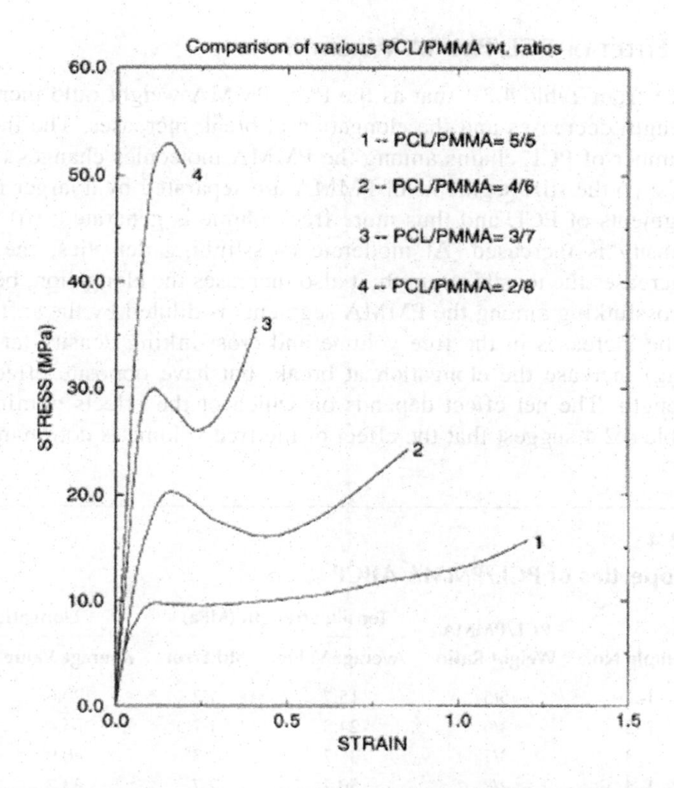

FIGURE 4.2.2 Stress-strain curves of ABCP made from VTPCL1250. PCL/PMMA weight ratio: 1, 5/5; 2, 4/6; 3, 3/7; 4, 2/8.

TABLE 4.2.3
Effect of Free Ends on Tensile Properties of PCL/PMMA ABCP

Sample No.	VT PCL2000/ UVTPCL2000 Weight Ratio	Tensile Strength (MPa) Average Value	Tensile Strength (MPa) Std Error	Elongation (%) Average Value	Elongation (%) Std Error
1	0/1	27.0	1.5	90.7	4.1
2	0.25/0.75	31.5	0.9	72.1	5.2
3	0.5/0.5	28.3	1.1	48.7	3.3
4	0.75/0.25	21.8	2.4	22.2	3.0
5	1/0	18.3	2.2	12.1	1.7

Notes: PCL/PMMA weight ratio = 4/6.

4.2.3.5 Effect of PCL/PMMA Weight Ratio

One can see from Table 4.2.4 that as the PCL/PMMA weight ratio increases, the tensile strength decreases and the elongation at break increases. The inclusion of a larger number of PCL chains among the PMMA molecules changes the system in two ways: (i) the stiff segments of PMMA are separated by a larger number of flexible segments of PCL and thus more free volume is generated; (ii) the crosslinking density is increased. At moderate crosslinking densities, the chemical bonding increases the tensile strength; it also increases the elongation, because the physical crosslinking among the PMMA segments is diluted by the soft segments of PCL. The increases in the free volume and crosslinking density (at moderate crosslinking) increase the elongation at break, but have opposite effects on the tensile strength. The net effect depends on which of the effects dominates. The data of Table 4.2.4 suggest that the effect of the free volume is dominant. At high

TABLE 4.2.4
Tensile Properties of PCL/PMMA ABCP

		PCL/PMMA	Tensile Strength (MPa)		Elongation (%)	
VTPCL	Sample No.	Weight Ratio	Average Value	Std Error	Average Value	Std Error
1250	1–1	5/5	15.7	1.2	120.6	4.4
	1–2	4/6	24.3	1.7	85.4	5.0
	1–3	3/7	35.7	2.3	40.9	3.2
	1–4	2/8	50.7	2.7	23.2	2.2
2000	2–1	5/5	18.5	0.9	130.5	7.7
	2–2	4/6	27.0	1.5	90.7	4.1
	2–3	3/7	37.2	2.8	43.3	2.7
	2–4	2/8	54.6	2.0	30.8	2.5
3250	3–1	5/5	21.6	1.5	67.7	4.2
	3–2	4/6	32.8	1.9	33.1	3.0
	3–3	3/7	42.1	2.5	17.0	1.1
	3–4	2/8	49.7	3.6	11.3	1.2
4000	4–1	5/5	25.9	1.2	110.8	6.3
	4–2	4/6	34.3	0.4	59.9	2.1
	4–3	3/7	41.1	2.6	30.6	2.3
	4–4	2/8	49.7	2.3	19.3	1.7
5250	5–1	5/5	22.2	1.4	51.1	4.3
	5–2	4/6	32.1	1.2	27.8	2.0
	5–3	3/7	38.6	2.1	15.6	1.1
	5–4	2/8	46.7	3.2	10.2	0.8
6000	6–1	5/5	17.2	0.8	77.6	5.4
	6–2	4/6	23.0	1.1	47.9	3.2
	6–3	3/7	32.5	1.6	29.4	1.3
	6–4	2/8	38.8	1.8	15.5	1.4

crosslinking (such circumstances appear to be outside the range considered in Table 4.2.4), it is expected that tensile strength increases and elongation at break decreases with increasing crosslinking.

4.2.3.6 Effect of the MW of the PCL Chains

Table 4.2.4 shows that, for the range of MWs of VTPCL employed, both the tensile strength and the elongation at break pass through a maximum as the MW of PCL changes. The maximum elongation at break is exhibited by PCL2000. For any given PCL/PMMA weight ratio, the increase in the MW of PCL decreases the crosslinking density. At large crosslinking densities, the free volume in the system is expected to grow as the density decreases. At low crosslinking densities, the free volume is expected to decrease as the density decreases, since the PMMA and VTPCL molecules can arrange in a more compact manner. This explains the occurrence of the maximum in the elongation at break. The samples based on VTPCL530 have a high crosslinking density, since the PCL chains are short; as a result, the material lacks flowability and could not be tested.

The effect on the tensile strength is somewhat more complex. At a given MW of VTPCL, the tensile strength increases as the ratio of PCL/PMMA decreases from 5/5 to 2/8 because the dominance of PMMA increases the strength. At a given ratio PCL/PMMA and low MW PCL, the tensile strength is expected to decrease as the MW of the VTPCL increases, because the number of crosslinks decreases; it attains a minimum (the minimum was not observed in our experiments because it is probably located below MW = 1250) and increases again, because the fewer PCL molecules allow for stronger interactions between the PMMA chains. At moderately high MW of VTPCL, the tensile strength is expected to decrease again with increasing MW because of the fewer crosslinks and increasing free volume among the PMMA chains.

Not only the MW, but also the MWD of PCL plays an important role in the elongation at break. Table 4.2.4 shows that the elongation at break of the samples based on PCL3250 or PCL5250 is unusually low. From Table 4.2.2 one can note that the polydispersity coefficients (M_w/M_n) of PCL3250 and PCL5250 are 2.5 and 3.0, respectively, which are much larger than those of the other PCLs (1.5–2.0). The broader distribution of the molecular weights results in non-uniformities in the network which may be the reason for the low elongation at break.

4.2.3.7 Effect of the Self-Crosslinking of PCL

The effect of self-crosslinking of PCL is presented in Table 4.2.5. Samples nos 1–4 (Table 4.2.5) were prepared as follows. Different percentages of OH groups present in the PCL diol 2000 were vinylized first, the remaining OH groups being crosslinked with TMDI, which is a trifunctional crosslinker. The crosslinked VTPCL was copolymerized with MMA molecules in a concentrated emulsion. Sample no. 5 constitutes an extreme case, representing a semi-IPN (linear PMMA penetrating a network of PCL) prepared by simultaneous bulk polymerization, in which the combination

TABLE 4.2.5
Effect of Self-Crosslinking on Tensile Properties of PCL/PMMA ABCP

Sample No.	Percentage of the OH Groups Vinylized (%)	Tensile Strength (MPa)		Elongation (%)	
		Average Value	Std Error	Average Value	Std Error
1	100	27.0	1.5	90.7	4.1
2	50	32.5	2.0	55.7	3.3
3	33	29.8	2.4	44.0	2.8
4	25	28.4	1.7	42.9	3.6
5	0	29.3	1.9	45.2	3.2

Notes: PCL/PMMA weight ratio = 4/6.

of PCL and PMMA is achieved through semi-interpenetration. Sample no. 1 is the same as the sample no. 2–2 of Table 4.2.4. It is supposed to be a full ABCP with all chain ends bound to PMMA chains. Samples nos 2, 3, 4 and 5 represent systems with decreasing percentage of OH groups vinylized, hence with increasing PCL self-crosslinking. The increase in the self-crosslinking increases the interpenetration of PMMA in the PCL network. One may note that sample no. 1 has the highest elongation. As the percentage of the vinyl groups decreases, the tensile strength remains essentially the same, but the elongation at break decreases rapidly. This indicates that the crosslinking via the vinyl groups plays a more important role in the elongation than the interpenetration. For compatible systems there are no major differences between crosslinking of PMMA with PCL and interpenetration. However, for incompatible systems, the chemical crosslinking between PMMA and PCL plays a greater role than interpenetration. It was reported that for the poly(propylene glycol)/PMMA composites, the compatibility of the IPN is poorer than that of the corresponding ABCP.[14,15] One can also see from Table 4.2.5 that the tensile properties of samples nos 3, 4 and 5 are comparable. This means that, when the density of chemical crosslinking between PMMA and PCL is lower than some critical level, its effect is no longer important.

4.2.3.8 Particle Morphology

The particle morphologies of the PCL/PMMA ABCP latexes are presented in the scanning electron micrographs of Figure 4.2.3 for two PCL/PMMA weight ratios. The samples of Figure 4.2.3a and c are based on VTPLC2000 and those of Figure 4.2.3b and d on VTPCL4000. Figure 4.2.3c and d show that the size of the latexes is in the range 0.5–2 μm. Because of the soft, flexible PCL, the latexes are bound to each other. The greater the amount of PCL, the stronger the latex binding. The particles prepared with VTPLC4000 are somewhat coarser than those prepared with VTPCL2000.

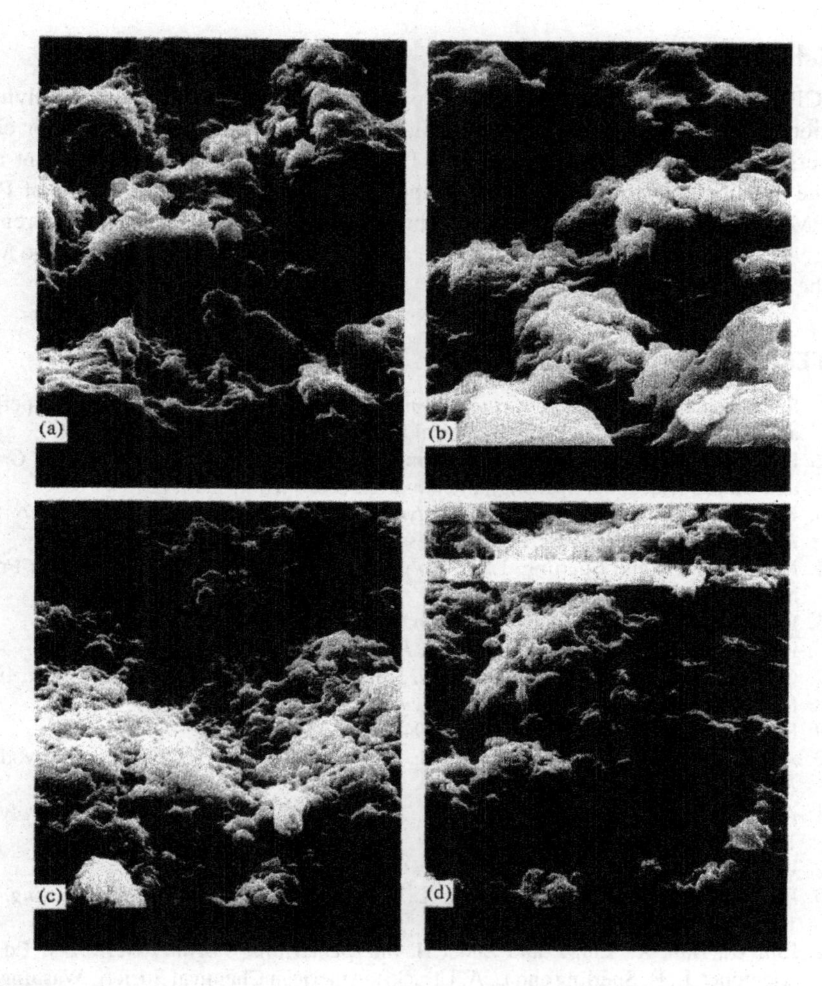

FIGURE 4.2.3 Particle morphology of PCL/PMMA ABCP. PCL/PMMA weight ratio: (a), (b), 5/5: (c), (d), 3/7.

4.2.3.9 A Comment on Flowability

As already noted in the Introduction, the use of the concentrated emulsion method instead of bulk polymerization provides latexes which have flowability and can therefore be used in melt-processing. A sufficiently high molecular weight of the PCL constitutes another important factor that affects the flowability of the final material. Since sufficiently long PCL chains are flexible, they can be easily deformed during processing. This flexibility also affects the toughness of the material. Short PCL chains will result in low flowability and toughness. The samples based on VTPCL530 constitute an example. Since the PCL chains are in this case short, the material had no flowability during the thermopressing and a sheet could not be generated.

4.2.4 CONCLUSION

ABCP latexes of PCL/PMMA were prepared via concentrated emulsion polymerization method. Depending on composition, they are either tough plastics or elastomers. The chemical bonding between PMMA and PCL plays an important role in the mechanical properties of the composites. As the relative proportion of PCL in PMMA increases, elongation at the break point increases, but tensile strength decreases. The best tensile properties are associated with an optimum average MW of the PCL chains.

REFERENCES

1. Sperling, L. H. in *Interpenetrating Polymer Networks* (Eds O. Klempner, L. H. Sperling and L. A. Utracki), American Chemical Society, Washington, DC, 1994, p. 1.
2. Bamford, C. H. and Eastmond, G. C. in *Recent Advances in Polymer Blends, Grafts and Blocks* (Ed. L. H. Sperling), Plenum Press, New York, 1974, p. 165.
3. Klempner, D. and Frisch, K. C. *Polymer Alloys II: Blends, Blocks, Grafts and Interpenetrating Networks*, Plenum Press, New York, 1980, p. 185.
4. Manson, J. A. and Sperling, L. H. *Polymer Blends and Composites*, Plenum Press, New York, 1976, Ch. 8.
5. Ruckenstein, E. and Park, J. S. *J. Polym. Sci., Chem. Lett. Edn* 1988, **26,** 529.
6. Ruckenstein, E. and Kim, K. J. *J. Appl. Polym. Sci.* 1988, **36,** 907.
7. Odian, G. *Principles of Polymerization*, 3rd edn, John Wiley & Sons, New York, 1993.
8. Ruckenstein, E. and Li, H. *J. Appl. Polym.***Sci.** 1994, **52,** 1949.
9. Ruckenstein, E. and Li, H. *Polymer* 1994, **35,** 4343.
10. Allen, G., Bowden, M. J., Blundell, D. J., Hutchinson, F. G., Jeffs, G. M. and Vyvoda, J. *Polymer* 1973, **14,** 597.
11. Allen, G., Bowden, M. J., Blundell, D. J., Jeffs, G. M., Vyvoda, J. and White, T. *Polymer* 1973, **14,** 604.
12. Djomo, H., Widmaier, J. M. and Meyer, G. C. *Polymer* 1983, **24,** 1415.
13. Frisch, K. C., Klempner, D., Xiao, H. X., Cassidy, E. and Frisch, H. L. *Polym. Eng. Sci.* 1985, **25,** 12.
14. Liu, W., Han, X., Liu, J. and Zhou, H. *Interpenetrating Polymer Networks* (Eds O. Klempner, L. H. Sperling and L. A. Utracki), American Chemical Society, Washington, DC, 1994, p. 571.
15. Liu, J., Liu, W., Zhou, H., Hou, C. and Ni, S. *Polymer* 1991, **32,** 1361.
16. Brandrup, J. and Immergut, E. H. (Eds) *Polymer Handbook*, 3rd edn, John Wiley & Sons, New York, 1989.

4.3 Concentrated Emulsions Pathways to Polymer Blending*

Eli Ruckenstein and Jun Seo Park

Department of Chemical Engineering, State University of New York at Buffalo, Buffalo, NY 14260, USA

ABSTRACT A new method of preparation of polymer composites involving the concentrated emulsion polymerization is described. In this kind of emulsion, the volume fraction of the dispersed phase is very large (as large as 0.99), while the volume fraction of the continuous phase is very small. In the present case, a monomer containing an appropriate initiator constitutes the dispersed phase and a dilute solution of surfactant in water constitutes the continuous phase. In a first step, two such concentrated emulsions containing different monomers were prepared and each of them was subjected to heating at 40°C for partial polymerization. Subsequently, the two partially polymerized systems were mechanically mixed, and the mixture was subjected to additional polymerization, drying, and sintering by heating at various temperatures for various time intervals. Partially polymerized concentrated emulsions of polystyrene, poly(butyl methacrylate), poly(butyl acrylate), and crosslinked polystyrene, whose conversions were less than 5%, were employed. Conversions higher than 5% led to large increases in the viscosity of the concentrated emulsions, making their mixing difficult. NMR spectroscopy was used to obtain information about the extent of copolymerization between the two monomers. Electron microscopy examination of the surfaces obtained by the fracture of the composites revealed that the latex particles aggregated with relatively slight changes in size and shape.

4.3.1 INTRODUCTION

Multiphase polymer blends of two or more structurally dissimilar polymers often exhibit a synergistic behaviour due to their different chemical structure and to the existence of various phase domains.[1–3] Because of their superior mechanical

* *POLYMER*, 1990, Vol 31, 2397–2401, (1990).

properties, such as high impact strength, the combination of dissimilar polymers by chemical or physical methods has been frequently exploited. The chemical methods include the copolymerization,[4,5] the interpenetration of two networks,[6] and the concentrated emulsion[7] and colloidal[8] pathways, while the physical method[9] involves mechanical blending, without covalent bonding, between the different species. The preparation method affects the morphology of the multiphase polymer composite and plays an important role in its physical properties. Of course, the properties of the polymer composites are also dependent on the properties of the component polymers.

In this paper, a new approach to composites, involving partially polymerized concentrated emulsions and their subsequent blending, is described. Partially polymerized concentrated emulsions of polystyrene (PS), poly(butyl methacrylate) (PBMA), poly(butyl acrylate) (PBA), and crosslinked polystyrene (PS) were prepared by the concentrated emulsion polymerization method.[7,10] In concentrated emulsions, which have the appearance of gels, the volume fraction of the dispersed phase (in the present case a monomer containing an appropriate initiator) can be as high as 0.99 and the continuous phase (in the present case water containing surfactant) is in the form of a network of thin liquid films that separate polyhedral cells of the dispersed phase. In a first step, the concentrated emulsion is partially polymerized by heating at 40°C until a conversion of about 5% is achieved. The length of this heating time depends upon the nature of the monomer. A greater conversion should be avoided because it increases the viscosity of the system, making the next step, the blending, very difficult. A composite can be prepared by blending two or more different partially polymerized concentrated emulsions and then subjecting the mixture to additional polymerization, followed by drying and sintering. The additional polymerization involves some copolymerization, which might occur because of the coalescence of the latex particles, and/or because of copolymerization near the surfaces of contact between the latex particles. Information about copolymerization was obtained by NMR spectroscopy. The microstructure of the polymer composites was investigated by scanning electron microscopy. The mixing of the gels without preliminary partial polymerization, followed by polymerization, led to the formation of almost only copolymer and not polymer composites.

4.3.2 EXPERIMENTAL

4.3.2.1 Materials and Preparation

Concentrated emulsions of styrene (Aldrich), butyl methacrylate (Aldrich), and butyl acrylate (Aldrich) were prepared at room temperature by dropwise addition of vacuum distilled monomers that contained recrystallized azobisisobutyronitrile (AIBN, Alfa) to stirred distilled and deionized water containing sodium dodecyl sulphate (SDS, Aldrich). For the preparation of crosslinked PS latexes, a mixture of distilled divinyl benzene (Polysciences) and styrene was employed. A concentrated emulsion mildly packed by centrifugation into a 15 mL capacity tube was immersed into a 40°C water bath for partial polymerization, in the presence of air. Two different partially polymerized gels were mixed by magnetic stirring, in the presence of air, at room temperature. The mixture of gels thus obtained was further heated at 40°C for 24 h for completing the polymerization. Finally, the polymer composite thus obtained was slowly dried at 80°C for 24 h and then sintered at 120°C for 6 h in a temperature controlled oven.

4.3.2.2 NMR Measurements

The solutions for NMR measurements were prepared by dissolving 0.15 g of purified and well dried polymer in 3 mL chloroform-d ($CDCl_3$, Aldrich). The polymer solutions were introduced into pyrex NMR tubes. A Varian Gemini-300, 300 MHz proton *FT* NMR instrument was employed.

4.3.2.3 Molecular Weight Measurements

Gel permeation chromatography (GPC Waters) was used to determine the molecular weight of the polymer with methylene chloride (Aldrich) as the mobile phase. Very dilute solutions (0.2 g L^{-1}) of polymer in chloroform (Aldrich) were prepared and injected in the g.p.c. The calibration curve of the GPC was obtained by using polystyrene standards (Polymer Laboratories) with molecular weights in the range of 10^4–12×10^6.

4.3.2.4 Electron Microscopy Investigation

Polymer latexes, obtained by dispersing small amounts of partially polymerized gels in water, were examined with a SEM (Amray 100A). The specimens were prepared by placing drops of highly diluted latex solutions on clean cover glasses. After drying they were coated with gold. Surfaces obtained by fracturing polymer composites of crosslinked PS and PBMA were contacted for 12 h with acetone (Aldrich), which is a good solvent for PBMA. Both the fractured and solvent treated surfaces were coated with gold before examination.

4.3.3 RESULTS AND DISCUSSION

Figure 4.3.1, in which the rate of the concentrated emulsion polymerization of PS, PBMA, and PBA is compared with that of the bulk polymerization, shows that the former rate of polymerization is higher than the latter. The molecular weights of the obtained polymers, listed in Table 4.3.1, are greater for the concentrated emulsion polymerization method. Both the higher molecular weight and the higher reaction

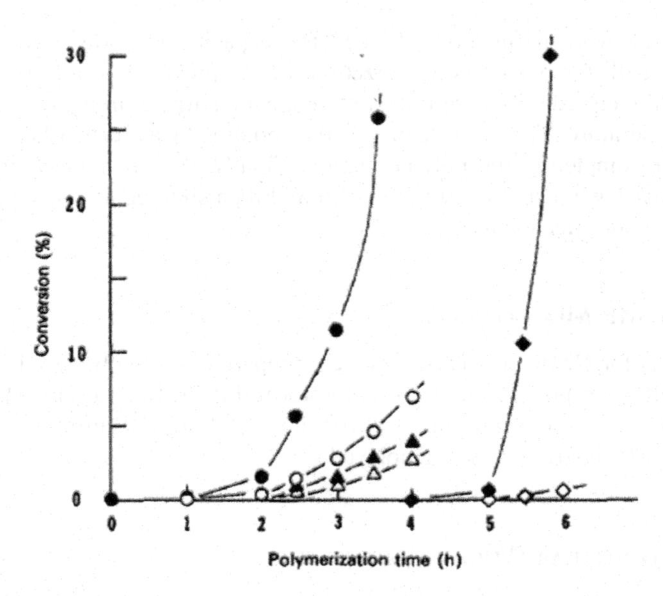

FIGURE 4.3.1 Polymer conversion at 40°C against polymerization time. ●, Styrene; ▲, butyl methacrylate; and ◆, butylacrylate in concentrated emulsion polymerization (monomer 50 mL, AIBN 0.30 mol L^{-1} monomer, SDS 0.35 g, water 4 mL). ○, Styrene; Δ, buthyl methacrylate; and ◊, butylacrylate in bulk polymerization (monomer 40 mL, AIBN 0.30 mol L^{-1} monomer).

TABLE 4.3.1
Molecular Weights Obtained by Concentrated Emulsion Polymerization (monomer 50 mL, AIBN 0.30 mol L^{-1} monomer, SDS 0.35 g, water 4 mL) and by Bulk Polymerization (monomer 50 mL, AIBN 0.30 mol L^{-1}) at 40°C for 8 h of Polymerization

	Molecular Weight	
	Bulk	Concentrated Emulsion
Polystyrene	6.7×10^5	4.2×10^6
Poly(butyl methacrylate)	2.1×10^4	6.1×10^4
Poly(butyl acrylate)	3.2×10^4	8.5×10^4

rate are probably due to a gel like effect induced by the higher rigidity of the monomer in the emulsion cells.[10] The concentrated emulsion becomes more viscous with increasing conversion and transforms into a solid at high conversion. A too high viscosity produces difficulties in the uniform mixing of two different partially polymerized concentrated emulsions.

Two different partially polymerized concentrated emulsions were blended under high shear. The partially polymerized latex particles rearrange and aggregate during the blending process. During the additional heating of the mixture of the two partially polymerized gels at 40°C for 24 h, polymerization inside the latex particles, copolymerization inside the coalesced particles, and/or copolymerization near the surfaces of contact between particles take place. The extent of copolymerization in the polymer composite was investigated by employing high resolution NMR spectroscopy. Figure 4.3.2 presents the proton NMR spectra of the polymers. Figure 4.3.2a and b are for the PS and PBMA, respectively, while Figure 4.3.2c is for a polymer blend of

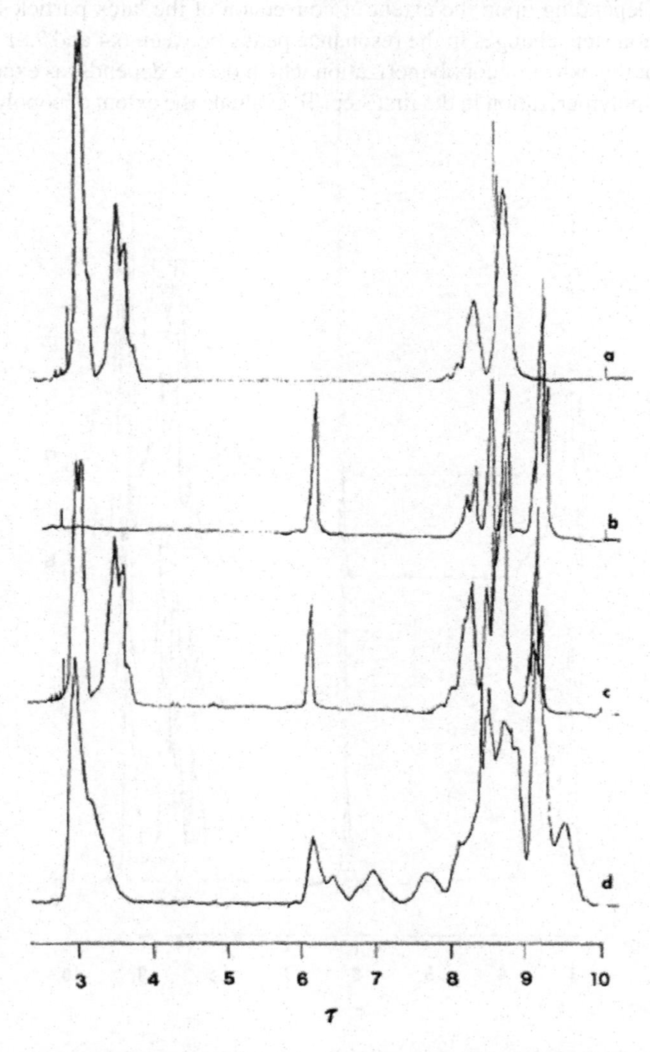

FIGURE 4.3.2 Proton NMR spectra of curve a, polystyrene; curve b, poly(butyl methacrylate), curve c, polymer blend of equal weight of polystyrene and poly(butyl methacrylate) prepared by solution blending, and curve d, copolymer of equal weights of styrene and butyl methacrylate.

equal weights of PS and PBMA prepared by solution blending. The resonance peaks between 2.8 and 3.8τ in Figure 4.3.2a are due to the phenyl protons of the styrene units. The proton NMR spectrum of the copolymer of equal weights of styrene and butyl methacrylate, is given in Figure 4.3.2d, where the resonance peaks between 6.4 and 7.8τ are due to the butoxyl protons. Similar complexities (several peaks) in the methoxyl region of the spectra of the copolymers styrene-methyl methacrylate[11] and styrene-methylacrylate have been reported.[12]

Figure 4.3.3 presents the proton NMR spectra of the polymer composites prepared by blending equal weights of two different partially polymerized concentrated emulsions. Depending upon the extent of conversion of the latex particles in the first polymerization step, changes in the resonance peaks between 6.4 and 7.8τ occur. This indicates that the extent of copolymerization which occurs depends, as expected, upon the extent of polymerization in the first step. To evaluate the extent of copolymerization

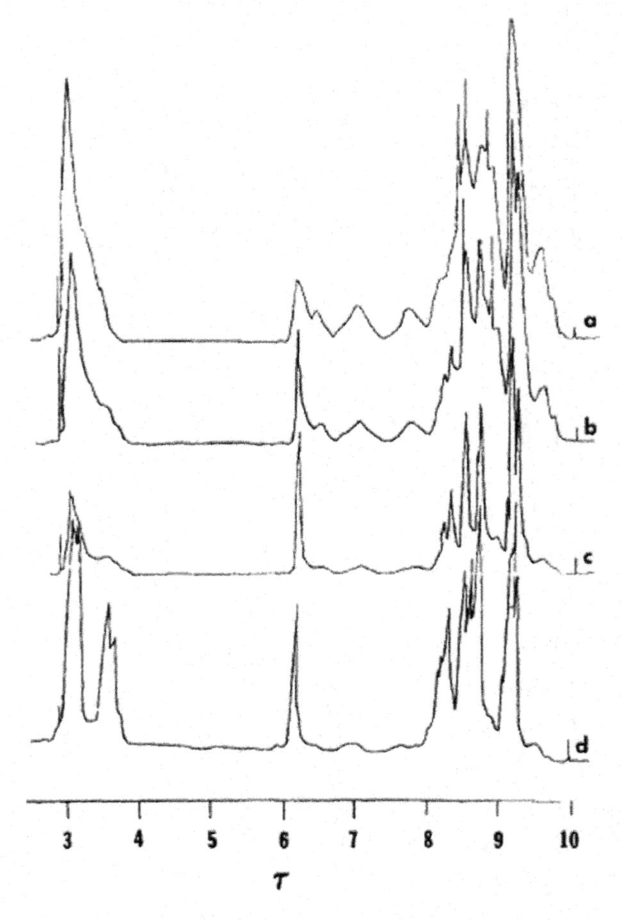

FIGURE 4.3.3 Proton NMR spectra of polymer composite of polystyrene and poly(butyl methacrylate), prepared from equal weights mixtures of partially polymerized concentrated emulsions. Curve a, 0%; curve b, 1%; curve c, 3.5%; curve d, 5% conversions of the monomers.

TABLE 4.3.2
The Ratio of the NMR Peaks Area of Buthoxyl Protons Resonance to Phenyl Protons Resonance

	Peaks Area between 6.4 and 7.8τ/ Peaks Area between 2.8 and 3.8τ
Copolymer, Figure 4.3.2d	0.41
Polymer composite, Figure 4.3.3a	0.38
Polymer composite, Figure 4.3.3b	0.32
Polymer composite, Figure 4.3.3c	0.27
Polymer composite, Figure 4.3.3d	0.10
Polymer blend prepared by solution blending, Figure 4.3.2c	0.0

of the mixed partially polymerized concentrated emulsions, the ratio of the areas of the peaks between 6.4 and 7.8τ of the butoxyl protons to those between 2.8 and 3.8τ of the phenyl protons of the styrene units was calculated and the results are listed in Table 4.3.2. As expected, the amount of copolymer in the polymer composites decreases with increasing conversion in the first step. As the conversion of the latex particles increases over 5%, the blending of the partially polymerized concentrated emulsions by mixing becomes difficult, due to the increase of their viscosities.

Figure 4.3.4 presents some scanning electron micrographs of the latex particles prepared by the concentrated emulsion polymerization, and shows that the

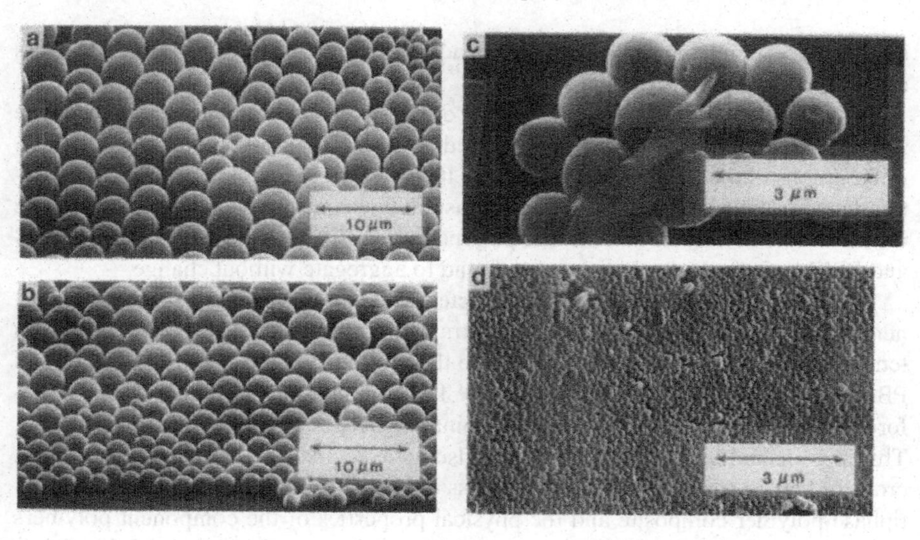

FIGURE 4.3.4 Electron micrographs of polymer latexes (monomer 50 mL, AIBN 0.30 mol L^{-1} monomer, SDS 0.35 g, water 4 mL, at 40°C). (a) Polystyrene, (b) poly(butyl methacrylate), (c) poly(butyl acrylate), (d) cross-linked polystyrene (40 mL of styrene and 10 mL of divinyl benzene.

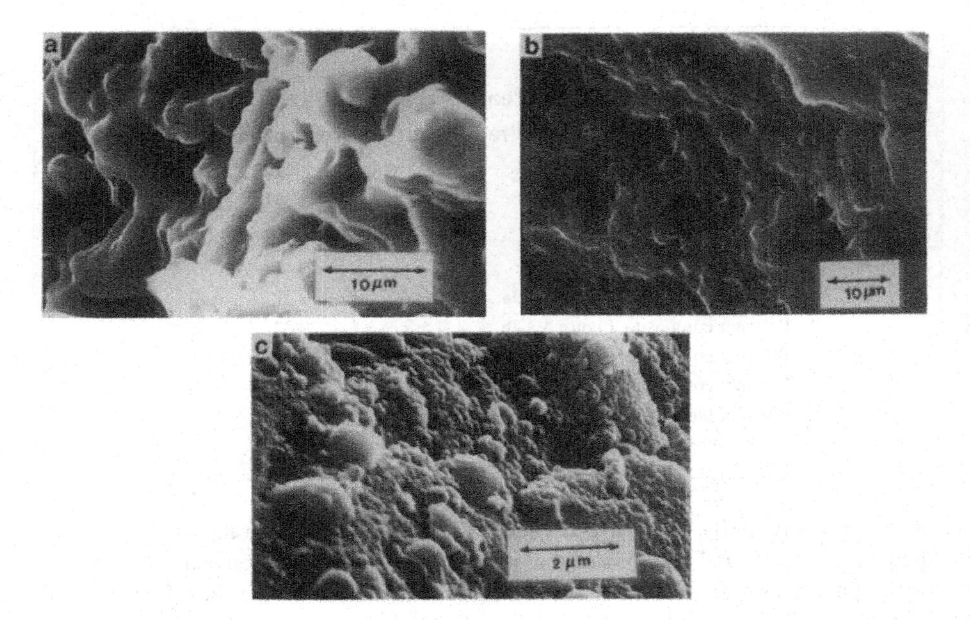

FIGURE 4.3.5 Electron micrographs of surfaces obtained by fracturing the polymer composite, which was prepared by blending two different 5% conversion concentrated emulsions. (a) polymer composite of polystyrene (80 wt%) and poly(butyl acrylate) (20 wt%); (b) polymer composite of polystyrene (50 wt%) and poly(butyl methacrylate) (50 wt%); (c) polymer composite of crosslinked polystyrene (50 wt%) and poly(butyl methacrylate) (50 wt%).

concentrated emulsion polymerization leads to spherical latexes whose diameters range from submicrons to microns.

In order to examine the microstructure of the two phase polymer composites, the sintered polymer composites were fractured. Figure 4.3.5 presents scanning electron micrographs of the surfaces thus obtained. It shows that the polymer composites contain latex particles whose sizes are not drastically changed when compared to those of the latexes in Figure 4.3.4. In the polymer composite containing cross-linked PS and PBMA, the polymer latexes were found to aggregate without change.

The toughness of a composite is expected to be higher when one of the components has a high and the other a low glass transition temperature. The glass transition temperatures of the polymers employed in the present experiments: PS, PBMA, and PBA are 100, 20, and –55°C respectively.[13] The combination of PS and PBA is therefore expected to be tough, while the combination between PS and PBMA less tough. This was indeed observed. In addition, also as expected, the combination between cross-linked polystyrene and PBMA is less tough. Both the morphology of the multiphase polymer composite and the physical properties of the component polymers play important roles in determining its mechanical strength. Figure 4.3.6 presents a solvent treated surface obtained by fracturing a composite formed of crosslinked PS and PBMA. The solvent dissolved the PBMA; the remaining cross-linked PS particles are seen to be in contact and well interconnected.

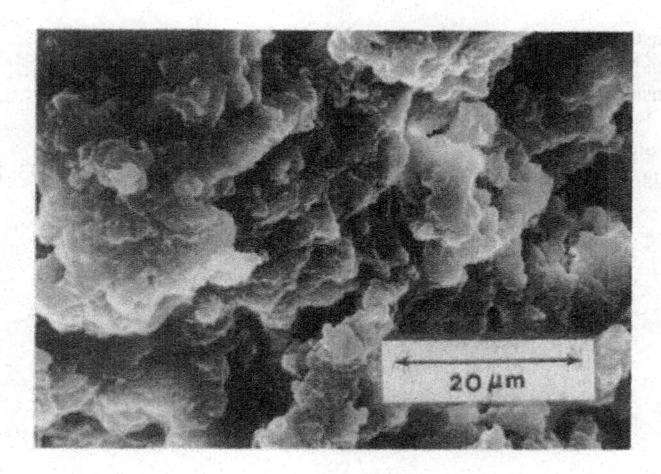

FIGURE 4.3.6 Electron micrograph of a solvent treated surface obtained by fracturing the polymer composite of crosslinked polystyrene and poly(butyl methacrylate) of Figure 4.3.5c.

4.3.4 CONCLUSION

Two different partially polymerized latex particles prepared by the concentrated emulsion method can be blended by the mixing of the corresponding concentrated emulsions, if their conversion is lower than about 5%. For larger conversions, the viscosities of the concentrated emulsions become high and the mixing difficult.

Aggregation of the two kinds of latex particles takes place after heating the mixture at 40°C for 24 h and further at 80°C for 24 h and at 120°C for 6 h. During these additional heating periods, homopolymerization inside the particles and copolymerization inside the coalesced particles and at the contact between particles take place. A high resolution NMR spectroscopy was employed to determine the extent of copolymerization in the polymer composites. Electron microscopy examination reveals that the microstructure of the two phase polymer composite is determined by the nature of the latex particles employed and that the aggregation during the blending and sintering processes occurs with slight changes in size and shape.

REFERENCES

1. Paul, D. R. and Newman, S. Eds, *Polymer Blends*, Vol I and II, Academic Press, New York (1978).
2. Manson, J. A. and Sperling, L. H. *Polymer Blends and Composites*, Plenum, New York (1976).
3. Walsh, D. J., Higgins, J. S. and Maconnachie, A. *Polymer Blends and Mixtures*, Martinus Nijhoff, Dordrecht, the Netherlands (1985).
4. Battaerd, H. A. J. and Tregear, T. W. *Graft Copolymers*, Interscience, New York (1967).
5. Henderson, J. F. and Swarc, M. *Macromol. Rev.* 1968, 3, 317.
6. Sperling, L. H. and Friedman, D. W. *J. Polym. Sci. A-2* 1969, 7, 425.
7. Ruckenstein, E. and Park, J. S. *J. Polym. Sci., Polym. Lett. Edn* 1988, 26, 529.

8. Ruckenstein, E. and Park, J. S. *Chem. Mat.* 1989, 1, 343.
9. Matsuo, M. *Japan Plastics* 1968, 2, 6.
10. Ruckenstein, E. and Kim, K. J. *J. Appl. Polym. Sci.* 1988, 36, 907.
11. Bovey, F. A. *J. Polym. Sci.* 1962, 62, 197.
12. Ito, K. and Yamashita, Y. *J. Polym. Sci., B* 1965, 3, 637.
13. Sperling, L. H., Chiu, T. W. and Thomas, D. A. *J. Appl. Polym. Sci.* 1973, 17, 2443.

4.4 Self-Compatibilization of Polymer Blends via Concentrated Emulsions*

Eli Ruckenstein and Hangquan Li

Department of Chemical Engineering, State University of New York at Buffalo, Amherst, NY 14260, USA

ABSTRACT A novel approach to the formation and compatibilization of polymeric blends is suggested, namely the self-compatibilization via concentrated emulsions. In this method, two concentrated emulsions are prepared from different monomers and subjected to partial polymerization; at least one concentrated emulsion contains also some divinyl-terminated macromonomer. A blend is formed by mixing the partially polymerized latexes and subjecting the mixture to complete polymerization. Some AB networks, with the macromonomer as chains A and the homo- and copolymers generated from the other monomers as chains B, are formed, which ensure the self-compatibilization of the resulting blend. Blends of styrene-co-methyl methacrylate and poly (vinyl acetate) were prepared and investigated as a model system.

4.4.1 INTRODUCTION

Polymer blending is one of the most important pathways to the development of new polymeric materials. However, in such blends, the problem of compatibility may arise, since satisfactory physical and mechanical properties are related to a fine dispersion of one phase in the other one and to the resistance to gross phase segregation. For this reason, the compatibilization of immiscible polymers has attracted significant interest.[1–3]

The compatibility of a blend can be promoted through copolymers which have segments capable of specific interactions and/or chemical reactions with the blend components.[4–7] The copolymer compatibilizers can be either added to or formed *in situ* in the blend during melt-mixing and processing. Being a convenient and fast way to produce polymer blends, the reactive extrusion[8,9] has been widely used in commercial manufacturing. However, it has some limitations,[10] such as the short residence time in the equipment, heat generation, and the need for special

* *Polymer Bulletin*, 35, Vol. 35, 517–524, (1995).

design for reactant feeding and volatile removing.[11] For these reasons, an alternative approach to the formation and compatibilization of blends of immiscible polymers, namely self-compatibilization[12,13] via concentrated emulsion polymerization,[14,15] was developed by this group. In the present paper, two concentrated emulsions of different monomers or monomer mixtures, in which at least one concentrated emulsion contains a small amount of divinyl-terminated macromonomer (DVTM), are prepared and subjected to partial polymerization. A blend is formed by mixing the partially polymerized concentrated emulsions instead of the melts. Further, the mixture is subjected to complete polymerization. Both before and after the two concentrated emulsions have been mixed, AB networks will be formed, with the DVTM as chain A and the homo- or copolymers of the other monomers as chain B. Such AB networks constitute good compatibilizers between the two components. Since the formation of the network occurs independently in each latex particle, the crosslinking is limited to each latex, and the latexes have flowability. A distinctive feature of this method is that the polymers which are blended and the compatibilizers are generated simultaneously. In other words, a compatibilized blend is produced via a single polymerization.

In this paper, blends from poly(styrene-co-methyl methacrylate) (PSM) and poly(vinyl acetate) (PVAc) were chosen as model systems. They were chosen not only because the flexible PVAc can toughen the brittle PSM, but also because the copolymerization between styrene (St) and vinyl acetate (VAc) and between methyl methacrylate (MMA) and VAc occurs with difficulty and hence little copolymer is generated. Indeed their copolymerization parameters are[16]: $r_1 = 0.01$ and $r_2 = 56.0$ for VAc and styrene and $r_1 = 0.03$, and $r_2 = 26.0$ for VAc and MMA. Vinyl-terminated polycaprolactone (VTPCL) was selected as the precursor of the network compatibilizer, because it possesses some compatibility with MMA.[17]

The effect of compatibilization is evaluated in this paper via the measurement of tensile properties. For a partially compatible blend, the higher the compatibility, the better the tensile properties. If the compatibility is lower than a certain level, a catastrophic fall in those properties will occur. Since both components, PSM and PVAc, are transparent materials, the transparency of the products can be used as a criterion of compatibility as well.

4.4.2 EXPERIMENTAL

4.4.2.1 MATERIALS

Azobisisobutyronitrile (AIBN, Kodak) was recrystallized from methanol. Other chemicals were purchased from Aldrich. Styrene (St), methyl methacrylate (MMA), and vinyl acetate (VAc) were filtered through an inhibitor removal column before use. The other compounds were used as received.

4.4.2.2 PREPARATION OF VINYL-TERMINATED PCL (VTPCL)

Preparation of VTPCL was described in a previous paper.[17] In this paper, it was used in the form of a solution (0.2g/mL toluene).

4.4.2.3 Concentrated Emulsion Polymerization

A mixture was first prepared from a monomer (styrene/MMA or VAc), an initiator AIBN (0.005 g/g of monomer) and a VTPCL solution. The mixture thus prepared was used as the dispersed phase of a concentrated emulsion. An aqueous solution of SDS (10 wt.%) was first placed in a flask provided with a magnetic stirrer. The mixture was added dropwise with vigorous stirring into the flask with a syringe, until the volume fraction of the SDS aqueous solution became 0.2. The whole addition process lasted for about 15 minutes, and took place at room temperature. The paste-like concentrated emulsion thus formed was additionally stirred for 15 min under a flow of nitrogen. Two concentrated emulsions of different monomers were prepared and introduced into a water bath at 50°C to carry out the partial polymerization of each until just 20% conversion was achieved (about 3 h). Subsequently, the two concentrated emulsions were mixed with magnetic stirring. The mixture of concentrated emulsions thus obtained was further heated at 60°C for 48 h for completing the polymerization. The product thus obtained was washed three times with 2-propanol and dried in a vacuum oven for 24 h. The blend product of the concentrated emulsion polymerization was obtained as a white powder, which will be denoted as concentrated emulsion blend (CEB). The ratio of the weight of the powder to the original weight of the reactants was considered as the conversion.

For comparison purposes, PSM, PVAc and copolymers from styrene/MMA/VAc were also prepared via the concentrated emulsion polymerization.

4.4.2.4 Preparation of Solution Blends

By "solution blend" we denote a blend obtained via casting a solution containing both polymers. In this paper a PSM/PVAc blend was obtained by casting the solution of PSM and PVAc in chloroform on a glass plate and evaporating the solvent.

4.4.2.5 The Tensile Testing

The powders of the CEBs or of the solution blends were thermo-pressed with a Laboratory Press (Fred S. Carver INC.) at 150°C for 3–5 min, and then cooled to room temperature. The sheets thus obtained were cut to the size required by the ASTM D.638–58T. The tensile testing was conducted at room temperature, with an Instron Universal Testing Instrument (Model 1000). The elongation speed of the instrument was 20 mm/min.

4.4.2.6 Solubility Measurements

A pre-weighed blend sheet prepared as for tensile testing was washed with methanol in an extractor for 12 h. The remaining sheet was retrieved and dried in a vacuum oven for 24 h to remove the solvent. The ratio of the weight lost during washing to that of the original sheet represents the content of methanol-soluble species (CMSS) and was taken as the content of VAc homopolymer (in percentage). The sheet remaining after methanol washing was further washed with chloroform for 12 h. The weight percentage of the remaining sheet to the original weight was considered as the gel content (the crosslinked part of the blend).

4.4.3 RESULTS AND DISCUSSION

4.4.3.1 Solution Blends

Data for the solution blends and pure PSM (styrene/MMA wt ratio = 1/1) and PVAc are listed in Table 4.4.1. The solution blends are milk opaque materials. Compared with the pure PSM and PVAc, the mechanical properties of these blends exhibit a dramatic decrease. Both the appearance and the mechanical properties indicate a complete incompatibility. The solution blends represent an extreme case in which no copolymers are present. From the content of methanol-soluble species one can conclude that the method employed in this paper to evaluate the content of VAc homopolymer is a suitable one.

4.4.3.2 Copolymers

The properties of the copolymers are listed in Table 4.4.2. As mentioned in the Introduction, the copolymerization between either styrene and VAc or MMA and VAc occurs with difficulty. For this reason the "copolymerization" of VAc with either styrene or MMA does not lead to copolymers only, but to mixtures of homopolymers and copolymers. This is clearly shown by the content of PVAc in Table 4.4.2, which indicates that only about 50% of VAc monomer was combined with styrene or MMA as copolymers, which do not dissolve in methanol. One can also note that the conversions are relatively low. A large fraction of the monomers remain unreacted or form only olig omers after 48 h of polymerization. However, because of the presence of the copolymer chains, the mechanical properties are generally better than those of the solution blends.

TABLE 4.4.1
Properties of the Solution Blends

wt. Ratio of PSM/PVAc	Tensile Strength (MPa)		Elongation at Break (%)		Content of Methanol-Soluble Species (%)
	Average	Std. Error	Average	Std. Error	
1/0	65.5	3.2	10.2	1.3	0
1/0.5	16.2	1.0	10.3	0.4	64.6
1/1	7.2	0.5	26.7	1.6	48.9
1/2	6.0	0.4	65.4	5.4	31.7
0/1	15.1	1.6	350	19.5	100

TABLE 4.4.2
Properties of the Copolymers

wt. Ratio of St/MMA/VAc	Tensile Strength (MPa) Average	Tensile Strength (MPa) Std. Error	Elongation at Break (%) Average	Elongation at Break (%) Std. Error	Conversion (%)	Content of Methanol-Soluble Species (%)
1/1/1	21.1	2.4	13.2	0.8	84	15.6
1/1/2	12.5	0.6	29.8	2.8	83	24.5
1/1/4	8.8	0.6	61.4	5.9	81	34.0

4.4.3.3 Concentrated Emulsion Blends without VTPCL

The data for the concentrated emulsion blends (CEBs) without VTPCL are presented in Table 4.4.3. One may note that the contents of homo-PVAc in the CEBs are higher than those in solution blends. This is a result of the preparation methodology. As described in the Experimental section, in the preparation of CEBs, each concentrated emulsion was first partially polymerized until just 20% conversion was achieved and subsequently the two were mixed. At that conversion, each concentrated emulsion acquires a high viscosity, which delays the transfer of the monomers and oligomers from one latex to another, thus limiting copolymerization. However, still about 10% of the VAc monomer was involved in copolymerization. One should also note that the conversions of the CEBs are higher than those of the copolymers; this probably occurs because of the homopolymerization of VAc. The mechanical properties of the CEBs are comparable to those of the copolymers; because of the higher contents of homo-PVAc in the CEBs, the tensile strengths are somewhat lower but the elongations at break are somewhat higher. Both the copolymers and the CEBs are translucent, which indicates that the compatibility in both systems is somewhat improved.

4.4.3.4 Concentrated Emulsion Blends with VTPCL

In order to strengthen the compatibilizing effect, small amounts of VTPCL were introduced. The reactions among VTPCL and styrene, MMA and VAc generate AB

TABLE 4.4.3
Properties of the Concentrated Emulsion Blends without VTPCL

wt. Ratio of St/MMA/VAc	Tensile Strength (MPa) Average	Tensile Strength (MPa) Std. Err	Elongation at Break (%) Average	Elongation at Break (%) Std. Err	Conversion (%)	Content of Methanol-Soluble Species (%)	Content of Chloroform-Insoluble Species (%)
1/1/1	20.5	0.5	14.1	2.5	92	27.3	0
1/1/2	11.4	0.8	32.7	1.7	91	41.7	0
1/1/4	8.5	0.7	62.2	3.8	89	57.6	0

networks, with PCL as chains A and the copolymer and homopolymers of those monomers as chains B. Because the network contains chains of both PSM and PVAc, it compatibilizes the two. This function of the network is similar to that of a block copolymer. It should be emphasized that the network compatibilizer is not added, but is formed during polymerization. For this reason we call such a methodology self-compatibilization. All the CEBs with VTPCL possess a transparent appearance, which indicates an improved compatibility. The effect of self-compatibilization is also made clear by Table 4.4.4. The tensile strength and elongation at break increase simultaneously, in contrast to what happens in Tables 4.4.2 and 4.4.3. Of course, the PCL chains themselves can improve to some extent the mechanical properties. Indeed, it was noted in a previous paper[17] that PCL can toughen PMMA when the weight ratio of PCL/PMMA is higher than 20/100. However, this kind of improvement differs from that caused by compatibilization. Comparison between different contents of VTPCL indicates that a content of 0.05 g/g of the other monomers is enough for self-compatibilization. A higher content of VTPCL results in a lower tensile strength and a higher elongation at break.

Another advantage brought by the network is a higher conversion. One can see from Table 4.4.4 that the higher the content of VTPCL, the higher the conversion.

4.4.3.5 Effect of the Distribution of VTPCL

In the preparation of the samples of Table 4.4.4, VTPCL was equally distributed between the two concentrated emulsions. CEBs in which only one concentrated emulsion contained VTPCL were also prepared, and the results are listed in Table 4.4.5. One can see that if VTPCL is introduced only in VAc, comparable mechanical properties are obtained as when VTPCL is introduced in both concentrated emulsions. When, however, VTPCL is added to styrene/MMA only, lower mechanical properties are achieved. This can be ascribed to the decrease in compatibility. Indeed, to

TABLE 4.4.4
Properties of the Concentrated Emulsion Blends with VTPCL

wt. Ratio of St/MMA/VAc	Tensile Strength (MPa)		Elongation at Break (%)		Conversion (%)	Content of Methanol-Soluble Species (%)	Content of Chloroform-Insoluble Species (%)
	Average	Std. Err	Average	Std. Err			
VTPCL = 0.05 g/g of other monomers							
1/1/1	46.6	5.8	20.7	4.1	93	19.6	18.2
1/1/2	42.7	3.7	43.0	1.6	94	30.3	17.1
1/1/4	35.1	2.6	61.4	5.4	91	42.7	15.2
VTPCL = 0.1 g/g of other monomers							
1/1/1	36.8	3.3	23.6	3.1	94	18.0	32.7
1/1/2	30.4	1.6	50.9	5.4	92	28.8	30.6
1/1/4	26.5	3.6	67.5	3.3	91	33.4	29.3

TABLE 4.4.5
Properties of the Blends[a] with Different VTPCL Distributions

VTPCL Containing Conc. Emul.	Tensile Strength (MPa)		Elongation at Break (%)		Conversion (%)	Content of Methanol-Soluble Species (%)	Content of Chloroform-Insoluble Species (%)
	Average	Std. Err	Average	Std. Err			
St/MMA	23.3	0.5	32.1	2.5	92	32.4	16.6
VAc	40.7	1.9	42.2	3.9	94	29.6	16.2
both	42.4	3.7	43.0	1.6	94	30.3	15.0

[a] St/MMA/VAc wt ratio = 1/1/2, Content of VTPCL = 0.1 g/g of other monomers

be a good compatibilizer, both PSM and homo-PVAc chains should be present in the network. Because of the copolymerization parameters, styrene and MMA are more easily included in the network than VAc after the concentrated emulsion containing styrene, MMA and VTPCL is mixed with that containing VAc alone. The homo-PVAc has the opportunity to combine with VTPCL only when VTPCL is present in the VAc containing concentrated emulsion.

4.4.3.6 Effect of the Styrene/MMA wt Ratio

The nature of the system affects the self-compatibilization. Indeed, Table 4.4.6 shows the effect of the styrene/MMA wt ratio on the mechanical properties. The elongation at break decreases with increasing styrene/MMA ratio, while the tensile strength exhibits a maximum. This behavior can be explained on the basis of the difference in the polarities of the components, which are reflected in the solubility parameters. The solubility parameters[16] of PS, PMMA and PVAc are 17.5–18.5,

TABLE 4.4.6
Properties of the Blends with Various St/MMA wt Ratios

wt. Ratio of St/MMA	Tensile Strength (MPa)		Elongation at Break (%)		Conversion (%)	Content of Methanol-Soluble Species (%)	Content of Chloroform-Insoluble Species (%)
	Average	Std. Err	Average	Std. Err			
100/0	24.5	0.6	28.9	3.6	81	22.1	33.3
25/75	38.2	0.8	43.3	2.7	85	24.9	30.4
50/50	30.4	1.6	50.9	5.4	92	28.3	29.4
75/25	24.5	0.8	67.6	4.1	95	27.6	32.2
0/100	19.5	0.7	100.5	6.8	97	28.4	31.7

VTPCL = 0.1 g/g of other monomers

18.5–19.5 and 19–21 $(MPa)^{1/2}$, respectively. The polarity of PVAc is closer to that of PMMA than to that of PS. In other words, the compatibility between PVAc and PMMA is higher than that between PVAc and PS. For this reason, PVAc has a stronger effect on the blends with higher MMA content. Therefore, the higher the MMA content, the higher the elongation at break and the lower the tensile strength. For the samples free of MMA, compatibility is very low, and both elongation at break and tensile strength are poor.

4.4.4 CONCLUSION

While the solution blends of PSM and PVAc are incompatible, the compatibility of the blends generated from the same components via concentrated emulsions is greatly improved because of self-compatibilization. The former are opaque and the latter are transparent; the mechanical properties of the former are poor while those of the latter are excellent. This study shows that the methodology of self-compatibilization constitutes an alternative approach to the reactive melt-processing for the preparation of blends from incompatible polymers.

REFERENCES

1. Utracki LA (1987) *Inter Polym Proc* 2:3.
2. Paul DR (1992) Phase coupling and morphology generation in Engineering polymer alloys. In Choy CL, Shin FG (ed) *Proc Inter Symp on Polym Alloys & Composites.* Hongkong, p. 21.
3. Keskkula K, Paul DR (1994) Toughening agents for engineering polymers. In Collyer AA (ed) *Rubber Toughened Engineering Plastics.* Chapman & Hall, New York, p. 136.
4. Xanthos M, Dagli SS (1991) *Polym Eng Sci* 31:929.
5. Xanthos M (1988) *Polym Eng Sci* 28:1392.
6. Gaylord NG (1989) *J Macromol Sci Chem* A26:1211.
7. Kreisher K(1989) *Plastics Technol* 35:67.
8. Brown SB (1989) Reactive Extrusion, course notes, Polymer Processing Institute, Hobken NJ.
9. Tzoganakis C (1989) *Adv Polym Technol* 9:321.
10. Todd DB (1988) *Am Chem Soc Polym Prep* 29:563.
11. Lambla M (1987–88) *Polym Proc Eng* 5:297.
12. Ruckenstein E and Park JS. (1990) *Polymer* 31:2397.
13. Ruckenstein E and Li H. (1994) *Polymer* 35:4343.
14. Ruckenstein E and Park JS. (1988) *J Polym Sci Chem Lett Ed* 26:529.
15. Ruckenstein E and Kim KJ. (1988) *J Appl Polym Sci* 36:907
16. Brandrup J, Immergut EH (ed) (1989) *Polymer Handbook.* 3nd Ed. John Wiley & Sons, New York.
17. Li H, Ruckenstein E (1995) *Polymer* 36:2281.

4.5 Composites via Heterogeneous Crosslinking of Concentrated Emulsions*

Eli Ruckenstein and Hangquan Li

Department of Chemical Engineering, State University of New York at Buffalo, Amherst, New York 14260, USA

ABSTRACT Polymer composites based on an epoxy resin oligomer, an amine terminated oligomer or maleic anhydride, and one or several monomers among styrene, methyl methacrylate, butyl methacrylate, glycidyl methacrylate and acrylonitrile have been prepared using the concentrated emulsion pathway. Each of the components of the composite was first transformed in a concentrated emulsion in which water with a low volume fraction was the continuous phase. The concentrated emulsions containing monomers were partially polymerized (to ensure their stability during the subsequent heating) before being mixed with one or two concentrated emulsions, each containing one of the above mentioned oligomers. The mixture of concentrated emulsions was subjected to heating for the complete polymerization of the monomer and for the reaction between the epoxy resin and the amine terminated oligomer or the maleic anhydride to occur. The reaction generates a crosslinking among the cells of the concentrated emulsions (heterogeneous crosslinking) and thus achieves compatibilization among various incompatible components. The composites were characterized using Fourier transform infrared spectroscopy, differential scanning calorimetry, swelling, as well as solubility measurements. Tensile measurements showed that the components of the composites were compatibilized and that the composites possess excellent mechanical properties.

* *Polymer,* Vol. 37, 3373–3378, (1996).

4.5.1 INTRODUCTION

Several methods have been developed in recent decades to prepare polymer composites. In addition to melt blending and graft copolymerization, the blending of two or more types of latexes by mechanical mixing was also suggested.[1,2] Because of the "low viscosity" of the latex components, intimate dispersion could be achieved by employing a relatively low mechanical energy for mixing.[3] When the latexes were further crosslinked, a network or interpenetrating latex networks was (or were) obtained.[4,5] The mixing and crosslinking were so far exclusively conducted in a multi-step way, i.e., the latexes were sequentially formed, coagulated and crosslinked. In this paper, we propose a novel procedure, in which each of the polymers involved is first transformed as such, or as monomers in a concentrated emulsion in water. The concentrated emulsion of monomers is partially polymerized before it is mixed with one or two concentrated emulsions of polymer or oligomer. At least one of the concentrated emulsions contains a component which can react with the components of the other concentrated emulsions. During heating, the monomers polymerize completely and the above reaction occurs. The final result is a composite in which the various incompatible polymers are made compatible, due to the reaction.

To illustrate the procedure, we prepared two kinds of composites. One is based on an epoxy resin (poly[bisphenol A-*co*-epichlorohydrin], PBPAE), an amine terminated oligomer (poly[butadiene-*co*-acrylonitrile]), denoted PBAn, and at least two monomers out of styrene (St), methyl methacrylate (MMA), butyl methacrylate (BMA) and glycidyl methacrylate (GMA), one of them being GMA. Concentrated emulsions in water were first prepared from each of the three components: PBPAE, PBAn and the monomers. After the concentrated emulsion containing monomers was partially polymerized (to ensure the stability of the concentrated emulsion in the subsequent heating), the three concentrated emulsions were mixed and subjected to heating. During heating, complete polymerization of the monomers takes place and the amine terminated oligomer reacts with the epoxy resin and the GMA. In what follows, the above composite will be called the epoxy/amine system. The other composite is based on PBPAE, maleic anhydride (MAn) and one or several monomers among St, BMA and acrylonitrile (An). Two concentrated emulsions in water were prepared, one from the epoxy resin, and the other from the mixture between the monomers and MAn. After the partial polymerization of the latter, the concentrated emulsions were mixed and subjected to heating. The MAn can react with the epoxy resin and can also copolymerize with the monomers. This provides the compatibilization between the epoxy resin and the formed polymer. This composite will be called the epoxy/anhydride system. In both composites, the components do not dissolve one in another. PBPAE and PBAn separate even in the presence of their common solvent chloroform. The best way to mix these components is via concentrated emulsions.

A concentrated emulsion has the appearance of a paste, and differs from the conventional emulsion in that the volume fraction, ϕ, of the dispersed phase is greater than 0.74 (which represents the volume fraction of the most compact arrangements of spheres of the same size), and may be as high as 0.99. At high volume fractions, the concentrated emulsions are composed of polyhedral cells less than 1 μm in size, separated by thin films of continuous phase. Its use in the preparation of latexes and composites has recently been suggested.[6–8]

Since the components cannot dissolve one in another, in the mixture of concentrated emulsions suggested in the present paper, the reactions between the functional groups are likely to take place at the interface of the cells. The compact packing of the cells in a concentrated emulsion provides the best opportunity for inter-cell surface reactions, i.e., heterogeneous crosslinking,[9] to take place. Unlike the homogeneous crosslinking, the heterogeneous one involves particularly the surface region of the latexes. The use of the concentrated emulsion method has another advantage. To acquire toughness, the flexible component of the composite must be in the size range 0.5–2 μm.[10,11] In the other processing methods, strict conditions must be employed to achieve such a size range. By using the concentrated emulsion mixing method, the size of the cells just falls in that favoured range, and hence toughening can be easily ensured.

4.5.2 EXPERIMENTAL

4.5.2.1 MATERIALS

Azobisisobutyronitrile (AIBN, Kodak) was recrystallized from methanol. All the other chemicals were purchased from Aldrich. St, BMA, MMA and GMA were filtered through an inhibitor removal column before use. MAn, PBPAE (epoxy resin, molecular weight = 4000, 1800, which are denoted as E4000 and El800, respectively), PBAn (molecular weight = 4900, amine equivalence = 1200), dodecyl sulfate sodium salt (SDS, 70%), chloroform (99%) and methanol (99%) were used as received.

4.5.2.2 PREPARATION OF THE CONCENTRATED EMULSION

An aqueous solution of SDS (10 wt%) was first placed in a flask provided with a magnetic stirrer. The flask was sealed with a rubber septum, after which the air inside was replaced with nitrogen. The dispersed phase was added dropwise with vigorous stirring into the flask with a syringe, until the volume fraction of the SDS aqueous solution became 0.2. The whole addition process lasted about 15 min and took place at room temperature.

4.5.2.3 PREPARATION OF THE EPOXY/AMINE COMPOSITES

Three concentrated emulsions in water were first prepared separately from a solution of PBPAE (0.5 g per g chloroform), a solution of PBAn (1 g per g chloroform) and a monomer mixture as dispersed phases, respectively. The monomer mixture containing an initiator (AIBN, 0.005 g per g of monomers) was partially polymerized at 50°C for 1 h, before preparing the concentrated emulsion, in order to moderately increase the viscosity, and thus to ensure the stability of the concentrated emulsion during subsequent heating. The concentrated emulsion based on monomers was kept at 50°C for 30 min in order to generate incipient latexes which can keep their identity to a high extent after mixing. Subsequently, the three concentrated emulsions were mixed, and the mixture was further heated at 50°C for 16 h to complete the polymerization of the monomers, and to allow the reactions of the amine groups with the epoxy resin and GMA units to take place. The product was washed with methanol and dried in a vacuum oven. The dried product was treated for ⩽12 hours, at 75°C to carry out the post-curing reaction.

4.5.2.4 Preparation of the Epoxy/Anhydride Composites

The preparation was similar to the one above, but only two concentrated emulsions in water were mixed: one based on PBPAE and another one based on a partially polymerized monomer mixture containing MAn. After the two concentrated emulsions were mixed, the mixture was kept in a water bath at 50°C for 24 h. No post-curing was employed, for reasons which are explained later.

4.5.2.5 Tensile Testing

The powders (fine and/or coarse) were thermopressed with a Laboratory Press (Fred S. Carver Inc.) at 150°C for 3–5 min, and then cooled to room temperature. The sheets thus obtained were cut with a die to the size required by ASTM D.638. The tensile testing was conducted at room temperature, using an Instron Universal Testing Instrument (Model 1000). The elongation speed of the instrument was 20 mm min^{-1}.

4.5.2.6 Measurement of the Insoluble Species Content and of Swelling

A pre-weighed sample sheet prepared as for tensile testing was immersed in chloroform in a test tube at room temperature, using 20 mL of solvent for each g of sheet. The solvent in the tube was changed with a fresh one every 8 h, until all the soluble species had been removed (less than 48 h). The swollen sheet was first weighed and subsequently dried in a vacuum oven for 24 h to remove the solvent. The ratio of the remaining weight to that of the original sheet is denoted as the insoluble species content, and the ratio of the weight of the swollen sample to that of the dried one is the swelling ratio.

4.5.2.7 Differential Scanning Calorimetry

The thermal behaviour of both the intermediate and final products was determined by differential scanning calorimetry (DSC), with a DuPont DSC instrument. Each sample was scanned twice. The first scanning was from −80°C to 200°C. After the first scanning, the specimen was cooled to below −80°C and then heated for a second time from −80°C to 140°C. The heating rate was 10°C min^{-1}.

4.5.2.8 Infra-Red Spectra

The Fourier transform infrared (FT-IR) absorption spectra were obtained with a Perkin-Elmer (1720) FT-IR instrument.

4.5.3 RESULTS AND DISCUSSION

4.5.3.1 Insoluble Species Content

The insoluble species content of the samples is listed in Tables 4.5.1 and 4.5.2. One can see that the wt% of the crosslinked structure of the final product is near or above 90%. In the ideal case, all the chains should be linked to a complete network and the insoluble species content should be 100%. However, in actual reactions, several non-ideal situations may occur: (i) the curing may be incomplete; and (ii) some linear polymer chains may not be linked to the network. It is reasonable to consider that most of these linear chains are copolymers of the monomers. Because these copolymers are statistical, most of the functional groups, GMA or MAn units, are not located at the end and, therefore, their reactivity is lower. Some copolymer chains may contain fewer reactive units. Indeed, Tables 4.5.1 and 4.5.2 show that the larger the amount of monomers introduced in the initial system, the lower the insoluble species content in the final product. With comparable molar ratios between the curing agent and the

TABLE 4.5.1
Insoluble Species Content and Glass Transition of Epoxy/Amine Systems

Amounts of the Components (g)							Time of Post-Curing (h)	Insoluble Species Content × 10^2	Swelling Ratio × 10^2	T_g (°C)
E4000	E1800	PBAn	St	BMA	MMA	GMA				
5		5	5	5	–	0.4	0	84.0	1569	–
							4	87.9	1466	–
							8	90.3	1204	70–105
							12	90.7	1179	70–105
	5	5	5	5	–	0.4	0	84.6	1449	–
							4	86.5	1211	–
							8	89.6	1070	75–110
							12	90.1	1028	75–110
5	–	5	10	–	–	0.4	8	92.0	1203	77–105
5		5	5	–	5	0.4	8	90.8	1188	80–105
	5	5	5	–	5	0.4	8	91.2	1050	83–100
5	–	5	5		–	0.4	8	95.3	1044	70–110
–	5	5	5	–	–	0.4	8	95.2	978	75–115
5	–	5	25	2.5	–	0.4	8	94.2	1052	65–100
–	5	5	25	2.5	–	0.4	8	95.2	992	70–105
5	–	3	25	2.5	–	0.4	8	89.8	1152	70–95
–	5	3	25	2.5	–	0.4	8	90.2	1040	75–105

TABLE 4.5.2
Insoluble Species Content and Glass Transition of Epoxy/Anhydride Systems

Amounts of the Components (g)						Insoluble Species Content × 10²	Swelling Ratio × 10²	T_g (°C)
E4000	E1800	BMA	St	An	MAn			
5	–	5	–	–	0.25	82.2	1180	83
5	–	5	2.5	–	0.25	80.5	1266	84
5	–	5	2.0	0.5	0.25	78.2	1232	87
–	5	5	–	–	0.25	86.1	1022	87
–	5	5	2.5	–	0.25	82.9	1176	88
–	5	5	2.0	0.5	0.25	80.5	1147	90
–	5	5	–	–	0.375	88.2	852	90
–	5	5	–	–	0.125	70.4	1354	–

epoxy groups, the insoluble species content for epoxy/amine systems is generally higher than those for the epoxy/anhydride systems, because in the latter case the amount of monomer employed is larger. However, the presence of the uncrosslinked soluble chains has a positive effect because it improves the flowability of the materials.

For the epoxy/amine systems, the post-curing time has an important effect on the insoluble species content. A longer curing time results in a higher fraction of chains linked to the network. Indeed, Table 4.5.1 shows that the insoluble species content increases initially with the curing time. However, the insoluble species content no longer increases for curing times longer than 8 h. This happens because the highest possible crosslinking is achieved in 8 h.

Obviously, the molar ratio of epoxy/curing agent also plays a role; the systems with a higher curing agent content having a higher insoluble species content.

One can note that the molecular weight of PBPAE has an effect on the insoluble species content of the epoxy/anhydride systems, but a more moderate one on the epoxy/amine systems. For the same weight amount, the PBPAE of low molecular weight possesses a larger number of epoxy groups, hence the curing should be more advanced in the latter case. However, in epoxy/amine systems, there is a competition between the epoxy groups of PBPAE and of the copolymers for the reaction with the amine groups. The increase in the crosslinking of PBPAE is offset by the decrease in the crosslinking of the copolymers in such cases. As a result the effect is moderate. In contrast, in the epoxy/anhydride systems, the PBPAE is cured by the MAn groups of the copolymer and the crosslinking increases as the number of epoxy groups increases.

4.5.3.2 Swelling Ratio

The swelling ratio, which is a measure of the extent of crosslinking, is listed in Tables 4.5.1 and 4.5.2. As expected, Table 4.5.1 shows that the swelling ratio decreases with increasing time of post-curing. However, the swelling ratios for 8 and

12 h post-curing are comparable. This again indicates, as noted above, that the highest possible curing extent is achieved in 8 h. For both systems, a lower swelling ratio is associated with a higher amount of curing agent employed. For the same post-curing time and the same amount of curing agent, the system with a greater amount of monomer possesses a higher swelling ratio, because of the lower extent of crosslinking.

4.5.3.3 Fourier Transform Infrared Determinations

The extent of curing of the epoxy/anhydride systems was also assessed on the basis of the FT-IR spectra (Figures 4.5.1 and 4.5.2). The spectrum in Figure 4.5.1 is for MAn, and the three spectra in Figure 4.5.2 are for the pure E4000 and the composites after 12 and 24 h of reaction at 50°C, respectively. MAn can be characterized by three consecutive sharp peaks between 1300 and 1200 cm^{-1}, and one can see that these peaks became smeared after 12 h, and completely disappear after 24 h of reaction. The epoxy group can be characterized by the sharp peak near 3050 cm^{-1}, pointed by an arrow. It becomes weaker after 12 h and can hardly be detected after 24 h of reaction. The disappearance of the MAn peaks is partly due to the curing, but also to hydrolysis. However, the weakening and disappearing of the characteristic peak of the epoxy group provides indications about the extent of curing.

4.5.3.4 Differential Scanning Calorimetry

In this paper, the DSC measurements have two purposes: the assessment of the extent of curing of the epoxy/amine systems; and the determination of the glass transition temperatures T_gs for both the epoxy/amine and epoxy/anhydride composites.

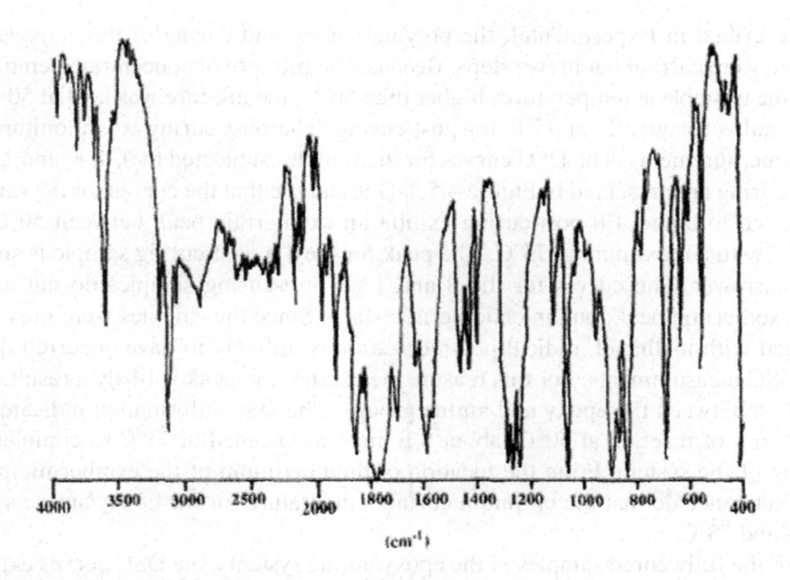

FIGURE 4.5.1 FT-IR spectrum of maleic anhydride.

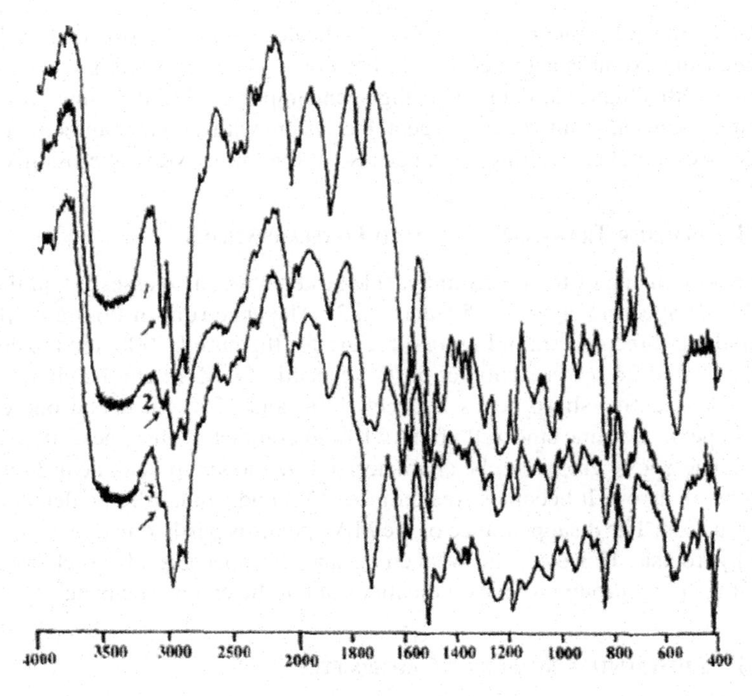

FIGURE 4.5.2 FT-IR spectra. (1) Poly(bisphenol A-*co*-epichlorohydrin), molecular weight = 4000. (2) Epoxy/anhydride composite after 12 h reaction. (3) Epoxy/anhydride composite after 24 h reaction. Amounts of the components (g) for the composite: E4000: 5; MAn: 0.25; BMA: 5.

As described in Experimental, the polymerization and curing of the epoxy/amine system were carried out in two steps. Because the mixture of concentrated emulsions became unstable at temperatures higher than 50°C, the mixture was kept at 50°C for 16 h and subsequently at 75°C for post-curing. The post-curing was monitored by DSC measurements. The DSC curves for the samples subjected to 0, 4, 8, and 12 h of post-curing are presented in Figure 4.5.3. One can see that the curves for the samples subjected to 0 and 4 h post-curing exhibit an exothermic peak between 50°C and 100°C with a maximum at 75°C. The peak for the 4 h post-curing sample is smaller and narrower. The curves for the 8 and 12 h post-curing samples do not exhibit any exothermic peak, but an endothermic stage. Since the samples were previously washed with methanol, radical polymerization is unlikely to have occurred during the DSC measurements. For this reason, the exothermic peak is likely a result of the reaction between the epoxy and amine groups. The DSC information indicates that after 16 h of reaction at 50°C, about 8 h more are needed at 75°C to complete the curing of the system. From the location of the maximum of the exothermic peaks, one can conclude that the optimum curing temperature for the epoxy/amine system is around 75°C.

For the fully cured samples of the epoxy/amine systems, the DSC curves exhibit a wide endothermic stage, which can be as wide as 40°C; no other stages or peaks are

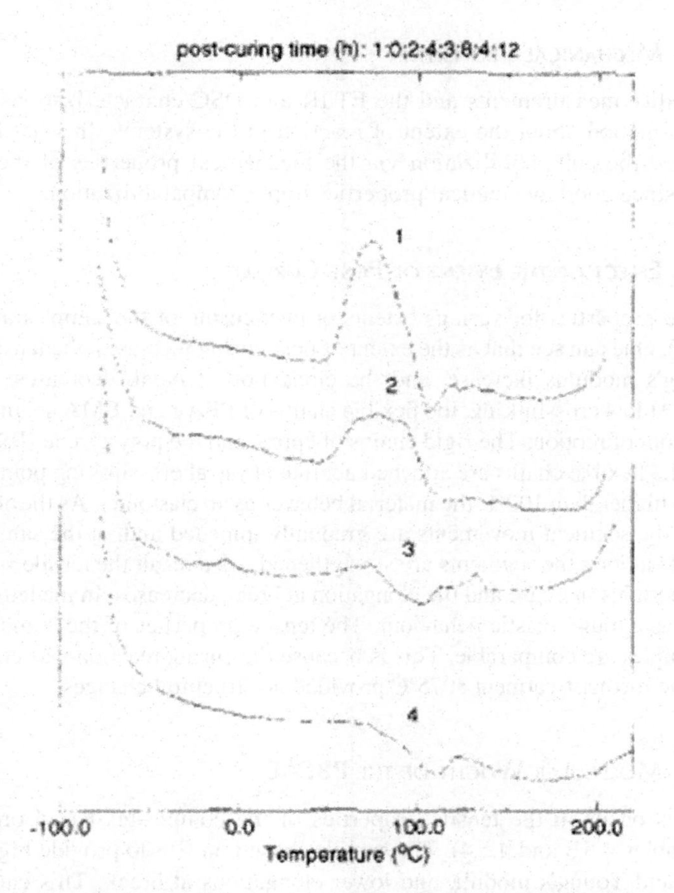

FIGURE 4.5.3 DSC diagrams for the composites. Post-curing time (h): (1), 0; (2), 4; (3), 8; (4), 12. Amounts of the components (g) for the composite: E4000: 5; PBAn: 5; St: 5; BMA: 5; GMA: 0.4.

present on the DSC curves. This indicates that the samples possess only one T_g range. A fully compatible homogeneous polymer blend usually has a single T_g,[12,13] which is not wide (3°C–5°C). Wide T_gs can be associated[14,15] with heterogeneous materials. The wide T_g of the epoxy/amine systems occurs because of the heterogeneity at the colloidal scale of the materials which were generated via the crosslinking of three types of latexes. A wide T_g range material can be used as a damping medium, since it can absorb vibrations in a wide range of frequencies.

No reaction peak was observed in the DSC measurements for the epoxy/anhydride system after 24 h of reaction. For this reason, the post-curing was not applied to these systems. In addition, no wide glass transition was exhibited. All the samples of epoxy/anhydride system provided a narrow T_g (Table 4.5.2). In contrast to the epoxy/amine systems which involved three concentrated emulsions, the epoxy/anhydride systems involve only two concentrated emulsions. Obviously, the latter systems should be less heterogeneous than the former.

4.5.3.5 Mechanical Properties

The solubility measurements and the FT-IR and DSC characterizations have provided information about the extent of reaction in the systems. In what follows we characterize the compatibilization via the mechanical properties of the materials obtained, since good mechanical properties imply compatabilization.

4.5.3.6 Effect of the Extent of Post-Curing

The tensile properties for various extents of post-curing of the samples are listed in Table 4.5.3. One can see that as the extent of post-curing increases, the tensile strength and Young's modulus increase, and the elongation at break decreases. This is as expected. At low crosslinking, the flexible chains of PBAn and BMA segments have a high freedom of motion. The rigid chains of epoxy and the polystyrene (PS) segments to which the flexible chains are attached act like physical crosslinking points. With an elongation higher than 100%, the material behaves as an elastomer. As the crosslinking increases, the segment movements are gradually impeded and, at the same time, the interactions among the segments are strengthened. As a result the tensile strength and Young's modulus increase and the elongation at break decreases. In the latter case, the material has a tough plastic behaviour. The tensile properties of the 8 and 12 h post-curing samples are comparable. This is because the curing reaction was completed in 8 h, and the further treatment at 75°C provided no structural changes.

4.5.3.7 Molecular Weight of the PBPAE

Differences occur in the tensile properties of the composites based on E1800 or E4000 (Tables 4.5.3 and 4.5.4). The samples based on E1800 provide higher tensile strengths and Young's moduli, and lower elongations at break. This can be easily explained as a result of the extent of crosslinking of the epoxy resin. For the same weight amounts, E1800 contains a larger number of epoxy groups. Consequently, E1800 can have a higher extent of crosslinking than E4000 and the composites based on E1800 can be stiffer and less flexible.

4.5.3.8 Amount of Curing Agent

For the epoxy/anhydride systems (see Table 4.5.4), the higher amount of curing agent (MAn) results in higher crosslinking and thus in a less flexible material. If the amount of curing agent is too low, the system cannot be properly cured, and the composite possesses hardly any mechanical strength. However for the epoxy/amine system (see Table 4.5.3), the situation is more complicated. PBAn is both a curing agent, because its amine groups can react with the epoxy groups, and a toughener, since its flexible segments toughen the glass matrix of epoxy and PS segments. In Table 4.5.3, the systems with the weight ratios PBAn/PBPAE = 5/5 and 3/5 are compared. The decrease of the PBAn/PBPAE weight ratio from 5/5 to 3/5 changes the amine/epoxy molar ratio from 1.5 to 0.94; this decreases the crosslinking extent and increases the flexibility of the system. On the other hand,

TABLE 4.5.3
Tensile Properties of Epoxy/Amine Systems

Amounts of the Components (g)								Tensile Properties (std error)			
E4000	E1800	PBAn	St	BMA	MMA	GMA	Time of Post-Curing (h)	Tensile Strength strength (MPa)	Elongation at Break (%)	Young's Modulus (MPa)	Toughness (MJ m^{-3})
5	–	5	5	5	–	0.4	0	24.5 (1.5)	126 (11.1)	613 (27)	15.4 (1.2)
							4	30.1 (1.4)	87 (4.2)	831 (21)	19.6 (1.1)
							8	37.6 (1.7)	65 (3.1)	922 (12)	21.6 (2.0)
							12	36.9 (2.2)	63 (4.7)	966 (35)	21.2 (1.7)
	5	5	5	5	–	0.4	0	27.5 (0.9)	108 (8.1)	634 (32)	14.8 (0.9)
							4	34.6 (1.8)	74 (5.5)	884 (15)	19.2 (1.3)
							8	42.6 (1.3)	58 (3.7)	984 (16)	22.2 (1.6)
							12	43.0 (2.0)	57 (2.6)	991 (34)	22.0 (1.3)
5	–	5	10	–	–	0.4	8	47.4 (2.5)	31 (1.8)	1022 (23)	12.9 (1.1)
5	–	5	5	–	5	0.4	8	48.6 (2.0)	21 (1.9)	1191 (25)	8.5 (0.4)
–	5	5	5	–	5	0.4	8	51.5 (3.7)	14 (1.1)	1224 (44)	6.1 (0.6)
5	–	5	5	–	–	0.4	8	41.7 (2.7)	55 (3.4)	966 (35)	20.6 (1.8)
–	5	5	5	–	–	0.4	8	45.3 (3.1)	51 (4.2)	1075 (21)	20.0 (0.7)
5	–	5	2.5	2.5	–	0.4	8	35.4 (1.1)	69 (3.5)	822 (10)	21.4 (2.2)
–	5	5	2.5	2.5	–	0.4	8	40.1 (2.0)	62 (4.2)	916 (29)	21.6 (1.3)
5	–	3	2.5	2.5	–	0.4	8	39.7 (2.8)	56 (3.3)	917 (17)	17.9 (0.8)
–	5	3	2.5	2.5	–	0.4	8	44.6 (3.1)	51 (2.5)	1016 (26)	18.2 (1.4)

TABLE 4.5.4
Tensile Properties of Epoxy/Anhydride Systems

Amounts of the Components (g)						Tensile Properties (std error)			
E4000	E1800	BMA	St	An	MAn	Tensile Strength (MPa)	Elongation at Break (%)	Young's Modulus (MPa)	Toughness (MJ m^{-3})
5		5			0.25	48.5 (1.4)	66 (3.3)	1048 (96)	23.8 (1.7)
5		5	2.5		0.25	53.2 (2.1)	21 (1.6)	1224 (54)	8.9 (0.9)
5		5	2.0	0.5	0.25	55.6 (2.4)	25 (1.3)	1322 (47)	11.1 (0.8)
	5	5			0.25	51.2 (1.7)	57 (2.7)	1152 (32)	22.1 (1.9)
	5	5	2.5		0.25	55.4 (2.5)	15 (1.0)	1304 (27)	6.7 (0.5)
	5	5	2.0	0.5	0.25	58.1 (2.7)	19 (0.7)	1414 (36)	8.7 (1.1)
	5	5			0.375	55.8 (3.2)	34 (1.5)	1378 (25)	14.4 (0.7)
	5	5			0.125				

a decrease in the PBAn content decreases the flexibility. The net effect depends which of the two effects prevails. The data in Table 4.5.3 show that the samples with less PBAn provide a higher tensile strength and Young's modulus but a lower elongation at break. This means that the rubber provides in this particular case the dominant effect.

4.5.3.9 Effect of the Monomers

Several monomers were used in different weight ratios. Tables 4.5.3 and 4.5.4 show that the tensile properties are sensitive to the nature(s) and amount(s) of monomer(s). St, MMA and An units increase the tensile strength and Young's modulus but decrease the elongation at break; the BMA units change the above properties in the opposite direction.

4.5.3.10 Toughness

The toughness of the samples, as measured by the area under the stress–strain curve, is listed in Tables 4.5.3 and 4.5.4, which show that most of them possess a high toughness near 20 MJ m^{-3} As shown by Table 4.5.3, a sufficient extent of crosslinking results in a high toughness. Longer post-curing times and higher contents of the curing agent usually provide higher toughness. However, a too high crosslinking extent may cause a decrease in toughness, because flexibility becomes too small. Rigid monomer units, such as those of St and MMA, are harmful to toughness.

4.5.4 CONCLUSION

Polymer composites based on epoxy resins are prepared via the mixing of concentrated emulsions of various components. This is followed by polymerization and by the heterogeneous crosslinking which occurs simultaneously with the transformation of the cells of the concentrated emulsion to latexes. All the samples exhibit a single T_g, which is wide for some systems and narrow for other systems. They possess excellent mechanical properties, indicating that the components of the composites were satisfactorily compatibilized. The curing time affects the extent of crosslinking for some systems and hence both the mechanical and thermal properties. Suitable compositions and curing lead to composites with a wide glass transition range, as wide as 40°C, which may be employed as energy-damping materials.

REFERENCES

1. Manson, J. A. and Sperling. L. H. in *Polymer Blends and Composites*, Plenum Press, New York. 1977. p. 89.
2. Ramera, J. B., Rajalingam, P. and Radhakrishnan. G. *J. Appl. Polym. Sci.* 1991, **43,** 23.
3. Rosen. S. L. *Polym. Eng. Sci.* 1967, **7,** 115.
4. Frisch, H. L. and Klempner. D. *Macromol. Rev.* 1970, **1,** 149.
5. Frisch. K. C. Klempner, D., Frisch, H. L. and Ghiradella. H. in *Recent Advances in Polymer Blends, Grafts and Blocks* (Ed. L. H. Sperling). Plenum Press, New York. 1974.
6. Ruckenstein. E. and Kim, K. J. *J. Appl. Polym. Sci.* 1988, **36,** 907.
7. Ruckenstein, E. and Park, J. S. *J. Polym. Sci., Chem. Lett. Edn* 1988, **26,** 529.
8. Ruckenstein. E. *Adv. Polym. Sci.* 1996, **128,** 1.
9. Ruckenstein, E. and Kim. K. J. *Chem. Mater.* 1989. **1,** 472.
10. Lovell. P. A. *Macromol. Symp.* 1995, **92,** 71.
11. Ortiz, C., McDonough, W., Hunston. D. and Hoffmann. D. *Polym. Mater. Sci. Eng.* 1993, **70,** 9.
12. Fox, T. G. *Bull. Am. Phys. Soc.* 1956, **1,** 123.
13. Gordon, M. and Taylor, J. S. *J. Appl. Chem.* 1952, **2,** 493.
14. Klempner, D. and Frisch. H. L. *J. Polym. Sci. Polym. Lett. Edn* 1970, **8,** 525.
15. Sperling, L. H. in *Interpenetrating Polymer Networks*, American Chemical Society, Washington DC, 1994, p. 1.

4.6 Self-Compatibilization of Polymer Blends Prepared via Functionalized Concentrated Emulsion Polymerization*

Hangquan Li[†], *Haohao Huang*[†], *and Eli Ruckenstein*[‡]

[†] Beijing University of Chemical Technology, Beijing, China 100029

[‡] Department of Chemical Engineering, State University of New York at Buffalo, Buffalo, NY 14260

ABSTRACT Two "functionalized" concentrated emulsions in water were prepared separately, one from a weakly polymerized mixture of styrene (S) and a small amount of acrylic acid (AA) and the other from a mixture of butyl acrylate (or butyl methacrylate) and a small amount of glycidyl methacrylate (GMA). After the two concentrated emulsions were polymerized partially, they were mixed and subjected to complete polymerization. During the latter polymerization, reactions between the carboxyl groups of the AA moieties of the S/AA copolymer and the glycidyl groups of the GMA-containing copolymer occurred, and copolymers and crosslinked structures were generated that constituted the compatibilizers of the system. The blend materials thus obtained possessed excellent toughness compared to those without functional groups.

* *Journal of Polymer Science: Part A: Polymer Chemistry*, Vol. 37. 4233–4240, (1999).

4.6.1 INTRODUCTION

Polymer blends can be prepared in various ways, of which melt mixing is the most widely employed; solution mixing and latex mixing also are occasionally used. In general, a compatibilizer needs to be added to the system.[1,2] Novel methodologies based on concentrated emulsions recently were proposed.[3–6] In contrast to a conventional emulsion, a concentrated emulsion (with globules of uniform size) has a volume fraction of the dispersed phase that is larger than 0.74 (this represents the most compact arrangement of spheres of equal size) and can be as large as 0.99. In those procedures, two concentrated emulsions of different monomers in water first were prepared separately. Each of the concentrated emulsions was polymerized partially until a suitable viscosity was reached, and then they were mixed and further heated for completing the polymerization. During the final polymerization step, reactions occurred at the interface between the cells, and so compatibilizers were generated and the system could self-compatibilize. In an improved version,[7,8] a vinyl-terminated macromonomer was added to the mixture of concentrated emulsions to promote the formation of compatibilizers. Blends with high toughness were prepared via such self-compatibilization methods. As a further improvement, a "functionalized" concentrated emulsion method is suggested in this paper. As in the previous methods, two concentrated emulsions in water first were prepared separately. In one of the concentrated emulsions, styrene (S) mildly copolymerized with a small amount of acrylic acid (AA) constituted the dispersed phase, whereas in the other emulsion, a mixture of butyl acrylate (BA) or butyl methacrylate (BMA) and a small amount of glycidyl methacrylate (GMA) constituted the dispersed phase. After the two concentrated emulsions were subjected to partial polymerization, they were mixed and subjected to complete polymerization, and the polymer chains containing carboxyl and glycidyl groups reacted as follows:

$$\text{—COOH} + \underset{\diagdown\,\text{O}\,\diagup}{\overset{|}{\text{C}}\text{H—CH}_2} \rightarrow \text{—COO—}\overset{|}{\text{C}}\text{H—CH}_2\text{OH}$$

Copolymers thus were generated that compatibilized polystyrene (PS) and poly(butyl acrylate) (PBA) or poly(butyl methacrylate) (PBMA). As a result, the compatibilization and polymerization were achieved simultaneously. This process is examined in this article.

4.6.2 EXPERIMENTAL

4.6.2.1 Materials

Azobisisobutyronitrile (AIBN; Shanghai reagent manufacturer No. 3, China) was recrystallized from methanol. S, BA, BMA, and AA were purchased from Shanghai Central Reagent Manufacturer, China, and they were filtered through inhibitor removal columns before use. The other compounds, sodium dodecyl sulfate (SDS; 70%; Beijing QIUXIAN Chemical Manufacturer, China) and tetrahydrofuran (THF; 99%; Tianjin Reagent Manufacturer No. 6, China), were used as received.

4.6.2.2 Preparation of the Concentrated Emulsion Blends

4.6.2.2.1 Concentrated Emulsion I

A mixture of S and AA (in various weight ratios) containing an initiator (AIBN, 0.033 g/g monomers) was polymerized partially in a test tube at 60°C for 1 h to incorporate some carboxyl groups onto the PS chains. The remaining AA was extracted with water. The partially polymerized system thus obtained subsequently was used as the dispersed phase of a concentrated emulsion in water. In the preparation of the concentrated emulsion, an aqueous solution of SDS (0.08 g/g H_2O) first was placed in a flask provided with a magnetic stirrer. The flask was sealed with a rubber septum, and the air inside was replaced with nitrogen. The dispersed phase was added dropwise with vigorous stirring into the flask with a syringe until the volume fraction of the SDS aqueous solution became 0.2. The whole addition process lasted about 15 min and took place at room temperature.

4.6.2.2.2 Concentrated Emulsion II

A mixture of BA or BMA and GMA (in various weight ratios) containing an initiator (AIBN, 0.003 g/g of monomers) was used directly as the dispersed phase of a concentrated emulsion without partial polymerization. The preparation method of the concentrated emulsion was the same as that for the concentrated emulsion **I**.

The two concentrated emulsions thus prepared were introduced into a water bath at 60°C to carry out another partial polymerization. The durations of the partial polymerization of the concentrated emulsions **I** and **II** were 2 and 3.5 h, respectively, for reasons discussed later. Subsequently, the two concentrated emulsions were mixed, and the mixture was heated further at 60°C for 24 h to complete the polymerization. The product obtained was washed with methyl alcohol and dried in a vacuum oven for 24 h. The blend product of the concentrated emulsion polymerization was obtained as a white powder.

The various samples prepared are listed in Table 4.6.1.

4.6.2.3 Solid Content Determination

A concentrated emulsion polymerized for a particular time was mixed with a sufficiently large amount of methanol. The precipitated polymer was washed several times with methanol and dried. The ratio of the weight of the polymer obtained to that of the original monomers was taken as the solid content (wt%).

4.6.2.4 Measurements of Gel Content

A preweighed powder was immersed in THF at room temperature by using 10 mL of solvent for each gram of powder. The solvent was replaced with a fresh one every 8 h until all the soluble species were removed (96 h). The remaining sample sheet was dried in a vacuum oven for 24 h to remove the solvent. The weight ratio of the remaining sample to the original one was considered the gel content (wt%).

TABLE 4.6.1
Recipes for Various Systems

Sample Number	Amount of Acrylic Acid (g/100 g styrene)[a]	Amount of GMA (g/100 g BA)[a]	Amount of GMA (g/100 g MBA)[a]	Gel Content (wt%)
1a	0.0	0.0	—	0.8
2a	1.0	1.8	—	2.2
3a	2.0	3.5	—	3.9
4a	3.0	4.3	—	6.3
5a	4.0	7.0	—	8.1
1b	0.0	—	0.0	0.7
2b	1.0	—	1.8	1.9
3b	2.0	—	3.5	3.5
4b	3.0	—	4.3	5.9
5b	4.0	—	7.0	7.8

[a] A weight ratio of acrylic acid to GMA of 4/7 was employed for the mole ratio to be 1/1. The weight ratio of the two concentrated emulsions mixed was 1/1. Amount of SDS (g/gH_2O) = 0.08; fraction of dispersed phase = 0.8; polymerization temperature = 60°C; time of partial polymerization for styrene/acrylic acid = 2 h; time of partial polymerization for BA or BMA/GMA = 3.5 h.

4.6.2.5 Fourier Transform Infrared (FTIR) Spectroscopy

Infrared study was carried out with a FTIR instrument (F-1000, Shimatsu, Japan) by using a sample film prepared by thermal pressing.

4.6.2.6 Scanning Electron Microscopy

The fractured surface of the specimen after impact testing was coated with a thin film of gold. The surface morphology was examined by scanning electron microscopy (S-250MKIII, Bam-bridge Instr., UK).

4.6.2.7 Impact Testing

The powders of the concentrated emulsion blends were thermopressed at a temperature of 140°C and a pressure of 8 MPa for 3 to 5 min to the size described by the ASTM standard D 256. The impact testing was conducted at room temperature (about 25°C) with a Charpy Xoj-4 (Wu Zhong Testing Instrument Co., China) impact testing instrument. Some specimens were tested at 30°C for comparison purposes.

4.6.3 RESULTS AND DISCUSSION

Polymer blends were prepared via the polymerization of a mixture of two partially polymerized concentrated emulsions. The purpose of the partial polymerization was to maintain the identity of the cells of the concentrated emulsions by increasing the viscosity of the cells. After the two concentrated emulsions were mixed, the diffusion of the polymer chains and unreacted monomers could be restricted to the outer shells of the cells. However, the viscosity of the cells should not be allowed to become too high because this would impede the uniform mixing of the concentrated emulsions. Because the viscosity of the cells depends on the solid content in the concentrated emulsion, it can be controlled easily through the time of partial polymerization. The solid contents of the concentrated emulsions of S, BA, or BMA for various polymerization times are presented in Figure 4.6.1. Figure 4.6.1 shows that after the concentrated emulsion of S was polymerized for 2 h or that of BA (or BMA) was polymerized for 3.5 h, a solid content of 70 wt% could be achieved, which provided suitable viscosities for the mixing of the concentrated emulsions. For solid contents higher than 70 wt%, the mixing was difficult. Also, the introduction of a small amount of AA or glycidal methacrylate did not affect the polymerization rate much. For these reasons, the durations of the partial polymerization of the concentrated emulsions were chosen to be 2 h for S and 3.5 h for BA or BMA.

After the two partially polymerized concentrated emulsions were mixed, diffusion of monomers and oligomers among the cells took place. Because the cells possessed relatively high viscosities, the diffusion was slow and was restricted to thin shells near the surface of the cells. In these thin shells, various reactions took place:

1. Different monomers randomly copolymerized.
2. The glycidyl and carboxyl groups reacted via eq 1. The double bonds thus inserted in the polymer chains could participate further in the polymerization process, leading to various kinds of copolymers.
3. The glycidyl groups of the PBA or PBMA chains reacted with the carboxyl groups present in the PS chains, resulting in graft copolymers and crosslinked structures.

The three kinds of reactions generated compatibilizers, the product of the third reaction being most likely the most effective one.

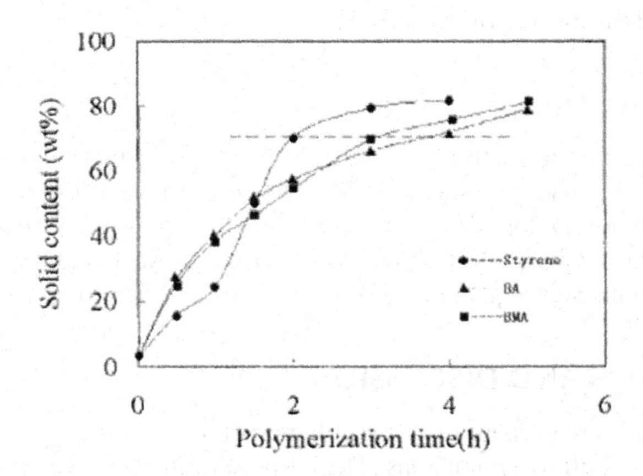

FIGURE 4.6.1 Solid contents of the concentrated emulsions of three monomers.

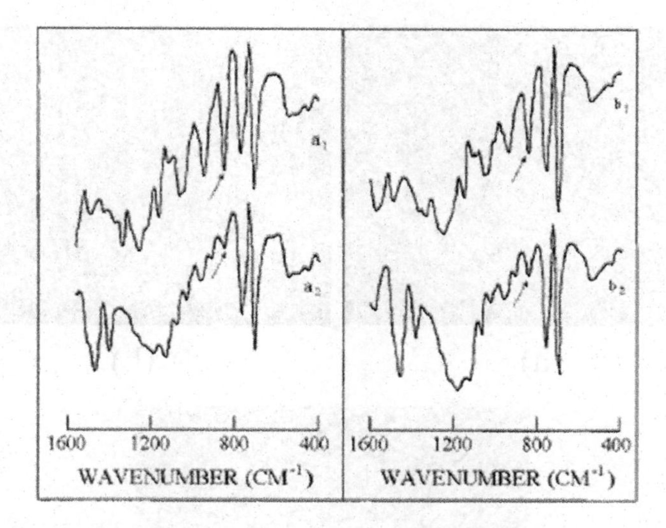

FIGURE 4.6.2 IR spectra of different systems: (a_1) sample from styrene/(butyl acrylate + glycidyl methacrylate) system, (b_1) sample from styrene/(butyl methacrylate + glycidyl methacrylate) system, (a_2) sample from (styrene + acrylic acid)/(butyl acrylate + glycidyl methacrylate) system, and (b_2) sample from (styrene + acrylic acid)/(butyl acrylate + glycidyl methacrylate) system.

The reaction between glycidyl and carboxyl groups was confirmed by infrared spectroscopy. In Figure 4.6.2, the IR spectra of the compatibilized products (a_2 and b_2) are compared with those of two reference products (a_1 and b_1). The latter were prepared as their counterparts, except no AA was present. In the spectra of a_1 and b_1, the peak at 845 cm^{-1}, which can be attributed to the glycidyl groups, is very sharp. In contrast, in the spectra of a_2 and b_2, that peak becomes smeared and unclear. For a_1 and b_1, no functional groups reacted with the glycidyl groups, whereas for the samples a_2 and b_2, most glycidyl groups were consumed in the reaction with the carboxyl groups, accounting for why its peak in the IR spectrum is much weaker.

The homopolymers and copolymers of S and BA (or BMA) free of functional groups completely dissolved in THF. However, for the compatibilized products, some insoluble species remained after 96 h of washing. This indicated that the reaction between the functional groups resulted not only in copolymers but also in some crosslinked structures (gel).[9–15] The gel content of the products is listed in Table 4.6.1, which shows that the greater the amount of the functional groups introduced, the higher the gel content was. As a combination of different chain segments, the crosslinked structures played an important role in compatibilization. However, they could affect the processibility. The preparation of the test bars for impact testing showed that a gel content lower than 10 wt% did not affect the processibility.

The compatibilizing effect was revealed clearly by the morphologies of the fractured surfaces. Figures 4.6.3 through 4.6.5 are micrographs of the fractured surfaces

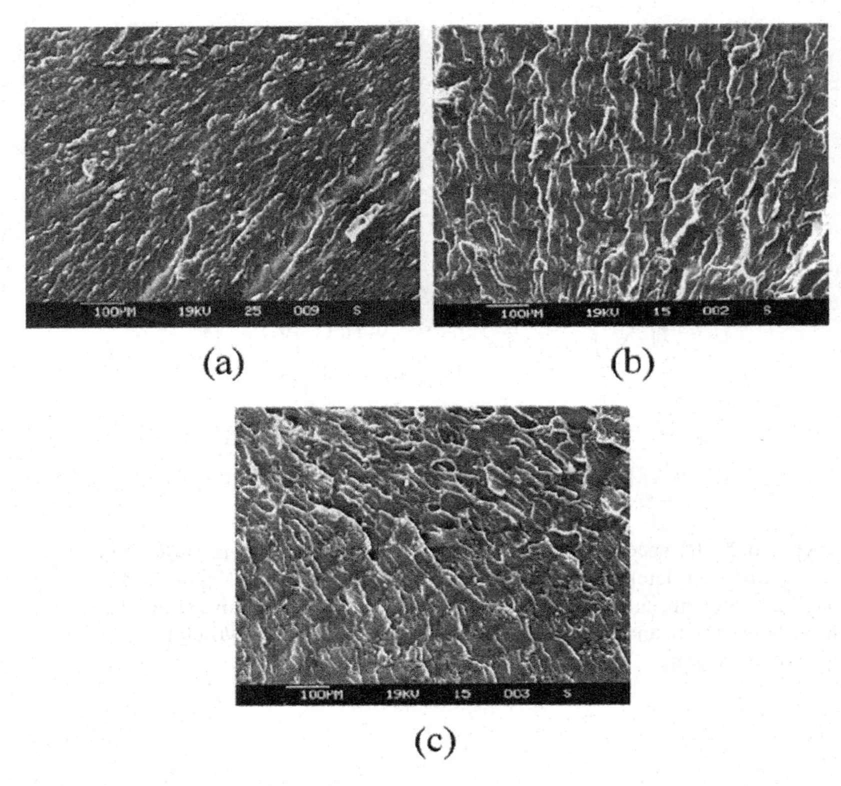

FIGURE 4.6.3 Micrographs of PS/PBA blends for various amounts of functional monomers: (a) None, (b) AA 2.0 g/100 g styrene/GMA 3.5 g/100 g BA, and (c) AA 4.0 g/100 g styrene/GMA 7.0 g/100 g BA.

obtained by scanning electron microscopy, from which one can extract information about the smoothness of the surface and finesse of the wrinkles. The former is a measure of the brittleness of a sample, and the latter provides indications about the domain size of the phases. When a brittle sample breaks, there is hardly any deformation, and the fractured surface is smooth. However, a tough sample undergoes relatively large deformations before break, and the fractured surface is coarse. It is clear from Figures 4.6.3 and 4.6.4 that, when no functional groups were present, the fractured surfaces were relatively smooth (Figures 4.6.3a and 4.6.4a), whereas the fractured surfaces of the systems containing functional groups were much coarser (Figures 4.6.3b, c and 4.6.4b, c). Figures 4.6.3 and 4.6.4 also show that the greater the amount of functional groups, the finer the wrinkles were. It is obvious that the deformation before break was provided by the more flexible component (PBMA or PBA) and that the finer wrinkles indicated smaller domain sizes. Therefore, one can conclude that the compatibilization resulted in a fine dispersion of the components. At a

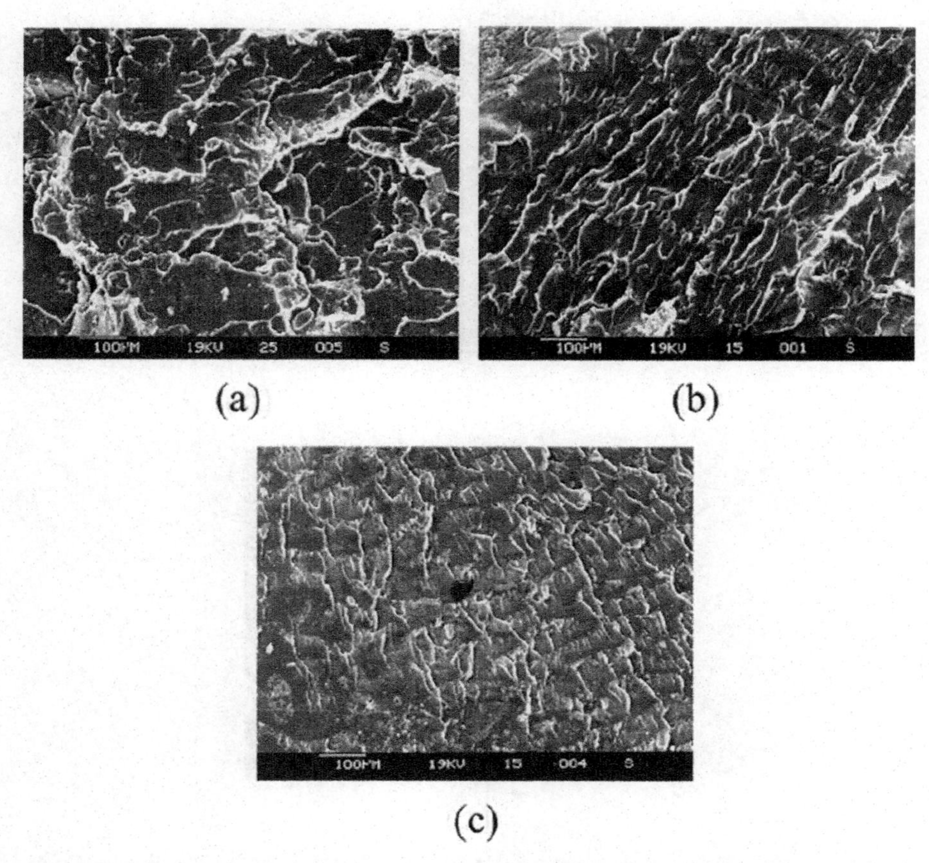

FIGURE 4.6.4 Micrographs of PS/PBMA blends for various amounts of functional monomers: (a) None, (b) AA 2.0 g/100 g styrene/GMA 3.5 g/100 g BMA, and (c) AA 4.0 g/100 g styrene/GMA 7.0 g/100 g BMA.

higher magnification (Figure 4.6.5), one can see more clearly that, as the amount of functional groups increased, the fractured surface became rougher.

The effect of compatibilization can be characterized by the impact strength of a sample, which constitutes a measure of its toughness. The relationship between impact strength and the amount of AA is presented in Figure 4.6.6, which shows that the greater the amount of functional groups, the higher the impact strength was. This increase occurred because a higher content of functional groups resulted in a greater amount of compatibilizer. However, the content of functional groups should be controlled. If it is too high, too much gel will be generated, and the product will not be melt processible.[16]

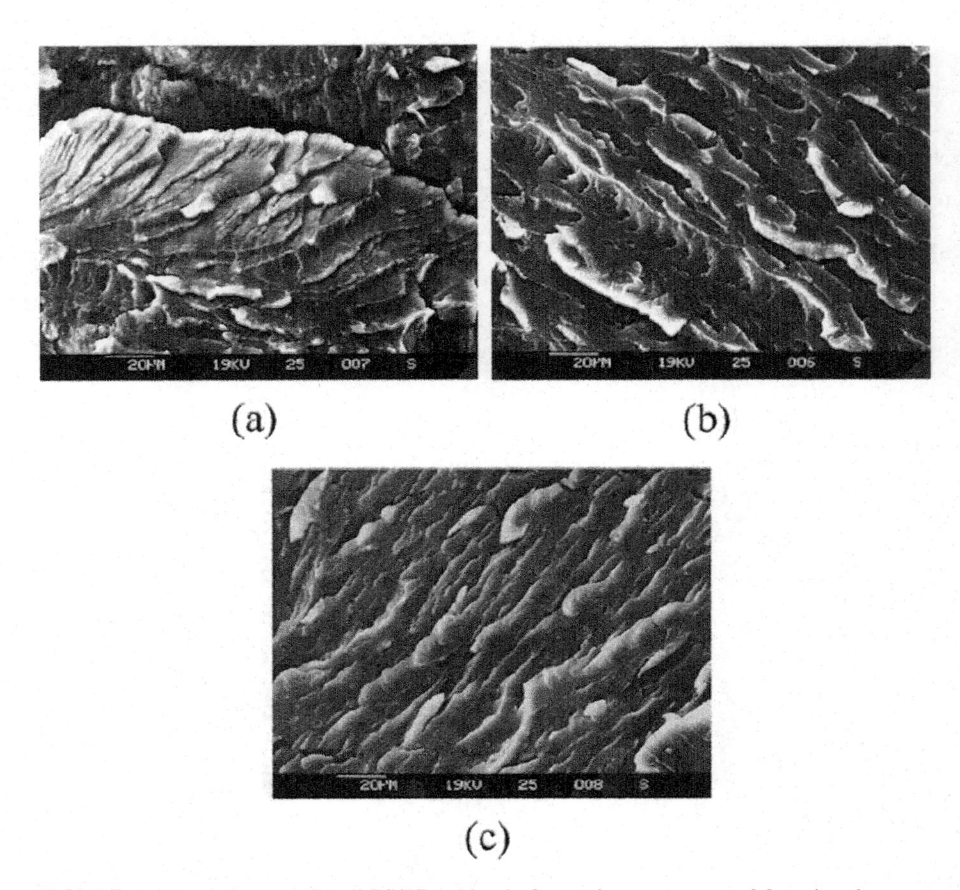

FIGURE 4.6.5 Micrographs of PS/PBA blends for various amounts of functional monomers (high magnification): (a) None, (b) AA 2.0 g/100 g S/GMA 3.5 g/100 g BA, and (c) AA 4.0 g/100 g S/GMA 7.0 g/100 g BA.

The toughening effects of PBA and PBMA on PS are presented in Figures 4.6.7 and 4.6.8. Because PBA has a glass-transition temperature of –35°C, it had a profound toughening effect on PS. The samples containing more than 50 wt% of PBA were too soft to be tested with the Charpy instrument. The toughening effect of PBMA, which has a glass-transition temperature of 27°C (which is a little higher than the room temperature), was more moderate. When the sample was tested at room temperature, PBMA was at its glass-transition temperature, and only a moderate toughening was observed.

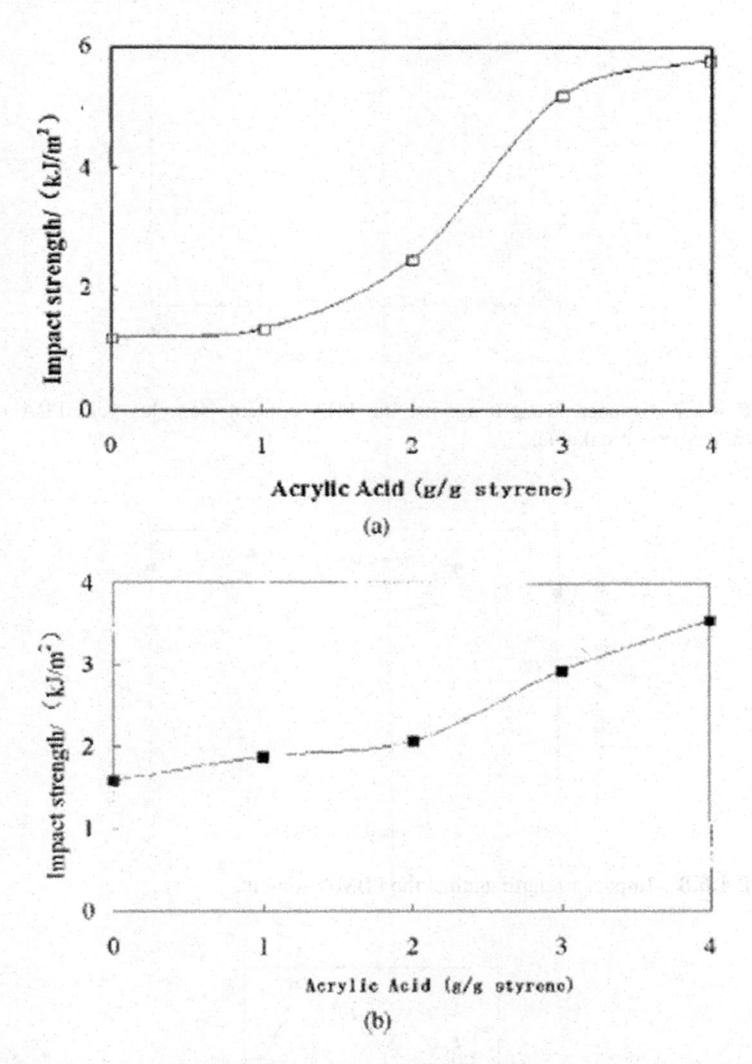

FIGURE 4.6.6 Impact strength against the functional monomer content: (a) PS/PBA system for a weight ratio of PS/PBA = 1 and a mole ratio of AA/GMA = 1/1 and (b) PS/PBMA system for a weight ratio of PS/PBMA = 1 and a mole ratio of AA/GMA = 1/1.

The impact bar was prepared via injection molding. To acquire some information about the reactions that occurred during processing, the IR spectra of the samples before and after melt processing are compared in Figure 4.6.9. The relative strengths of the peak at 845 cm^{-1} are essentially identical, from which one can conclude that the reaction between carboxyl and glycidyl groups was completed substantially during the concentrated emulsion polymerization.

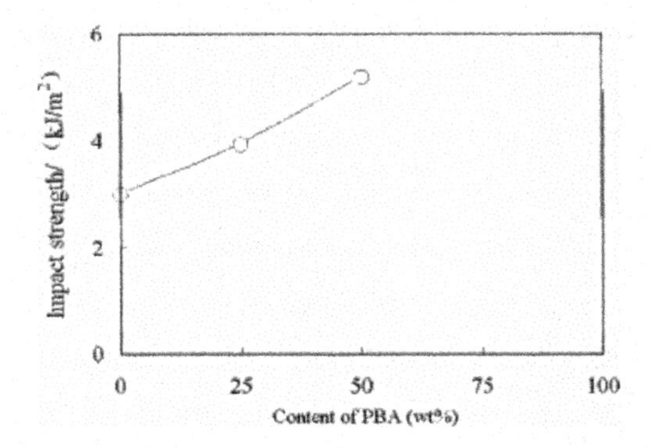

FIGURE 4.6.7 Impact strength against the PBA content (samples with PBA contents > 50 wt% were not breakable).

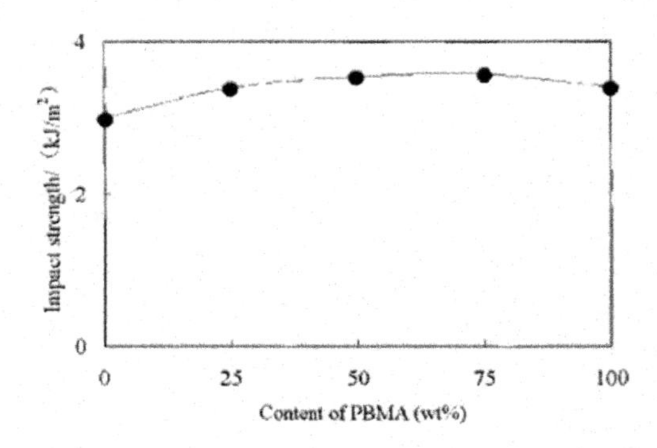

FIGURE 4.6.8 Impact strength against the PBMA content.

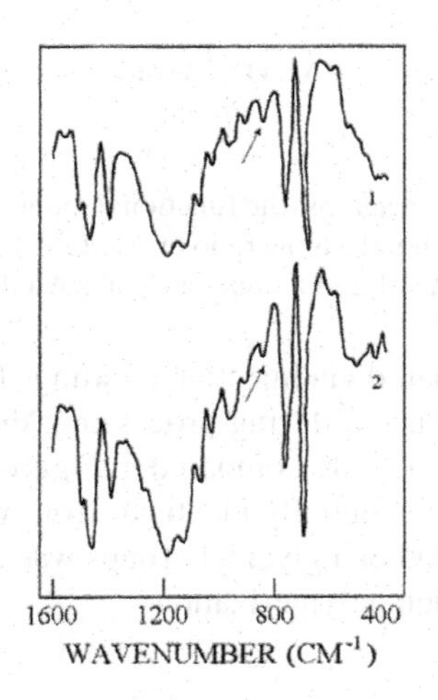

FIGURE 4.6.9 IR spectra of the samples before (1) and after (2) melt processing.

4.6.4 CONCLUSION

When two functionalized concentrated emulsions, styrene/acrylic acid (**I**) and butyl (meth)acrylate/glycidyl methacrylate (**II**) in water were polymerized partially and subsequently mixed and subjected again to polymerization, some copolymers and crosslinked gels were generated near the interfaces of the cells of the emulsions. The system thus was self-compatibilized by the compatibilizers generated *in situ*. The compatibilization resulted in a fine dispersion of the two phases. The impact strengths of the functional groups containing blends could be about four times higher than those without the functional groups.

This project was funded by the Natural Science Foundation, China, and the National Science Foundation, the United States of America.

REFERENCES AND NOTES

1. Coleman, M. M.; Pehlert, G. J.; Painter, P. C. *Macromolecules* 1996, 29, 6820.
2. Tzur, A.; Narkis, M.; Siegemann, A.; Seadan, M.; Lambla, M. *Polym Networks Blends* 1995, 5, 107.
3. Ruckenstein, E.; Park, J. S. *Polymer* 1990, 31, 2397.
4. Ruckenstein, E.; Li, H. *Polymer* 1994, 35, 4343.
5. Li, H.; Ruckenstein, E. *J Appl Polym Sci* 1996, 61, 2285.
6. Ruckenstein, E.; Li, H. *Polym Composites* 1997, 18, 320.
7. Ruckenstein, E.; Li, H. *J Appl Polym Sci* 1994, 52, 1949.
8. Li, H.; Ruckenstein, E. *Polymer* 1995, 38, 2281.
9. Vazquez, F.; Cartier, H.; Landfester, K.; Hu, G. H.; Pith, T.; Lambla, M. *Polym Adv Technol* 1995, 6, 309.
10. Scranton, A. B.; Peppas, N. K. *J Polym Sci Part A: Polym Chem* 1990, 28, 39.
11. Bansil, R.; Hermann, H. J.; Stauffer, D. *Macromolecules* 1984, 17, 998.
12. Tobita, H.; Hamielec, A. E. *Polymer* 1991, 32, 2641.
13. Durand, D.; Bruneau, C. M. *Makromol Chem* 1982, 183, 1007.
14. Tobita, H.; Hamielec, A. E. *Macromolecules* 1989, 22, 3098.
15. Boots, H. M.; Kloosterboer, J. G.; van de Hei, G. M. M. *Br Polym J* 1985, 17, 219.
16. Nie, L.; Narayan, R.; Grulke, E. A. *Polymer* 1995, 36, 2227.

4.7 Room Temperature Initiated and Self-Heating Polymerization via Concentrated Emulsions: Application to Acrylonitrile Based Polymers*

Eli Ruckenstein and Hangquan Li

Department of Chemical Engineering,
State University of New York at
Buffalo, Amherst, NY 14260, USA

ABSTRACT A novel concentrated emulsion polymerization procedure, in which the polymerization is initiated at room temperature and the heat generated by the reaction accelerates the process, is proposed. The polymerization of acrylonitrile (AN) and its copolymerization with vinylidene chloride (VDC) are used as examples. AN (alone or with a comonomer) containing an oxidant was first dispersed in water to generate a concentrated emulsion. The polymerization of the monomers was initiated at room temperature by introducing an aqueous solution containing a mixture of reductants (ferrous sulfate and sodium metabisulfite) into the concentrated emulsion. The heat generated in the system increased its temperature and accelerated the polymerization. Under optimal conditions, the polymerization could reach >90% conversion within one hour. The small volume of the continuous phase in a concentrated emulsion constitutes an advantage of the procedure, since only a small amount of the produced heat is used for its heating. In addition, because the reductant,

* *Polymer Bulletin*, Vol. 37, 43–50, (1996).

which is present in the water phase, together with the oxidant, which is present in the oil phase, constitute the initiator, the large oil-water interfacial area of the concentrated emulsion constitutes an additional advantage.

4.7.1 INTRODUCTION

In the last few years, concentrated emulsions have been employed in this laboratory as precursors in the preparation of homopolymers, copolymers and polymer composites.[1–4] A concentrated emulsion is however stable only if one of the phases is sufficiently hydrophobic and the other one sufficiently hydrophilic. There are cases in which a concentrated emulsion can be generated at room temperature, but loses its stability at the polymerization temperature, which must be higher than 50°C for most initiators. In such cases, concentrated emulsions cannot be used as polymerization precursors. The goal of this paper is to use a redox initiator, which can initiate polymerization at room temperature. In addition, the heat evolved during polymerization is employed to accelerate the process. Because of the initial rapid polymerization, the concentrated emulsion, which would have lost its stability at high temperatures, does not segregate. While the present room temperature initiated, self-heating polymerization has some generality, the basic ideas will be illustrated using the homo- and co-polymerization of acrylonitrile (AN) with vinylidene chloride (VDC) as examples.

In order to achieve a high conversion of AN with the conventional emulsion polymerization method, either high initial temperature (usually >90°C)[5–9] or long polymerization time (>4 h)[10] had to be employed. To initiate polymerization, the system had to be heated to an elevated temperature. Because of the highly exothermic nature of the polymerization of AN, the heat generated had to be removed. The initial heating and the subsequent cooling make production highly energy-consuming.

In the present paper, a novel procedure based on concentrated emulsions is proposed, which provides a faster and energy-saving preparation of AN-based materials. AN (or AN with VDC) containing an oxidant was employed as the dispersed phase of a concentrated emulsion in water. An aqueous solution of reductants (ferrous sulfate and sodium metabisulfite) was subsequently introduced into the concentrated emulsion. The polymerization started at room temperature as soon as the reductants were added. Since the reactor was insulated, the heat generated by the polymerization reaction rapidly increased the temperature. The increased temperature accelerated the polymerization, which was finished within one hour. Since in this method the system is heated by itself, the procedure will be called "self-heated" polymerization. The heat of reaction is no longer a problem, but constitutes an advantage. Since both the oxidant (which is present in the oil phase) and the reductants (which are introduced in the water phase) constitute the initiator, the polymerization is initiated at the oil-water interface. The large interfacial area of the concentrated emulsion accelerates the polymerization and a high temperature is achieved. "Self-heating" polymerization is also stimulated by the fact that the volume fraction of the continuous phase of the concentrated emulsion is small and hence only a small amount of the generated heat is consumed for its heating.

A concentrated emulsion has the appearance of a paste, and differs from a conventional emulsion in that the volume fraction of the dispersed phase is greater than 0.74 (which represents the volume fraction of the most compact arrangements of spheres of the same size), and can be as high as 0.99. At high volume fractions, the concentrated emulsions are composed of polyhedral cells less than 1 μm in size, separated by thin continuous phase films. A large oil-water interfacial area can be achieved in concentrated emulsions.

The effects of the wt ratio of the monomers, the kind and composition of the reductant, the type of dispersant (surfactant), etc. on the polymerization and the properties of the products were investigated.

4.7.2 EXPERIMENTAL

4.7.2.1 Materials

Hexadecyltrimethylammonium bromide (CTAB), polyethylene glycol dodecyl ether (Brij35) and polyoxyethylene sorbitanmonolaurate (Tween20) were purchased from Fluka. All the other chemicals were purchased from Aldrich. Acrylonitrile (AN) and vinylidene chloride (VDC) were filtered through an inhibitor removal column before use. Cumene hydroperoxide (tech, 70%), ferrous sulfate heptahydrate ($FeSO_4 \cdot 7H_2O$, 99%), sodium metabisulfite (SMBS, 97%), polyvinyl alcohol (PVA; 80% hydrolyzed, MW 9000–10000), dodecyl sulfate sodium salt (SDS, 70%), dimethyl formamide (DMF, 99%) and methanol (99%) were used as received.

4.7.2.2 Preparation Method

An aqueous solution (1.6 mL) of CTAB (3 wt.%) and a very small amount of aqueous PVA solution (0.018 g of wt% solution per lg monomer) were first placed in a 50 mL flask provided with a magnetic stirrer. The flask was sealed with a rubber septum and insulated. The dispersed phase [monomer or monomers (8 g) containing an oxidant (cumene hydroperoxide, 0.06 g/g)] was added dropwise with a syringe to the flask with vigorous stirring, until the volume fraction of the aqueous solution became 0.17. The addition process lasted about 5 min at room temperature. A concentrated emulsion was thus generated. Subsequently, an aqueous solution of reductants (0.5 g/g water of $FeSO_4$ and $Na_2S_2O_5$ in various proportions) was introduced (0.16 g solution per g monomer) with a syringe. As soon as the reductant solution was uniformly dispersed in the concentrated emulsion, stirring was stopped. A glass thermometer was inserted through the rubber septum into the concentrated emulsion. One hour after the reductant was introduced, the product was retrieved, washed with methanol and dried in a vacuum oven. The percentage of the weight of the final product with respect to the original weight of the monomers is taken as the conversion. It is important to note that the surfactant CTAB had to be combined with PVA to ensure the formation of a stable concentrated emulsion, and that two reductants had to be used in appropriate proportions to maintain the stability of the concentrated emulsion and thus to achieve high conversions.

4.7.2.3 Product Characterization

The glass transition of the products was determined by differential scanning calorimetry, with a Perkin-Elmer DSC7 instrument. Elemental analysis was performed by Quantitative Technologies Inc. (Whitehouse, NJ). Molecular weights were determined from intrinsic viscosities using the relation: $[\eta] = KM^{\alpha}$, in which, because the constants K and α are not available for the copolymers, a set of constants for the AN homopolymer, $K = 52 \times 10^{-3}$ and $\alpha = 0.69$,[11] were used.

4.7.3 RESULTS AND DISCUSSION

In the "self-heating" polymerization, the polymerization starts at room temperature because of the initiator employed. The heat generated increases the temperature of the system, thus accelerating the process. Because of insulation, the temperature of the concentrated emulsion increased initially (3–5 min) rapidly to a maximum, after which it decayed very slowly. The decay is due to the heat loss and the increasingly lower amount of heat generated. The maximum temperature is listed in Table 4.7.1.

The maximum temperature plays an important role. The rate of polymerization at room temperature is low. It can be increased by raising the temperature. In order to achieve the highest possible rate of polymerization, the flask has to be insulated and the stirring has to be stopped as soon as the aqueous solution of the reductants is uniformly dispersed. If the flask would not have been insulated or the stirring would have been maintained, most of the heat generated in polymerization would have been dissipated, and the temperature in the flask would have risen only a little (our experiment has indicated a rise from 25°C to 30°C).

To achieve a high temperature by "self-heating," the concentrated emulsion method provides several advantages. The more concentrated the emulsion, the higher the interfacial area between the phases and the higher the conversion and the maximum temperature, for reasons already explained in the Introduction. Of course, a too highly concentrated emulsion may cause difficulties in stirring and subsequent handling. For this reason, a volume fraction of the dispersed phase of 0.83 was employed in most of the present experiments. For comparison purposes, polymerizations were also carried out in more dilute emulsions and the results are listed in Table 4.7.2, which shows that the more dilute the emulsion, the lower the maximum temperature and the lower the conversion.

It should be emphasized that only suitable surfactants and suitable reductants ensure the stability of the concentrated AN emulsions. A concentrated emulsion is very stable when the monomer dispersed in water is extremely hydrophobic. Because AN is somewhat polar, the corresponding concentrated emulsion in water is stable at room temperature, but becomes unstable at 50°C. In this paper, besides the cationic surfactant CTAB, an anionic surfactant (SDS) and two non-ionic surfactants (Brij 35 and Tween 20) were also tried. They all generated stable concentrated emulsions at room temperature only when coupled with PVA. However, all the concentrated emulsions broke when the temperature was raised to 50°C. In addition, when the reductants were introduced at room temperature, the concentrated emulsions based on the anionic and non-ionic surfactants broke immediately and almost no polymerization

TABLE 4.7.1
Characteristics of Polymerization and Products[a]

Wt Ratio of AN/VDC	Wt Ratio of FS/SMBS	Maximum Temperature (°C)	Conversion (%)	T_g (°C)	MW × 10^{-3}	State of the Concentrated Emulsion
7/3	pure FS	45	35.0			stable
9/1		58	69.4		–	
8/2	9/1	54	65.7		–	
7/3		51	47.9		160	stable
6/4		46	39.4		–	
8/2		74	90.9		–	
7/3		72	86.3		103	stable
6/4	3/1	66	81.0		–	
5/5		65	75.7		–	
4/6		61	74.0		–	
10/0		110	98.8	125	–	
9/1		90	97.7	109	–	
8/2		81	95.1	85	–	
7/3	2/1	76	92.0	69	81.5	stable
6/4		73	87.0	59	–	
5/5		68	76.1	52	–	
4/6		64	73.4	48	–	
3/7		60	70.6	45	–	
8/2	1/1	84	94.1		–	partial
7/3		81	90.3		51.8	segregation
7/3	1/2	87	85.2		49.4	partial segregation
7/3	1/3	76	82.2		51.2	complete segregation
7/3	pure SMBS	67	62.4		64.9	complete segregation

[a] Amount of the reductant solution: 0.16 g/g monomer, concentration of the reductant solution: 0.5 g/g water, polymerization time: 1 hr

occurred. Those based on the cationic surfactant started to polymerize immediately and this polymerization stabilized the concentrated emulsion. When the temperature became high, a large fraction of the monomer was already converted and the stability was no longer a problem.

One can see from Table 4.7.1 that the AN/VDC wt ratio has a major influence on polymerization. When the other conditions are kept the same, the higher the AN content, the higher the maximum temperature and the conversion. This result can be attributed to: (i) The polymerization of AN is more exothermic than that of VDC and (ii) since the boiling point of VDC is 32°C, the VDC in the vapor phase remains

TABLE 4.7.2
Effect of Emulsion Concentration[a]

Vol. Ratio of Continuous/Dispersed Phases	Maximum Temperature (°C)	Conversion (%)	MW × 10^{-3}
1/5	76	92.0	81.5
1/3	72	76.3	45.0
1/2	63	54.5	39.7
1/1.5	61	47.4	34.8
1/1	55	28.7	28.6

[a] Amount of monomers = 8 g, wt ratio of AN/VDC = 7/3, wt ratio of reductants FS/SMBS = 2/1, amount of the reductant solution: 0.16 g/g monomer, concentration of the reductant solution: 0.5 g/g water, CTAB concentration: 3 wt%, PVA: 0.018 g of 20 wt% aqueous solution per g monomer, polymerization time: 1 hr

TABLE 4.7.3
Results of Elemental Analysis

No.	Initial wt ratio of AN/VDC	Maximum Temperature (°C)	N wt%	Cl wt%	Wt Ratio of AN/VDC in Copolymer
1	4/6 = 0.67	64	10.06	40.86	0.68
2	6/4 = 1.50	73	13.77	27.13	1.40
3	7/3 = 2.33	76	17.69	18.71	2.62
4	9/1 = 9.00	90	22.77	7.28	8.71

[a] Amount of the reductant solution: 0.16 g/g monomer, concentration of the reductant solution: 0.5 g/g water, wt ratio of reductant FS/SMBS = 2/1, polymerization time: 1 hr

unpolymerized. To clarify this issue, VDC polymerization was also carried out. No temperature increase was observed when the reductant solution was introduced, and the polymerization was so slow that after 24 h only 10% conversion was achieved. This indicates that the heat generated by VDC polymerization is negligible, and that only the reactions involving AN generate significant heat. The compositions of AN/VDC copolymers were determined via elemental analysis, and the results are listed in Table 4.7.3. They show that the content of the VDC units in copolymers is comparable to that in the initial mixture. The glass transition temperatures (T_g) of the copolymers measured by differential scanning calorimetry (DSC), which are listed in Table 4.7.1, show that T_g decreases with increasing VDC. The T_gs of AN and VDC homopolymers are 125°C and −18°C,[12–14] respectively. The presence of VDC units is expected to decrease the T_g of the copolymers, and indeed this happens.

An aqueous solution of a mixture of ferrous sulfate (FS) and sodium metabisulfite (SMBS) was selected as reductant. Comparing the samples with the same wt ratio of AN/VDC = 7/3 in Table 4.7.1, one can see that the FS/SMBS wt ratio (for the same total weight) has a major effect on polymerization.

When FS was added alone, the concentrated emulsion maintained its stability, the polymerization started immediately and the viscosity of the system increased, but the conversion after one hour was low (35%) and the maximum temperature was 45°C. The stability of the concentrated emulsion is a result of the electrical double layer repulsion (due to the charge generated by the surfactant molecules adsorbed upon the oil-water interface) and of the steric repulsion (caused by the PVA molecules co-adsorbed upon the same interface) which compensate for the van der Waals attraction among the polyhedral droplets. The amount of FS added (0.06 g) increases the ionic strength in the water solution to a large value (2 mol/L). At these high ionic concentrations, the free ions and the ion pairs compete for the water molecules and one can no longer identify individual hydrated ions; the water and ions become more collectively organized, probably acquiring a quasi-liquid crystalline structure. The double layer repulsion is annihilated, but is replaced by a repulsion generated by the "dipoles" of the ion pairs formed between the anions present and the cations of the surfactant molecules at the oil-water interface. The concentrated emulsion can maintain its stability when the steric repulsion due to the adsorbed PVA molecules and the ion pair repulsion because the adsorbed surfactant molecules compensate for the van der Waals attractive interactions. The higher viscosity of the quasi-liquid crystalline structure of water may provide an additional stabilizing effect. The conversion is low probably because of the barrier to the redox initiation process generated by the surfactant molecules present as ion pairs at the oil-water interface.

In contrast, just the addition of SMBS has broken the concentrated emulsion, since the viscosity of the system decreased abruptly, but the conversion after one hour was 62% and the maximum temperature was 67°C, higher than in the previous case. The concentrated emulsion is not stable, either because the condition mentioned above is not satisfied, or, more likely, because of the salting out effect [caused by the high ionic strength (3 mol/L in this case)], which pushes the surfactant from the water to the oil phase. The latter effect occurs because the organization of water by the electrolyte decreases the compatibility between the head group of the surfactant and water. As a result, the oil-in-water concentrated emulsion is transformed to a water-in-oil emulsion. While the interfacial area between oil and water is decreased, the interfacial barrier to the redox initiation process is probably greatly decreased, and, as a result the conversion is higher than in the previous case.

By combining FS with SMBS one can avoid the breaking of the concentrated emulsion and achieve high conversion. Indeed, the conversion is 92% for a weight ratio FS/SMBS = 2 [for the same total amount of reductant (0.16 g/g monomer) and for the wt ratio AN/VDC = 7/3].

The maximum temperature changes with the FS/SMBS wt ratio, but has surprisingly the largest value at FS/SMBS about 1/2. One expects, however, the highest conversion to be associated with the highest maximum temperature. The maximum temperature is highest at another value of FS/SMBS, probably because another reaction, involving the reductants and the surfactant, occurs.

The molecular weight is affected by the volume fraction of the dispersed phase of the emulsion and by the wt ratio FS/SMBS of reductants. As shown in Table 4.7.2, the more dilute the emulsion, the smaller the molecular weight. In a concentrated emulsion, the polyhedral droplets are small and the presence of the surfactant at the oil-water interface decreases its mobility. The chains which grow inside the droplets have for this reason also a decreased mobility. This delays the termination and, as a result, higher MWs are achieved. Because the size of the droplets in a dilute emulsion is much larger, the bimolecular termination occurs more easily, since the presence of the interface does not decrease as much the mobility of the chains. Consequently, the resulting MW is lower. Table 4.7.1 shows that the MW of the AN/VDC copolymer changes with the FS/SMBS wt ratio and exhibits a minimum at FS/SMBS = 1/2. The higher the maximum temperature, the lower the MW, because the mobility of the growing chains is affected by the temperature.

4.7.4 CONCLUSION

Using the method proposed in this paper, both homo- and co-polymerization of acrylonitrile via the concentrated emulsion method can be initiated at room temperature. The heat generated increases the temperature and accelerates the reaction, and a high conversion can be achieved within one hour. In dilute emulsions, lower conversions are obtained. Only cationic surfactants coupled with PVA were suitable in the present cases as dispersants; when anionic and non-ionic surfactants were used, even when coupled with PVA, the concentrated emulsions broke when the reductant was introduced. An aqueous solution of FS and SMBS with a wt ratio of 2/1 constitutes the best reductant, since it leads to the highest conversion.

REFERENCES

1. Ruckenstein E, Park JS (1988) *J Polym Sci Chem Lett Ed* 26:529.
2. Ruckenstein E, Kim KJ (1988) *J Appl Polym Sci* 36:907.
3. Ruckenstein E, Li H (1995) *Polym Bull* 35:517.
4. Li H, Ruckenstein E (1995) *Polymer* 36:2281.
5. Kobashi T, Shiota H, Umetami H (1977) *Ger Offen* 2,709,503.
6. Koenig J, Sueling C, Boehmke G (1977) *Ger Offen* 2,604,630
7. Turner JJ (1975) US 3,873,508.
8. Iwata H, Otani T, Kobayashi T (1978) *Jpn Kokai Tokkyo Koho* 78 104,689.
9. Kohashi N, Shioda H, Umetami H (1977) *Japan Kokai* 77 107,045.
10. Sakai H, Hamada S, Inoue H, Hosaka S (1973) *Japan Kokai* 73 90,380.
11. Kamide K, Miyazaki Y, Kobayashi H (1985) *Polym J* 17:607.
12. Gupta AK, Vhand N (1980) *J Polym Sci Polym Phys Ed* 18:1125.
13. Zutty NZ, Whitworth SJ (1964) *J Polym Sci B* 2:709.
14. Beevers RB, White EFT (1960) *Trans Faraday Soc* 56:1529.

4.8 High-Rate Polymerization of Acrylonitrile and Butyl Acrylate Based on a Concentrated Emulsion*

Chen Zhang[†], *Zhongjie Du*[†], *Hangquan Li*[†], *and Eli Ruckenstein*[‡]

[†]School of Materials Science and Engineering, Beijing University of Chemical Technology, Beijing 100029, People's Republic of China

[‡]Department of Chemical Engineering, Furnas Hall, State University of New York at Buffalo, Buffalo, NY 14260, USA

ABSTRACT A mixture of acrylonitrile and butyl acrylate was polymerized starting from a concentrated emulsion in water. The polymerization was initiated at room temperature by a redox system, consisting of reductants dissolved in the water phase and an oxidant dissolved in the dispersed phase. The initial polymerization was carried out adiabatically with self-heating until a temperature of about 70°C was reached; this was followed by additional heating in a water bath at a higher temperature, for a total polymerization time up to 30 min. The conversion thus achieved was higher than that obtained via the adiabatic process alone. An optimum temperature during the additional heating was observed. Because of the gel effect, the molecular weight of the

* *Polymer* Vol. 43, (5391–5396), 2002".

product increased with time and the reaction rate became affected by diffusion. The additional heating enhanced the mobility of the species in the system, thus ensuring a final product with a composition near that of the feed.

4.8.1 INTRODUCTION

The acrylonitrile (AN) homopolymer could not be used as a plastic because it softens only slightly below its decomposition temperature of about 300°C. In addition, it does not dissolve in its monomer and hence cannot be shaped by bulk casting. However, the copolymers of AN with a number of comonomers could be used as plastics or rubbers. The copolymers with vinylidene chloride have been used for years to prepare films of low gas permeability, often as coatings on various materials [1]. Styrene–acrylonitrile (SAN) with styrene as the predominant unit has also been available for a long time [2]. More recently, many plastics containing high proportions of AN, with the comonomer being an acrylate or a methacrylate, have been developed. About 95% of the materials produced are used for packaging and refrigerator liners, because their barrier capabilities are superior to those of more conventional plastics, such as ABS (acrylonitrile/butadiene/styrene ternary copolymer) and SAN. The copolymers of AN and butyl acrylate (BA) combine superior mechanical properties with a low gas permeability. However, they have been prepared by graft copolymerization, which often lasts for more than 6 h [3]. The radiation-induced emulsion polymerization of AN and BA was also explored [4]; however, little was applied industrially because of the complexity of the equipment. The purpose of this paper is to suggest an alternate preparative methodology, which can tremendously shorten the time of synthesis.

A high-rate polymerization method based on a concentrated emulsion [5–7] was proposed by this group. A concentrated emulsion [8] is an emulsion whose volume fraction of the dispersed phase is higher than 74%, which corresponds to the close packing of spheres of the same size. At high-volume fractions, the concentrated emulsions are composed of polyhedral cells, separated by thin films of continuous phase. This kind of structure of the concentrated emulsion provides, because of its large interfacial area, high rates of initiation and interfacial reaction. In the high-rate polymerization, the initiation occurred at the interface between the droplets and the continuous phase. Redox initiators, consisting of reductants and an oxidant soluble in the continuous and dispersed phases, respectively, were employed and the polymerization was carried out under adiabatic conditions. Since the concentration of droplets is high, the heat generated during the reaction could be efficiently used to accelerate the polymerization, and a high conversion of above 80 wt% was reached in less than 1 h. This procedure was employed to polymerize AN and a comonomer, the high reaction rate of AN being responsible for the high rate of polymerization. The comonomers studied included vinylidene chloride [5] and vinyl acetate [6,7]. It was found that the reaction heat could maintain a high polymerization rate for less than about 15 min, after which the rate slowed down.

In this paper, an initial adiabatic self-heating was followed by an external heating to achieve a high rate of polymerization of AN with BA. The initial step was performed under adiabatic conditions until a temperature of 70°C was reached, whereas the latter one was carried out in a water bath. The adiabatic self-heating followed by the external heating ensured a higher conversion than the adiabatic process alone. The effect of various reaction conditions on the rate of the concentrated emulsion polymerization of AN/BA system was investigated, and a comparison was made between the adiabatic alone and the combined procedures.

4.8.2 EXPERIMENTAL

4.8.2.1 Materials

Acrylonitrile (AN, CP, 99%) and butyl acrylate (BA, AR, 99.5%) were provided by Shanghai No. 3 Reagent Manufacturer; they were distilled before use to remove inhibitors. Azobisisobutyronitrile (AIBN, CP) was provided by Shanghai No. 4 Reagent Manufacturer, and was recrystallized from methanol before use. Cumene hydroperoxide (CHPO, CP, >70%) and sodium metabissulfite (SMBS, CP, 90%) were purchased from the Shanghai Zhongxin Chemical Manufacturer, hexadecyltrimethylammonium bromide (HTAB) from Beijing Chemical Reagent Co., poly(vinyl alcohol) (PVA, grade 124) from Beijing Donghuan Chemical Manufacturer, ferrous sulfate (FS, AR, 99%) from Beijing Chemical Reagent Manufacturer, *N, N*-dimethylformamide (DMF) and ethyl acetate (EA) from Beijing Yili Chemical Co. The water was distilled and deionized.

4.8.2.2 Concentrated Emulsion Polymerization

Into a 50 mL flask equipped with a magnetic stirrer, 3 mL of an aqueous solution of HTAB (0.3 g/g H_2O) and about 0.04 mL of an aqueous solution of PVA (0.009 g/g H_2O) were first introduced. Then, with stirring, a mixture containing AN, BA and an oxidant (CHPO) was added dropwise with a syringe, which penetrated the rubber septum sealing the flask, until the volume fraction of the dispersed phase became 0.8. The addition lasted for about 20–30 min, and a paste-like concentrated emulsion was thus generated. After the flask containing the concentrated emulsion was properly insulated, an aqueous solution of reductants (0.167 g/g H_2O) was added with stirring using a syringe. After the solution of reductants was uniformly dispersed, the stirring could no longer be continued, because the rapid adiabatic polymerization that occurred tremendously increased viscosity. A thermometer that penetrated the rubber septum to the center of the concentrated emulsion was employed to measure the temperature. As soon as the system attained a temperature of about 70°C, external heating was provided. After a total of 30 min of polymerization (adiabatic plus monadiabatic), the system was taken out from the flask, washed with methanol and dried in a vacuum oven. The weight ratio of the dried product to the initial reactants represents conversion (in wt%).

4.8.2.3 Elemental Analysis

The elemental analysis was carried out using an Elemental Analyzer (Carloerba 1106, Italy).

4.8.2.4 Intrinsic Viscosity Measurements

Intrinsic viscosity was used as a measure of molecular weight. It was determined using an Ubbelhode viscometer with DMF as solvent. The samples were dissolved in DMF at a concentration of 20 mg/10 mL. The testing temperature was 30°C.

4.8.3 RESULTS AND DISCUSSION

4.8.3.1 Conversion at 30 min

The polymerization was carried out for only 30 min. In a preliminary work [6] it was found that during adiabatic polymerization one could attain a conversion of about 80–95 wt% in 1 h. However, most of the conversion (70–85 wt%) occurred in 30 min, after which the temperature of the system decreased slowly due to heat losses. The behavior in the first 30 min is, therefore, sufficient to represent the entire polymerization. In this paper, unless otherwise mentioned, the term "conversion" denotes exclusively the conversion after 30 min of reaction.

The conversion and some polymerization parameters of AN and BA for various wt ratios of AN and BA are listed in Table 4.8.1. One can note that the fraction of

TABLE 4.8.1
Conversion (wt%) at 30 min of the Concentrated Emulsion

	AN/BA Wt Ratio		
	8/2	**7/3**	**6/4**
Adiabatic system			
Maximum temperature (°C)	91	86	71
Temperature at 30 min (°C)	73	69	62
Conversion (wt%)	80.7	76.7	61.3
Additionally heated system			
Temperature of water bath (°C)	Conversion (wt%)		
70	86.9	84.9	70.1
80	87.4	86.0	75.0
90	83.7	77.8	73.8

Total weight of the monomers: 12 g; volume fraction of the dispersed phase: 0.8; HTAB: 0.09 g; PVA: 0.04 mL of its aqueous solution (0.009 g/g H_2O); oxidant: 0.4 g; reductants: 0.257 g; (SMBS/FS = 1.5/1 w/w). The reductants were introduced as an aqueous solution (0.167 g/g H_2O).

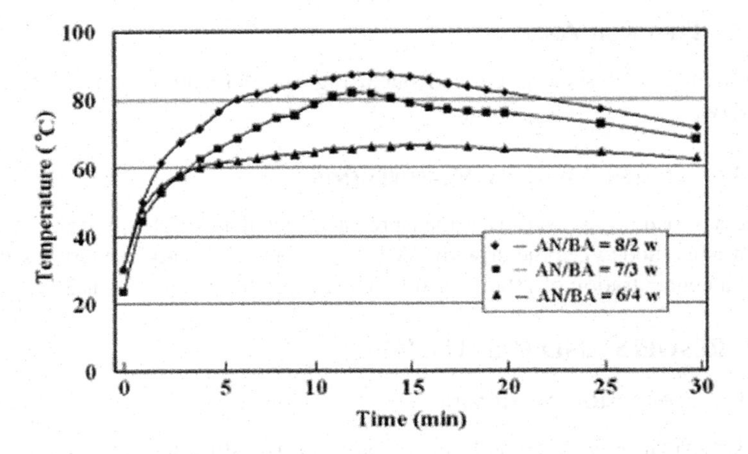

FIGURE 4.8.1 Temperature–time plots for adiabatic systems of various AN/BA wt ratios. The other polymerization conditions are as for Table 4.8.1.

AN in the mixture played an important role in the conversion: the greater the fraction of AN, the higher was the conversion. The temperature of the adiabatic systems is plotted against time in Figure 4.8.1, which shows that the greater the fraction of AN, the higher the temperature achieved. It is obvious that *high* temperature promotes polymerization rate. The heat of reaction for the homopolymerizations of AN and BA are 77.5 and 78 kcal/mol [9], respectively, hence the polymerization of both monomers is exothermic. However, the propagation constants for AN and BA are 28,000 and 2100 [10,11], respectively, hence the reaction rate of AN is by an order of magnitude higher than that of BA. For this reason, the polymerization of AN is much more exothermic per unit time. As a result, the polymerization was maintained mainly by AN. In fact, experiments indicated that when the fraction of AN was lower than 50 wt%, the polymerization could hardly be initiated at room temperature.

Figure 4.8.1 shows that during the adiabatic polymerization, the temperature inside the flask first increased rapidly, reached a maximum in less than 15 min, after which polymerization rate slowed down and temperature decreased slowly because of some heat losses. In order to enhance conversion, an additional external heating was employed. However, Table 4.8.2 indicates that when an external heating was employed from the beginning of the polymerization, the conversion has

TABLE 4.8.2
Conversion in Systems Additionally Heated from the Beginning of Polymerization

	Temperature of the Water Bath (°C)				
	50	60	70	80	90
Conversion (wt%)	68.6	72.9	60.8	55.5	70.7

AN/BA = 7/3 (wt); the other polymerization conditions are as for Table 4.8.1.

decreased instead of being enhanced. This may be attributed to the rate of decomposition of the organic initiator, which was larger at higher temperatures. When the polymerization was carried out under adiabatic conditions, the decomposition rate of the oxidant was enhanced slowly as the temperature increased, and thus the system could be provided steadily with fresh radicals. When, however, the system was subjected to a high temperature from the beginning, the initial decomposition rate of the organic oxidant became too high, and little organic oxidant remained available for further initiation. For these reasons, lower conversions were reached.

Another mechanism, involving the temperature enhancement of the reaction rate between the organic oxidant and reductants is, however, also possible. During the adiabatic polymerization, the rate of formation of the radicals is more uniform because, while the temperature gradually increases, the concentrations of the initiators decrease, and the system was provided continuously with radicals. In contrast, when the system was subjected to a high temperature from the beginning, a large number of radicals were formed initially and much fewer later, and the competition among the initial radicals decreased the conversion.

In order to avoid the above behavior, the concentrated emulsion was first subjected to an adiabatic polymerization (for about 9 min) until the temperature reached a value of about 70°C, after which the flask was introduced into a water bath at a temperature equal or higher than 70°C, the total polymerization time being 30 min. Table 4.8.1 shows that higher conversions were thus achieved than via the adiabatic polymerization alone. The enhancement of the conversion was dependent on the AN/BA wt ratio and on the temperature of the water bath. The lower the fraction of AN, the greater the increment of the conversion, because a smaller AN content made the adiabatic reaction less exothermic and the external heating had a greater effect. In addition, Table 4.8.1 shows that the change in temperature should not be too sharp. The conversion attained in a water bath of 90°C was, for reasons already noted, less than that at 80°C, and, in some cases, even below that at 70°C. One can conclude that 80°C represents an optimum temperature.

It should be pointed out that the polymerization in a water bath at a constant temperature is not necessarily an isothermal process. Indeed, the concentrated emulsion acquired a high viscosity and became a semi-solid for a conversion of about 20 wt%. At this conversion, stirring could no longer be employed. Under such conditions, the heat exchange with the water bath occurred only near the wall of the flask and the polymerization in the central region of the flask was still driven mostly by the self-heating due to the reaction. Therefore, the external heating accelerated essentially the region near the wall of the polymerization system.

Besides the monomer wt ratio and the external heating, the wt ratio between the two reductants also affected conversion. Two reductants (SMBS and FS) were employed to initiate the polymerization of the concentrated emulsion. When SMBS or FS was used alone, the polymerization could hardly be initiated. Table 4.8.3 shows that the highest conversion was attained when the SMBS/FS wt ratio was 1.5/1. For higher or lower SMBS/FS wt ratios, the conversions were lower. The temperature–time curves in Figure 4.8.2 show that for an SMBS/FS wt ratio of 1.5/1 the temperature was always higher than for the other systems. For this reason the SMBS/FS ratio of 1.5/1 constitutes an optimum ratio. This optimum was explained in a previous

TABLE 4.8.3
Effect of the Wt Ratio of the Reductants under Adiabatic Conditions

	SMBS/FS (w/w)				
	2/1	1.5/1	1/1	1/1.5	1/2
Maximum temperature (°C)	84	86	78	65	58
Temperature at 30 min (°C)	79	80	77	65	58
30 min conversion (wt%)	68.3	71.5	66.4	36.0	30.8
Intrinsic viscosity (dl/g)	85.7	124.4	115.2	97.1	67.0

AN/BA wt ratio = 7/3, other polymerization parameters are as for Table 4.8.1.

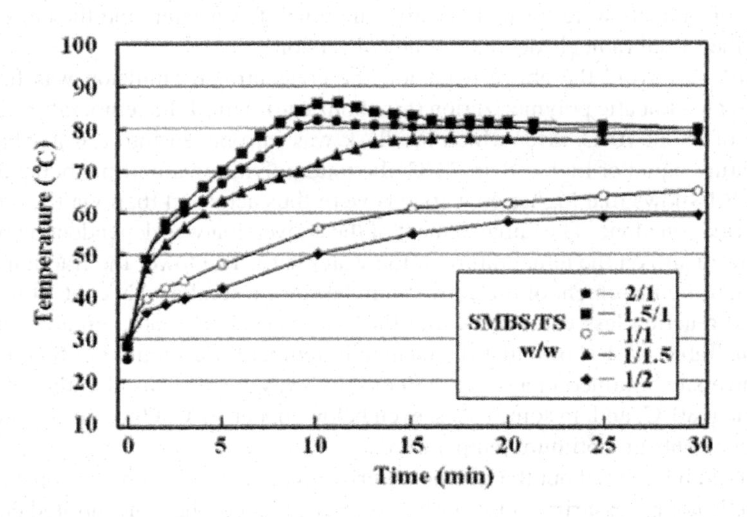

FIGURE 4.8.2 Temperature–time plots for adiabatic systems of various reductant wt ratios. The other polymerization conditions are as for Table 4.8.1.

paper [6] as being a result of the synergism caused by the cooperative reactions of the two reductants. The reaction between the oxidant CHPO and the reductant FS is as follows:

$$C_6H_5C(CH_3)_2OOH + Fe^{2+} \rightarrow C_6H_5C(CH_3)_2O^{\cdot} + Fe^{3+}$$

and leads to a radical and Fe^{3+}. Further, SMBS reduces Fe^{3+} to Fe^{2+}, thus stimulating the activity of FS. In addition, we found that the SMBS/FS wt ratio affected the stability of the concentrated emulsion. When the SMBS/FS wt ratio was 1/1.5, the concentrated emulsion was partially segregated, and became less stable as the fraction of FS increased. Because the valencies of Fe^{2+} and SO_4^{2-} are two, the doubling of the concentration of FS increased the ionic strength by four times. The high ionic strength was responsible for the instability of the concentrated emulsion, because of the shielding of the repulsive double layer force. A stable concentrated emulsion possesses a large surface area of the interface between the dispersed and the continuous phases where the initiation can occur; as a consequence, a higher initiation rate and thus a higher polymerization rate can be achieved.

4.8.3.2 Molecular Weight

The molecular weight in this paper was characterized by the intrinsic viscosity. Table 4.8.3 shows that under adiabatic conditions, as the SMBS/FS wt ratio increases, the molecular weight of the product passes through a maximum that coincides with that for the conversion. This indicates that there is a relationship between the polymerization rate and the molecular weight. Table 4.8.4 provides the intrinsic viscosities of the copolymers for various AN/BA wt ratios and temperatures of the water bath. It shows that high molecular weights are always associated with high polymerization rates. One can therefore conclude that the heat supply affected both the polymerization rate and the molecular weight. The greater the heat supply, the higher the polymerization rate and the molecular weight. Under adiabatic conditions, the only source of heat was the reaction heat. During the initial stage of polymerization, the higher the initiation rate, the higher was the polymerization rate, and an optimum wt ratio of the reductants provided a maximum polymerization rate. Using the same wt ratio of reductants, the larger the amount of AN, the greater was the heat generated. When additional heating was employed, an optimum temperature was found which provided a maximum polymerization rate. The gel effect was responsible for the high polymerization rate and the high molecular weight. Indeed, the higher the polymerization rate, the more rapid was the increase in the viscosity of the system. As the reaction mixture became more viscous, the radical ends of the polymer chains had an increasing difficulty in diffusing towards each other to terminate their growth.

TABLE 4.8.4
Intrinsic Viscosity (dl/g) of the Concentrated Emulsion with External Heating

Temperature of Water Bath (°C)	AN/BA Wt Ratio		
	8/2	7/3	6/4
70	165.2	134.3	123.7
80	181.6	159.4	125.6
90	163.6	131.2	116.9
Adiabatic system	154.4	124.4	102.8

The polymerization conditions are as for Table 4.8.1. For the additional heating, the flask was introduced into a water bath when the temperature inside became about 70°C.

On the other hand, the smaller monomer molecules had lesser difficulty in diffusing so that the propagation reaction could be continued. The decrease in the termination rate constant increased both the overall rate and molecular weight. In systems with high initiation and polymerization rates, the gel effect occurs earlier, and higher molecular weights can be achieved. Particularly, when the polymer is insoluble in its monomer, the radical end becomes entrapped in the coiled chains and the termination reaction rate becomes slower. Because the AN/BA copolymer is insoluble in either AN or BA, the gel effect became very active and lasted during almost the entire process of polymerization.

The conversion and molecular weight at various times for a particular system are presented in Table 4.8.5, which indicates an abrupt increase in molecular weight at about 2.5 min, for a conversion of about 12 wt%. This behavior was most likely caused by the starting of the gel effect. Afterwards the molecular weight increased with increasing conversion, hence, the gel effect continued to be significant. Comparing the results of the adiabatic and additionally heated systems one can note that after the system was introduced into the water bath, the molecular weight kept still increasing rapidly with time, whereas in the adiabatic system it increased only moderately. This occurred because in the later stages of polymerization, the system became a semi-solid and the species needed a higher temperature to polymerize at a high rate. While the self-heating alone could not provide such conditions, the external heating could provide them and hence could maintain a high polymerization rate and achieve a high molecular weight.

4.8.3.3 Composition

The change in the AN content of the product with time during the adiabatic polymerization is presented in Table 4.8.6, which shows that for each AN/BA wt ratio the AN fraction was lower than in the initial feed. The reactivity ratios reported for the copolymerization of AN and BA are comparable: 1.08 and 0.82 [12] (or 1.00 and 1.20 [13]). Consequently, the difference between the compositions of the product and the initial feed cannot be a result of their reactivity ratios alone. The AN contents for various polymerization times of a typical system are listed in Table 4.8.5, which shows that during the initial stages of polymerization the fraction of AN was much lower than that of the initial feed, but as the polymerization proceeded it increased with the conversion. In the additionally heated system, the AN fraction reached finally a value approximately equal to that in the initial mixture. However, during the adiabatic polymerization, the increase of the AN fraction was slower, being after 30 min by about 5 wt% lower than in the initial feed. The apparent higher reactivity of BA can be attributed to the gel effect, which became active a few minutes after the starting of the polymerization, when the viscosity of the droplets reached a sufficiently high level. The high molecular weight chains lost their mobility almost completely, whereas the monomers could still diffuse with some difficulty. To be incorporated into the copolymer chains, the monomers had to diffuse to become available to the radical ends. Under such circumstances, the reaction was no longer controlled by the reactivity alone, but was also affected by diffusion. Because the AN homopolymer is not soluble in either monomer, but the BA homopolymer is

TABLE 4.8.5
The Conversion and Intrinsic Viscosity of the Product at Various Polymerization Time

		Time (min)								
		0.5	**2.5**	**5**	**7.5**	**10**	**15**	**20**	**25**	**30**
Conversion (wt%)	Adiabatic	4.1	11.8	19.4	34.8	57.1	62.9	61.9	67.5	77.8
	Water bath					60.4	65.2	72.4	82.2	88.2
Intrinsic viscosity (dl/g)	Adiabatic	28.7	93.7	102.7	105.6	108.1	110.0	116.3	122.0	124.4
	Water bath					114.8	125.2	136.9	149.3	159.4
Content of AN (wt%)	Adiabatic	54.5	54.7	55.0	56.5	62.6	61.3	62.6	63.3	64.6
	Water bath					65.5	64.4	66.7	66.8	69.8

AN/BA wt ratio = 7/3. For additional heating, the flask was introduced into a water bath of 80°C after about 8 min when the temperature inside became about 70°C. The other polymerization conditions are as for Table 4.8.1.

TABLE 4.8.6
AN Content of the Products of Adiabatic Polymerization of Different AN/BA Wt Ratios

	AN/BA Wt Ratio in the Feed			
	9/1	8/2	7/3	6/4
AN fraction in the product (wt%)	84.3	75.1	64.6	57.5

The polymerization conditions are as for Table 4.8.1.

soluble in its monomer, BA had a higher affinity for the BA sequences. As a result, BA acquired a higher access to the polymer chains than AN and had a higher chance to be incorporated. In addition, because AN is more polar than BA, the former molecule has stronger interactions with the polar chain and had, therefore a lower mobility than the latter molecule. As the polymerization proceeded, the fraction of AN increased, and this enhanced its incorporation rate. More importantly, as the viscosity further increased, the system became semi-solid. The movement of all the species became slower and the polymerization had to occur in an in situ way. As a result, the AN content increased with time. Because the external heating enhanced the mobility, at the end of 30 min a product with the composition of the feed was obtained. However, for adiabatic conditions, the polymerization rate was lower because of the lower temperature, and the increase in the AN fraction was more moderate. A longer reaction time would have led to an AN composition in the product comparable to that in the feed.

4.8.4 CONCLUSION

The effect of an additional heating that followed an adiabatic one on the high rate of polymerization of AN/BA in a concentrated emulsion was investigated. It was found that the two steps in succession achieved a conversion higher than 80 wt% within 30 min. An optimum heating temperature was found. The high rate of polymerization was caused by the gel effect and by the additional heating. The latter enhanced the mobility of the species in the semi-solid system generated during the adiabatic polymerization. The two effects ensured the generation of a high molecular weight product with a composition near that of the initial feed.

REFERENCES

1. Kodani T, Sakai H, Takayama A, Nomura M. *J Appl Polym Sci* 1998; 69:573.
2. Schneider M, Pith T, Lambla M. *Polym Adv Technol* 1996;7:577.
3. Capek I, Juranicova V, Barton J. *Eur Polym J* 1999;35:691.
4. Duflot VR, Dobrov IV, Lukhovitskii VI. *High Energy Chem* 2001;35:23.
5. Ruckenstein E, Li HQ. *Polym Bull* 1996;37:43.
6. Li HQ, Du ZJ, Ruckenstein E. *J Appl Polym Sci* 1998;68:999.
7. Zhang C, Du ZJ, Li HQ, Ruckenstein E. *Polymer* 2002;43:2945.

8. Ruckenstein E. *Adv Polym Sci* 1997;127:1.
9. Brandrup J, Immergut EH. *Polymer Handbook*, 3rd ed. New York: Wiley; 1989.
10. Bengough WJ, Melville HW. *Proc R Soc* (*London*) *A* 1959;249:455.
11. Dainton FS. *J Polym Sci* 1959;39:313.
12. Tamikado T, Iwakura Y. *J Polym Sci* 1959;36:529.
13. Tatemichi H, Suzuki S. *Kogyo Kagaku Zassi* 1960;63:1843.

Index